VOLUME SIXTY EIGHT

THE PSYCHOLOGY OF LEARNING AND MOTIVATION

Current Topics in Language

Series Editor

KARA D. FEDERMEIER

Department of Psychology, Program in Neuroscience,
and the Beckman Institute for Advanced Science and Technology
University of Illinois Champaign, IL, United States

VOLUME SIXTY EIGHT

THE PSYCHOLOGY OF LEARNING AND MOTIVATION

Current Topics in Language

Edited by

KARA D. FEDERMEIER

Department of Psychology, Program in Neuroscience, and the Beckman Institute for Advanced Science and Technology
University of Illinois Champaign
IL, United States

DUANE G. WATSON

Department of Psychology and Human Development
Vanderbilt University, Nashville
TN, United States

ELSEVIER

ACADEMIC PRESS
An imprint of Elsevier

Academic Press is an imprint of Elsevier
125 London Wall, London EC2Y 5AS, United Kingdom
525 B Street, Suite 1650, San Diego, CA 92101, United States
50 Hampshire Street, 5th Floor, Cambridge, MA 02139, United States
The Boulevard, Langford Lane, Kidlington, Oxford OX5 1GB, United Kingdom

First edition 2018

Notices

Knowledge and best practice in this field are constantly changing. As new research and experience broaden our understanding, changes in research methods, professional practices, or medical treatment may become necessary.

Practitioners and researchers must always rely on their own experience and knowledge in evaluating and using any information, methods, compounds, or experiments described herein. In using such information or methods they should be mindful of their own safety and the safety of others, including parties for whom they have a professional responsibility.

To the fullest extent of the law, neither the Publisher nor the authors, contributors, or editors, assume any liability for any injury and/or damage to persons or property as a matter of products liability, negligence or otherwise, or from any use or operation of any methods, products, instructions, or ideas contained in the material herein.

ISBN: 978-0-12-815086-3
ISSN: 0079-7421

Publisher: Zoe Kruze
Acquisition Editor: Sam Mahfoudh
Editorial Project Manager: Teresa Pons-Ferrer
Production Project Manager: Vignesh Tamil
Cover Designer: Matthew Limbert

Typeset by TNQ Technologies

CONTENTS

Contributors ix

1. Perceptual Learning for Native and Non-native Speech **1**

Melissa Baese-Berk

1. Perceptual Flexibility of Native Phonological Categories 4
2. Acquisition of Novel Phonological Categories 8
3. Adaptation to Foreign Accented Speech 14
4. Conclusions and Speculation 19

References 23
Further Reading 28

2. A Common Representation of Serial Position in Language and Memory **31**

Simon Fischer-Baum

1. Position Representation Schemes 33
2. A General Method for Contrasting Schemes 36
3. Experimental Findings 39
4. Discussion 44

References 52

3. Neurocomputational Emergentism as a Framework for Language Development **55**

Arturo E. Hernandez, Juliana Ronderos and Hannah L. Claussenius-Kalman

1. Localization of Function 58
2. The Sensorimotor Hypothesis 61
3. Variability of Language Development 67
4. Neuroemergentism: Dynamic Interactions Across Systems 75
5. Conclusion 78

Acknowledgments 78
References 79

4. Syntactic Adaptation **85**

Edith Kaan and Eunjin Chun

1. Introduction 86
2. Evidence for Syntactic Adaptation in Comprehension 88
3. Mechanisms Underlying Syntactic Adaptation 90
4. Evidence for Syntactic Adaptation? 93
5. What Is the Relation Between Adaptation and Learning? 102
6. Concluding Remarks 110

Acknowledgments 110
References 111

5. Neural Indices of Structured Sentence Representation: State of the Art **117**

Ellen Lau

1. Linguistic and Non-linguistic Relations 119
2. Neural Correlates of Linguistic Relations: Three Candidates 121
3. Conclusion 138

References 139

6. Familial Sinistrality and Language Processing **143**

Chia-lin Lee

1. Introduction 144
2. Language-Related Processing as a Function of Familial Sinistrality 145
3. Theories of Familial Sinistrality and Language 148
4. New Considerations for Familial Sinistrality and Language Processing 153
5. Conclusions 168

References 169

7. How Special Is Language Production? Perspectives From Monitoring and Control **179**

Nazbanou Nozari

1. The Special Place of Language 180
2. Language Production and the Need for Control 191
3. Assessing the Need for Control: The Primary Job of a Monitoring System 195
4. Domain-Generality of Monitoring and Control in Language Production 202
5. Conclusion: How Special Is Language Production? 207

Acknowledgments 209
References 209

8. Alternatives and Inferences in the Communication of Meaning 215

Hannah Rohde and Chigusa Kurumada

1. Two Types of Alternatives 218
2. When Nothing Says Something: Roles of Alternative Forms 223
3. Intonational Phonology: Role of Alternative Meanings 227
4. Coreference: Combining Expectations for Meanings and Forms 235
5. Inferences Based on Alternative Forms and Meanings: A Bayesian Approach 240
6. Where Next? 246

Acknowledgments 252
References 252

9. Reading Ahead by Hedging Our Bets on Seeing the Future: Eye Tracking and Electrophysiology Evidence for Parafoveal Lexical Processing and Saccadic Control by Partial Word Recognition 263

Elizabeth R. Schotter

1. The Reading Speed Paradox 265
2. *Hedging Our Bets*: Coordinating Eye Movements and Language Processing 266
3. *Seeing the Future*: Obtaining Parafoveal Preview of Upcoming Words 267
4. *Reading Ahead*: Using Parafoveal Preview to Initiate Word Recognition 269
5. Prior Debates About Parafoveal Processing: *Serial Versus Parallel Lexical Processing* 276
6. Evidence for Parafoveal Processing From Electrophysiology Studies 280
7. Modeling Parafoveal Preview Effects 288
8. Summary 291

References 292

10. Individual Differences in the Real-Time Neural Dynamics of Language Comprehension 299

Darren Tanner, Maria Goldshtein and Benjamin Weissman

1. Introduction 300
2. Language-Related ERP Components 305
3. Some Quantitative Individual Differences 307
4. Qualitative Individual Differences in Non-native and Bilingual Morphosyntactic Processing 310
5. Monolingual Morphosyntactic, Semantic, and Thematic Processing and Some Psycholinguistic Conundrums 314
6. Higher-Order Inferencing and Multi-Modal Communication 319
7. Conclusions and Challenges for the Future 327

References 329

11. *He Gave My Nose a Kick* or *He Kicked My Nose*? Argument Structure Alternations and Event Construal **337**

Eva Wittenberg

1. The Linking Structures Between Grammar and Event Construal 339
2. Case Study I: Grammar and Event Roles, and What Happens When They Don't Align 342
3. Case Study II: From Grammar to Time 347
4. Loose Ends: Processing, Reference, and Acquisition 351
5. Conclusion: Linguistics Is Worth Your while 355

References 356

12. The Role of Discourse Context in Reference Production and Comprehension: Insights From the Lexical Differentiation Effect **361**

Si On Yoon

1. Introduction 362
2. Lexical Differentiation in Production 367
3. The Source of Lexical Differentiation 374
4. Lexical Differentiation in Comprehension 381
5. Future Directions 385
6. Conclusion 387

References 388

CONTRIBUTORS

Melissa Baese-Berk
Department of Linguistics, University of Oregon, Eugene, OR, United States

Eunjin Chun
Department of Linguistics, University of Florida, Gainesville, FL, United States

Hannah L. Claussenius-Kalman
Department of Psychology, University of Houston, Houston, TX, United States

Simon Fischer-Baum
Department of Psychology, Rice University, Houston, TX, United States

Maria Goldshtein
Department of Linguistics, University of Illinois at Urbana-Champaign, Champaign, IL, United States

Arturo E. Hernandez
Department of Psychology, University of Houston, Houston, TX, United States

Edith Kaan
Department of Linguistics, University of Florida, Gainesville, FL, United States

Chigusa Kurumada
Department of Brain and Cognitive Sciences, University of Rochester, Rochester, NY, United States

Ellen Lau
Department of Linguistics, University of Maryland, College Park, MD, United States

Chia-lin Lee
Graduate Institute of Linguistics, National Taiwan University, Taiwan; Department of Psychology, National Taiwan University, Taiwan; Graduate Institute of Brain and Mind Sciences, National Taiwan University, Taiwan; Neurobiology and Cognitive Neuroscience Center, National Taiwan University, Taiwan

Nazbanou Nozari
Department of Neurology, Johns Hopkins University, Baltimore, MD, United States; Department of Cognitive Science, Johns Hopkins University, Baltimore, MD, United States

Hannah Rohde
Linguistics and English Language, University of Edinburgh, Edinburgh, United Kingdom

Juliana Ronderos
Department of Psychology, University of Houston, Houston, TX, United States

Elizabeth R. Schotter
Department of Psychology, University of South Florida, Tampa, FL, United States

Darren Tanner
Department of Linguistics, University of Illinois at Urbana-Champaign, Champaign, IL, United States

Benjamin Weissman
Department of Linguistics, University of Illinois at Urbana-Champaign, Champaign, IL, United States

Eva Wittenberg
Department of Linguistics, University of California, San Diego, CA, United States

Si On Yoon
Beckman Institute for Advanced Science and Technology, University of Illinois, Urbana, IL, United States

CHAPTER ONE

Perceptual Learning for Native and Non-native Speech

Melissa Baese-Berk
Department of Linguistics, University of Oregon, Eugene, OR, United States
E-mail: mbaesebe@uoregon.edu

Contents

1. Perceptual Flexibility of Native Phonological Categories 4
2. Acquisition of Novel Phonological Categories 8
3. Adaptation to Foreign Accented Speech 14
4. Conclusions and Speculation 19
References 23
Further Reading 28

Abstract

In order to successfully perceive a language, a listener must be able to flexibly adapt to their input; however, a listener must also maintain the stability of their perceptual system, in order to maintain its integrity. Given the competing requirements of demonstrating flexibility while maintaining stability, it is critically important to understand under what circumstances listeners will demonstrate flexibility and under what circumstances they will not. In this chapter, I address this question using three types of perceptual learning for speech: perceptual flexibility of existing speech sound categories, acquisition of novel speech sound categories, and adaptation to unfamiliar accented speakers. These types of learning highlight the importance of balancing flexibility and stability in the perceptual system. In addition to highlighting previous results, their implications for language production and language learning are explored. Further, exploring these three types of learning together underscore the importance of investigating speakers and listeners from a variety of language backgrounds.

One hallmark of the human perceptual system is the remarkable flexibility of the system, a feature which allows individuals interpret new experiences and adapt to their surroundings. This flexibility, however, is coupled with stability, which maintains the integrity of the perceptual system. Within the subcategory of auditory perception, the human speech perception system is a foundational example of both the flexibility and stability of the perceptual system. In fact, speech perception provides a good testing ground for

Psychology of Learning and Motivation, Volume 68
ISSN 0079-7421
https://doi.org/10.1016/bs.plm.2018.08.001

examinations of the circumstances under which the human perceptual system is able to adapt to novel stimuli, and what the constraints on this flexibility are. In this manuscript, I outline three types of perceptual learning for speech: perceptual flexibility of existing speech sound categories, acquisition of novel speech sound categories, and adaptation to accented speakers. These types of perceptual learning demonstrate some important commonalities, suggesting foundational factors of perceptual flexibility more broadly. Further, the cases in which perceptual learning does not occur, or is disrupted, suggest some important limitations to the perceptual system.

Perception of speech sounds is characterized by categorical perception (see Repp, 1984 for a review). There are two critical behaviors linked with categorical perception of speech sounds. First, individuals tend to show a rather sharp categorization function for speech sounds. Rather than perceiving an acoustic continuum in a continuous fashion, listeners perceive such a continuum as being instances of one category and then sharply shift their perception to that of another category. Second, listeners are able to differentiate between speech sounds that cross a category boundary (i.e., the point of the shift in identification described above); however, they are typically less capable of differentiating between sounds within a single category. These sound categories are typically called "phonemes". That is, phonemes are the sounds that a language differentiates from one another. A phoneme can have multiple phonetic (i.e., acoustic) realizations. For example, the 'k' sound at the beginning of the word 'kit' is produced slightly differently than the 'k' sound that appears after an 's' at the beginning of the word, as in 'skit'. These multiple phonetic realizations are referred to as "allophones".

This categorical perception is language specific. That is, the category boundaries that determine a listener's perception are dictated by the phonemic categories of a particular language. Listeners whose native language is Hindi, which distinguishes between dental and retroflex contrasts (i.e., /ɖ/ and /d/), will demonstrate categorical perception for those phonemes. However, native English speakers will not demonstrate categorical perception for the same sounds, as English does not differentiate between dental and retroflex articulations for stops. Similarly, listeners of a language like English will differentiate between the phonemes /r/ and /l/. However, native speakers of Japanese will not demonstrate this same behavior, because they do not have two distinct categories for these sounds.

It is important to note that in spite of this categorical perception, listeners are still sensitive to subcategorical differences within a phoneme category.

While listeners demonstrate categorical behavior in their clicking responses to minimal pairs (i.e., words that differ only in a single sound; 'palm' and 'bomb'), they demonstrate gradient behavior in their eye movements (e.g., McMurray, Tanenhaus, & Aslin, 2002; 2009). That is, they look longer at competitor words when the difference between the two words is relatively small compared to when the difference is larger. This graded within-category sensitivity suggests that listeners are sensitive to small within-category differences even if they do not demonstrate such sensitivity in all tasks. Still, it is critically important that a listener learns how to group sounds that vary into categories, such that they understand a sound that is produced by one individual should be interpreted as the same sound as one produced by another individual, even if the acoustic properties of these sounds are likely not identical. Development of this skill is critically important for successful functioning in a language.

Much is known about how our perception of speech sounds develops over time. Very young infants are able to differentiate between a wide range of sound contrasts; however, before a child's first birthday, the perceptual system begins to shift from a language-general system to a more language specific one (Werker & Tees, 1984). For example, 6–8 months old infants raised in an English speaking environment are able to differentiate between sounds in Hindi and Salish, languages they have not been exposed to. However, by the time the children are 10–12 months old, they are unable to differentiate between these sounds. This is not due to general maturation, as native Hindi and Salish infants are able to differentiate between these sounds even at 12 months of age. This is not to say that every contrast behaves similarly. Some contrasts, which all infants have difficulty differentiating between demonstrate language-specific improvement over time (Kuhl et al., 2006).

While language specific patterns are evident very early in life, the perceptual system must retain flexibility for a variety of purposes, including adaptation to novel talkers. Speech is characterized by a great deal of variability, or the lack of invariance (Blumstein & Stevens, 1981). That is, each time an individual produces a word or phoneme, they produce it in a variable way. The speech sound is not identical to previous productions of the same word or segment by other speakers or even by the same speaker. Therefore, a listener's perception must be flexible enough to accommodate novel speakers and novel productions by familiar speakers.

Importantly, though, this flexibility must also be limited. Were it not, the structure of categories would be so flexible that a category itself would be quite difficult to define. Categories are clearly important for speech perception, as described above because they allow listeners to generalize over variability from a number of sources. Further, it is likely that flexibility in the perceptual system for speech comes at some cognitive cost.

Therefore, in the present paper, I examine flexibility in the speech perception system and limitations of this flexibility in three cases of perceptual learning and adaptation. First, I address perceptual adaptation for existing speech categories. Substantial literature has suggested that listeners shift the boundaries of their speech categories as a function of their exposure to speech (e.g., Kraljic & Samuel, 2005; Norris, McQueen, & Cutler, 2003). This type of adaptation is presumably quite important for adaptation to other native speakers. However, this work has also demonstrated that there are some limitations to this adaptation. Second, I address acquisition of novel phonological categories. While listeners demonstrate language-specific phonological perception very early in childhood, some evidence has suggested that adults with very little experience with non-native languages are able acquire novel categories both in the lab and through naturalistic exposure (e.g., Goto, 1971; Logan, Lively, & Pisoni, 1991; Strange & Dittman, 1984). Similar to adaptation within an existing phonological category, there are cases in which this learning is disrupted, which might suggest important limitations to perceptual flexibility. Third, I address adaptation to accented speech. Accented speech provides a substantial challenge to listeners; however, listeners are able to adapt to unfamiliar speech, including non-native speech, with exposure to this speech. Taken together, these three types of perceptual adaptation for speech provide critical test-cases for the flexibility and limitations of flexibility of the perceptual system more broadly.

1. PERCEPTUAL FLEXIBILITY OF NATIVE PHONOLOGICAL CATEGORIES

As noted above, by the time children are two years old, their perception of speech sounds is quite similar to perception by adult listeners in the same language community. This is not to say, however, that a listener's perception of a particular set of categories does not involve any flexibility. Rather, the specific boundary between two sounds is quite flexible, and depends deeply on the context in which the sound is presented. The Ganong effect (Ganong, 1980) suggests that the perception of a particular sound depends, in part, on whether it is part of a real word or a non-word. On a continuum from /d/ to /t/, native English listeners perceive more of the continuum as /d/ when it is presented in a lexical frame of '?ash' where 'dash' is a real word and 'tash' is a non-word. The reverse is true when the sound is presented in an '?ask' frame, as 'task' is a real word and 'dask' is a non-word. In this condition, listeners here more of the continuum as /t/ instead of /d/. Similar flexibility is seen in compensation for co-articulation (Elman & McClelland, 1988; Mann & Repp, 1981) and selective adaptation (Eimas & Corbit, 1973; Samuel, 1986; Vroomen, van Linden, De Gelder, & Bertleson, 2007). Co-articulation is a property of speech that results in speech sounds being produced differently depending on their context. For example, 'green boat' is often produced as 'greem boat', with the 'n' sound being produced as an 'm' sound which is more similar the 'b' in boat than 'n' is. Listeners compensate for this co-articulation when perceiving speech, allowing them to "repair" the 'm' sound to the intended 'n'. Similarly, selective adaptation can also shift perception a single speech sound. If a listener is exposed to repeated productions of 'ba', they will report fewer instances of 'ba' on a 'ba'-'da' continuum, even if the 'b' sound is not ambiguous. Taken together, these results demonstrate that a single speech sound can be perceived differently depending on the context in which the listener hears it.

The aforementioned shifts in perception presumably reflect the knowledge of the speaker about language-specific phonetic patterns which are integrated with lexical and syntactic knowledge. Interestingly, shifts in perception of speech sounds also occur with short periods of exposure, that are less clearly related to language-specific patterns and perhaps more closely attributable to talker-specific patterns. This shift is often described as "perceptual learning" (Norris et al., 2003), "perceptual recalibration" (Bertleson, Vroomen, & De Gelder, 2003), or "phonetic retuning" (Samuel, 2011). Of particular interest here is the perceptual learning that emerges as a function of lexical context, or lexically-guided phonetic recalibration.

Norris et al. (2003) were the first to investigate this phenomenon. Listeners were exposed to a single speaker, who produced tokens that were midway between a clear production of [f] and a clear production of [s], henceforth [?]. This ambiguous token was presented to listeners in one of three conditions. Listeners in the non-word condition heard the ambiguous token at the end of non-words in Dutch. Whether the word ended in [f] or [s], it did not make a real Dutch word. Listeners in the [f]-trained condition heard the ambiguous token at the end of Dutch words which end in [f] (e.g., *witlof* — chicory; *witlos* is not a Dutch word). Listeners in the [s]-trained condition heard the ambiguous token at the end of Dutch words which end in [s] (e.g., *naadlbos* – pine forest; *naadlbof* is not a Dutch word). After a brief exposure period, listeners were tested on an [ɛf]-[ɛs] continuum. Compared to listeners in the non-word condition, listeners in the [f]-trained condition showed a bias toward perceiving more of the continuum as [f]. Listeners in the [s]-trained condition similarly showed a bias, but in the opposite direction, perceiving more of the continuum as [s]. These findings suggest that within established phonological categories, listeners are able to adjust the boundaries of these categories as a function of experience.

Since this seminal finding, subsequent studies have examined how this perceptual learning emerges, and what constraints on this type of learning may exist. While the learning demonstrated above was triggered with lexical cues, non-lexical cues can trigger this learning as well, including visual cues (e.g., Bertleson et al., 2003) and orthographic cues (e.g., Mitterer & McQueen, 2009). This type of perceptual learning is remarkably robust. It generalizes to novel words (McQueen, Cutler, & Norris, 2006), novel syllable positions within words (Jesse & McQueen, 2011), and lasts for a substantial period of time beyond initial exposure (Eisner & McQueen, 2006). Further, this perceptual adjustment remains even when listeners have been exposed to a number of canonical tokens from a different speaker between the first and second post-test periods (Kraljic & Samuel, 2005), suggesting some amount of speaker specificity to this learning.

However, the results examining speaker specificity of learning have been mixed. Eisner and McQueen (2006) and Kraljic and Samuel (2005) found some evidence for speaker specificity of perceptual adjustment of category boundaries of fricatives. Interestingly, Kraljic and Samuel (2006) found no such speaker specificity for perceptual adjustment of stop category boundaries. In fact, they found that not only did perceptual learning generalize to novel talkers, but it also generalized to a novel contrast (i.e., another stop consonant pair that shared the voicing feature participants were exposed to during training; see also Kraljic & Samuel, 2007).

The issue of whether this type of perceptual learning is talker specific is made more complicated by the fact that listeners do not demonstrate this type of learning when it can be attributed to an external factor. Kraljic, Samuel, and Brennan (2008b) demonstrated that when a speaker was shown with a pen in their mouth during production of the ambiguous segments, listeners did not adjust the perceptual boundary of their phoneme categories. Further, this adjustment does not take place when the ambiguous production is attributable to features of a particular dialect (Kraljic, Brennan, & Samuel, 2008a).

Another critical question is whether listeners are truly adjusting their phoneme category boundaries, or are they instead showing decision biases that behave similarly to a true shift in category boundary? Clarke-Davidson, Luce, and Sawusch (2008) suggest the former, since exposure not only shifts categorization performance, but also discrimination abilities. Myers and Mesite (2014) used neuroimaging to ask a similar question. They demonstrate that neural activation initially emerges in right frontal and medial temporal areas typically activated during perceptual adjustments. This result could initially be interpreted as evidence for a shift in decision bias, since these areas process more low-level information. Over the course of the experiment, however, this activation shifts to left superior temporal areas, which are traditionally implicated in top-down processing. This shift suggests that listeners may be truly integrating category information in perception, and is interpreted as a shift toward phonemic processing over time. Using eye-tracking, Mitterer and Reinisch (2013) demonstrate a similar result. Listeners demonstrate an effect of phonetic recalibration very early in perception (200 msec after the onset of the target speech sound — as early as a listener could plan and implement an eye movement), suggesting that this effect is perceptual, not post-perceptual.

Given these findings, a natural question is what, exactly, is being learned by listeners? One line of research has asked what properties of the sounds are being learned. Mitterer, Scharenborg, and McQueen (2013) demonstrated that perceptual learning operates over position-specific allophones. That is, learning for word-final /r/ and /l/ in Dutch did not generalize to other, context-dependent allophones of these phonemes. Reinisch, Wozny, Mitterer, and Holt (2014) extend this finding, demonstrating that visually-guided phonetic recalibration occurs over context-sensitive allophones, not phonemes or acoustic cues more generally speaking. Interestingly, however, it does appear that phonetic recalibration can operate over phonemes across languages (see Reinisch, Weber, & Mitterer, 2013), suggesting that in some ways this re-calibration may be more general. More research, therefore, is needed to determine what, exactly, is being learned about speech sounds during phonetic recalibration.

However, interestingly, Drouin, Theodore, and Myers (2016) suggest that perceptual learning may not be identical for both categories at the boundary. Specifically, they demonstrate, using a goodness judgment task, that internal restructuring of the category occurred only for /ʃ/ and not

/s/ when listeners were exposed to ambiguous tokens between the two phonemes. While such an asymmetry may be surprising, other studies (e.g., Eisner & McQueen, 2006; Zhang & Samuel, 2013) demonstrate similar asymmetries, suggesting that acoustic properties of phonemes, including the variation listeners are typically exposed to within a particular may differ and influence this type of adaptation.

Taken together, these results demonstrate that, while listeners establish perceptual categories for speech sounds early on, these categories (or at least the boundaries within categories) are quite flexible. Listeners are able to adjust the boundaries of these categories with relatively little exposure, and these changes last beyond the initial period of exposure. Some evidence suggests that these shifts are truly shifts in phonological processing, not simply post-perceptual-decision oriented processes. However, in spite of this flexibility, there are also some limits to this flexibility. Specifically, listeners do not adjust the boundaries of these categories for all talkers in all cases. Taken together, these data suggest that perceptual categories are flexible later in life.

2. ACQUISITION OF NOVEL PHONOLOGICAL CATEGORIES

The learning described above primarily examines shifts within already established phonological categories. That is, the listener was presumed to have already acquired both categories of which the ambiguous sound could be a member. While this skill is critically important for adaptation to novel talkers, it is not the only case in which listeners' speech sound categories must be flexible. While listeners begin to show language-specific perception patterns very early in life, the acquisition of another language critically requires the listener to acquire novel sound categories. That is, as a later learner of a language, the listener must shift their perceptual system from one which is exquisitely tuned to one language to a system which is tuned to two languages. This flexibility in acquisition of novel categories also demonstrates some important limitations, which help us understand the processes underlying speech perception and the flexibility of the perceptual system more broadly. A wide body of work has examined perception and acquisition of non-native speech sound categories in adulthood. Below, I review these studies with special attention to both the flexibility and limitations on this type of learning.

It has long been known that non-native speakers have difficulties acquiring novel phonological categories. In speech production, this is most clear in the fact that most non-native speakers retain a non-native accent in the target language even when they are relatively proficient in the language. Evidence also suggested that non-native speakers faced similar challenges in speech perception as well (Briere, 1966; Goto, 1971). This was often attributed to differences in exposure. Many early studies demonstrated that naïve listeners did not show native-like perception of non-native categories (e.g. Beddor & Strange, 1982), and that listeners from different language backgrounds divide continua of speech sounds differently as a function of their native language (e.g., Keating, Mikós, & Ganong, 1981; Lisker & Abramson, 1964). In the early 1980s, researchers began to investigate whether listeners could better perceive new phonological categories after a brief period of in-lab exposure, initiating a flurry of research aimed at investigating both naïve perception of non-native contrasts and studies of acquisition of these contrasts after periods of training.

As noted above, our native language clearly shapes how we perceive speech sounds. Most research has centered around a primary finding: Individuals who do not have experience with, or exposure to, a set of sounds will find differentiating between those sounds to be hugely challenging. This has been demonstrated for many speech sounds (e.g., English /r/ and /l/ for Japanese listeners: Hindi dental-retroflex stops and Nthlakampx velar-uvular ejectives: Werker & Tees, 1984; Werker, Gilbert, Humphrey, & Tees, 1981). When a listener has tuned their perceptual system to their native language, they are able to differentiate among sounds from different native phonological categories. However, in doing so, they lose their ability to differentiate among non-native contrasts, as the variation between non-native categories is typically uninformative for perception in the listener's L1.

In spite of these robust findings, it is also the case that not all contrasts are equally challenging to all listeners. For example, native English listeners are quite good at differentiating among Zulu clicks, even though they have no exposure to these sounds in their everyday linguistic environment (Best, McRoberts, & Sithole, 1988). In fact, behavior by non-native listeners seems to vary quite widely. While some contrasts are very challenging for non-native listeners (e.g., English /r/ and /l/ for native Japanese listeners) and some are quite easy (e.g., Zulu clicks for native English listeners), other sound contrasts show intermediate behaviors (e.g., Best, McRoberts, & Goodell, 2001; Guion, Flege, Akahane-Yamada, & Pruitt, 2000; Polka, 1991; 1992). This pattern of results for naïve non-native listeners suggests that the perceptual system may be more flexible even without additional training. That is, the perceptual system of a naïve listener is language-specific in some ways, but this does not necessarily limit their perception.

A variety of theories have predicted which sound contrasts may be easier or more difficult for a naïve listener to perceive or for a non-native listener to acquire. Below, I summarize the most prominent theories: the Native Language Magnet Model (NLM; Kuhl, 1994), the Perceptual Assimilation Model (PAM Best, 1995; and PAM-L2; Best & Tyler, 2007), the Second Language Linguistic Perception Model (L2LP; Escudero, 2005), and the Speech Learning Model (SLM; Flege, 1995). Each of these models shares a critical component: They each predict that perception of particular sounds or sound contrasts by naïve or non-native listeners will vary as a function of the listener's native language and the relationship of the phonological system in the native language and the target language. How, specifically, this is implemented differs among each of these theories. While each theory focuses on the perception of non-native speech sounds, the specifics of their predictiosn differ. The NLM and the SLM focus primarily on perception of individual speech sounds, predicting how individual sounds may be perceived. The PAM and the L2LP Model focus, instead, on pairs of speech sounds, predicting an individual's ability to differentiate between those sounds.

The Native Language Magnet Model has examined a "perceptual magnet effect." This effect is typified by warped perception of an acoustic continuum. That is, they do not easily differentiate between tokens close to a "good" exemplar of a particular category but are better at differentiating between tokens near "poor" exemplars of this category. Kuhl and colleagues have demonstrated that this holds true within listener's native languages (i.e., perception of tokens within the /i/ category in English are perceived in a manner which is congruent with this perceptual magnet effect; Iverson & Kuhl, 1995; Kuhl, 1991; Kuhl, Williams, Lacerda, Stevens, & Lindblom, 1992). This seems to be true only for contrasts within the listener's native language. Kuhl et al. (1992) demonstrated that English speaking infants showed decreased sensitivity with an /i/ category but Swedish infants demonstrated similar reduced sensitivity within an /y/ category. Iverson & Kuhl (1996) demonstrate that not only is perception within a category influenced by this perceptual magnet effect, but it also impacts perception across categories (see Kuhl et al., 2008 for an extension of the model: NLM-e). While the NLM makes some accurate predictions about why contrasts are quite difficult for L2 listeners to perceive, it does not make clear predictions about why some contrasts are relatively easy for non-native listeners. The other aforementioned theories do make clear predictions about this, however.

The Speech Learning Model (SLM), for example, predicts that learners will show better production of "new" phonemes (e.g., French /y/ for native American English speakers) than "similar" phonemes (e.g., French /u/ for native American English speakers (Flege, 1987). This differences in performance is hypothesized to be driven by "equivalence classification." That is, when listening to French, native American English speakers perceive the French high front rounded vowel /y/ as being new, unlike any phonological category in English. Therefore, their perception of this phoneme, and their subsequent production is relatively good. However, these same listeners classify the French high back rounded vowel /u/ as being the same as the English back rounded vowel, even though production of the two vowels differs substantially. Given this equivalence classification, listeners are unable to perceive the phoneme correctly, and simply replace it with their native phoneme. SLM also asserts that a listener is not constrained by a critical period of acquisition. Instead, Flege predicts that age of acquisition will clearly impact learning in a linear way over time. This may, in fact, represent increased expertise in the learner's L1, which is a primary factor influencing perception and production of non-native speech sounds. The SLM proposes that L1 and L2 phonemes interact in a common space, and assimilation and dissimilation of phonemes between the two languages, as a function of their phonetic properties, will predict performance on any given set of phonemes. However, some researchers have argued that what will be assimilated or dissimilated is not well-predicted in this model.

The Perceptual Assimilation Model makes more fine-grained predictions about how speech sounds will be assimilated into the L1 phonological category space and makes specific predictions about how well pairs of non-native phonemes should be discriminated. For example, Best et al. (2001) lay out six possible assimilation types for pairs of non-native phonemes: two-category, where non-native phonemes are categorized as instances of two separate native phonological categories; single category assimilation, where both sounds are assimilated into a single native category; category goodness, where both categories are assimilated into a single native category, but one is viewed as a "better" exemplar of that category; uncategorized-categorized, where one sound is categorized into a native category, but the other is not; uncategorized—uncategorized, where neither sound is categorized into a native category; and non-assimilable, where neither sound is perceived as a speech sound. PAM provides predictions about how well (and how variable) individuals will perceive non-native sounds as a function of this categorization schema. For example, as mentioned above, native English listeners are quite good at discriminating Zulu clicks, in spite of a lack of experience with these sounds. Best and colleagues claim that this is because clicks are non-assimilable for native English listeners. Similarly, they predict that Zulu lateral fricatives, implosive stops, and ejective stops should each fall under a different type of classification, and that naïve listener performance can be predicted as a function of this classification. PAM was originally developed as a model for describing naïve perception of non-native speech sounds. However Best and Tyler (2007), extend PAM to second language learning, through the development of PAM-L2. In this model, predictions are made about the likelihood of forming a novel phonological category as a function of the perceptual phonetic similarity of a particular speech sound in the non-native language, as compared to the learner's native language.

Escudero et al. propose a computational model to predict performance by non-native learners, the Second language Linguistic Perception (L2LP model; Escudero, 2005; 2009). Like PAM, L2LP focuses primarily on sound contrasts instead of isolated sounds. Similar to both PAM and SLM, the initial perceptual state and developmental trajectory of the learner are highly influenced by the learner's native language. As in SLM, this model predicts that "new" and "similar" sounds will be perceived differently. However, rather than asserting that "similar" sounds will pose no challenge, L2LP predicts that adjustment of the phoneme boundaries required during acquisition of phonemes in an L2 (even if they are similar to the L1) will still provide substantial challenge for the learner; however this challenge will be smaller compared to the challenge of learning "new" phonemes. Some early results of this computational model demonstrated that the model did not closely predict human performance (e.g., Weiand, 2007). However, in a revision of the L2LP Model, van Leussen and Escudero (2015) demonstrate improved performance with the inclusion of lexical meaning in the model, something often not investigated in studies of acquisition of non-native phonemes, which is obviously crucial for real-world second language acquisition.

While substantial work has tried to predict circumstances under which learning would be more or less difficult for native speakers, other work has focused more on what particular training and learning circumstances might impact learning. That is, rather than focusing on properties of language, they focus on properties of learning. Below, I outline two primary types of investigations, the first about the role of variability during training and the second examining the role of sleep.

Researchers have long wondered how to define the "ideal" paradigm for training non-native speech sounds. Early investigations of perceptual learning of non-native speech sounds speculated that perhaps very distinct tokens are necessary initially in order to enhance the perceptibility of a contrast. However, Logan et al. (1991) demonstrated that variability in exposure correlated with enhanced generalization during a post-test. When listeners were exposed to multiple lexical items, they generalized to novel lexical items. Similarly, when listeners were exposed to multiple talkers during training, they generalized to novel talkers during test. This learning was found to be relatively long-lasting (Bradlow, Akahane-Yamada, Pisoni, & Tohkura, 1999), suggesting that the learning is relatively robust.

Some work has attempted to directly compare various training paradigms to determine which are most efficient for learners. Iverson, Hazan, and Bannister (2005) compared the high-variability phonetic training described above, which used all natural recordings, with three methods of alteration via signal processing: Enhancement, where the relevant cues were heightened; perceptual fading, where participants were exposed to the "best" tokens which then became more naturalistic and less distinct over the course of training; and secondary cue variation where the secondary cues to the contrast were varied over the course of training. All training types improved learning; however, because high variability phonetic training is seen as the most naturalistic, this has become the state of the art for the field.

However, a significant amount of recent research has suggested that increased variability may not be helpful for all listeners under all circumstances. For example, Perrachione, Lee, Ha, and Wong (2011) demonstrated that high-aptitude learners demonstrate benefits from high-variability phonetic training for Mandarin tone. However, individuals with low-aptitude performed less well with high-variability training than with low-variability training. It has been suggested this is perhaps in part because low-aptitude listeners may be less efficient at managing cognitive load (Antoniou & Wong, 2015). Further, varying multiple features of a stimulus (both relevant and irrelevant) disrupted perceptual learning (Antoniou & Wong, 2016). Fuhrmeister and Myers (2017) suggest that increasing variability before or after training can also disrupt learning. Therefore, the blanket assumption that variability improves perceptual learning of non-native speech sounds seems to be incorrect, given the complexity of these results.

An additional factor that may impact learning of this type is consolidation of learning during sleep. Previous work has suggested that sleep plays an important role in skill acquisition including both perceptual and motor learning. Fenn, Nusbaum, and Margoliash (2003) demonstrated consolidation of learning for phonological speech sounds. That is, participants who slept between training and post-test demonstrated more learning than participants who were awake for that period of time. In their review of this phenomenon, Earle and Myers (2013) suggest that sleep-mediated consolidation may be as critical for speech learning as it is for other types of skill learning. Their subsequent work has examined some factors of sleep in more detail, and has demonstrated that consolidation improves generalization to new talkers (Earle & Myers, 2015). Recently, they have examined individuals with Specific Language Impairment, a disordered marked by impoverished representations of speech sounds and have demonstrated that these listeners do not demonstrate consolidation overnight, unlike control participants, further implicating the critical role of sleep and consolidation in speech sound learning (Earle, Landi, & Myers, 2018).

In conclusion, the work described in the current section has demonstrated that individuals are able to acquire novel speech categories from non-native languages with exposure in the lab or in everyday life. However, while speech sound categories remain somewhat flexible, there are also limitations on this learning, including the type of training in which learners are exposed to the new categories. Finally, some recent work has highlighted some of the general learning factors that may directly impact perceptual learning of non-native speech sounds.

3. ADAPTATION TO FOREIGN ACCENTED SPEECH

A third type of learning bridges the two aforementioned types. While shifting single phonological categories is necessary to tune to a novel native talker, and acquisition of novel categories is necessary for acquisition of a non-native language, adaptation to the speech of non-native talkers is also a critically important and challenging skill in our increasingly global society. This adaptation is likely to require some of the same perceptual recalibration skills discussed above, but adaptation to novel non-native speakers likely requires recalibration of a wide number of phonological categories and the ability to interpret non-native productions of phonemes. Here, we review some of the evidence about the flexibility of listeners when perceiving non-native speech, how this adaptation differs from adaptation to other types of unfamiliar speech or challenging listening conditions, and what the limits are on this adaptation.

Before examining how native listeners adapt to nonnative speech, it is important to recognize that listening to non-native speech often provides a substantial challenge to listeners. Native listeners often find non-native speech harder to understand than native speech both in terms of objective measures (e.g., intelligibility, typically defined as how many words a listener is able to transcribe) and subjective measures (e.g., comprehensibility, typically defined as how difficult a listener find the speech to understand (Munro & Derwing, 1995). Recent work has demonstrated that the challenges underlying perception of non-native speech are in some ways similar to those faced by listeners when listening to other types of unfamiliar speech. For example, Bent, Baese-Berk, Borrie, and McKee (2016) demonstrated that perception of non-native speech (i.e., Spanish-accented English) correlates with an individual's ability to perceive both an unfamiliar native accent (i.e., Irish English) and the speech of dysarthric speakers. Perception of all three varieties was predicted by an individual's vocabulary size as measured by the Peabody Picture Vocabulary Test. However, while there are some similarities, it is not the case that the skills that underlie non-native speech perception help listeners understand speech in all challenging circumstances. McLaughlin, Baese-Berk, Bent, Borrie, and Van Engen (2018) examined perception of non-native speech and speech in noise. They demonstrate that listeners who succeed in perceiving one type of speech (i.e., non-native speech or speech in noise) are successful at understanding that speech in multiple circumstances; however, those listeners are not necessarily successful at understanding the other type of challenging listening condition. Further, while vocabulary predicted performance on all types of adverse listening conditions, working memory only correlated with perception of non-native speech, suggesting some important differences in the types of mechanisms underlying perception of non-native speech compared to other adverse listening situations.

Further, non-native listeners do not show similar challenges to native listeners in their perception of other non-native speakers. Bent and Bradlow (2003) demonstrated the "interlanguage speech intelligibility benefit" (ISIB). That is, non-native listeners were not only quite successful at understanding the non-native speech of those that shared their language background, they were also more successful at perceiving non-native speech from talkers who did not share their language background. In fact, for many listeners, non-native speech was more intelligible than native speech. However, subsequent work has suggested that this may be limited to certain proficiency levels of speakers. Hayes-Harb, Smith, Bent, and Bradlow (2008) demonstrated that the ISIB only holds for listeners who are relatively low proficiency, and only occurs when listening to relatively low proficiency speech. Xie and Fowler (2013) demonstrated a similar finding, suggesting that proficiency, and perhaps even experience in an English speaking country can modulate this benefit.

While baseline perception of non-native speech demonstrates some of the challenges listeners face, like perception of non-native phonemes, perception of unfamiliar non-native accents can improve with relatively little exposure. Clarke and Garrett (2004) demonstrated that a minute of exposure could increase processing speed for perception of non-native speech. Sidaras, Alexander, and Nygaard (2009) demonstrated a similar improvement in intelligibility of Spanish-accented English after very brief exposure.

While this learning is quite rapid, it does demonstrate some limitations. Specifically, Bradlow and Bent (2008) ask whether the type of exposure listeners received influenced their ability to generalize to novel talkers and novel accents. They examined exposure to native Mandarin speakers in five training conditions: a single talker who was also the talker used at test, a single talker who was different from the talker used at test, and five different talkers, none of whom was the test talker. These listeners were compared to two control groups. The first received no training, and the second received training on the task (i.e., sentence transcription), but were exposed to native English speakers, not native Mandarin speakers. At test, listeners were exposed to a native Mandarin talker and a native Slovakian talker. Listeners exposed to a single talker demonstrate talker-specific adaptation to Mandarin speech. That is, listeners who heard the test talker during training perform very well, while listeners who heard a different single talker at test and training do not perform better than the individuals who were trained on the task but with native English speakers. However, they also demonstrate that with increased variation in number of talkers, listeners demonstrate talker-independent adaptation for Mandarin. That is, listeners exposed to multiple talkers during training were able to generalize to a novel talker from the same language background at test. Interestingly, this adaptation was accent-dependent. That is, no group performed better than the task-trained controls on the speech of an unfamiliar accent (i.e., Slovakian). This suggests some important limitations to adaptation to non-native speech that appear to be driven by variation in exposure.

Support for this finding was strengthened by results of a study conducted by Baese-Berk, Baese-Berk, Bradlow, & Wright (2013). Using the same training paradigm as Bradlow and Bent (2008), they ask whether increasing variability further may result in accent-independent adaptation. That is, listeners were exposed to five talkers as in the multiple talker condition described above; however, instead of exposure to only native Mandarin speakers, listeners heard speakers from five different native language backgrounds (Mandarin, Hindi, Thai, Romanian, & Korean). Listeners generalized to a novel talker from a background they were exposed to during training (i.e., Mandarin) and a talker from a background they were not exposed to during training (i.e., Slovakian). This was interpreted as further evidence that variation in the exposure set impacted adaptation to non-native speech and to what extent that adaptation generalized to novel instances.

Interestingly, this effect also emerges for older adults with and without hearing loss. That is, older adults demonstrate accent-independent adaptation after exposure to multiple accents to a similar extent as younger listeners (Bieber & Gordon-Salant, 2017). The effect of accent general adaptation has been replicated, not only within the lab, but also with real world experience. Laturnus (2018) demonstrated that individuals who have more (and more varied) experience with unfamiliar speech in their everyday lives also demonstrate improved perception of non-native speech. Further, real-world experience can improve perception of specific accents. Witteman, Weber, and McQueen (2013) recruited participants with varying exposure to German-accented Dutch, and demonstrated that listeners with more naturalistic experience demonstrated improved performance when listening to stronger accents. However, in other work looking at perception of non-native speech by English as a Second Language teachers (who arguably have substantial experience with non-native speech) has shown mixed results, with some findings suggesting that they more accurately understand non-native speakers (Gass & Varonis, 1984; Kennedy & Trofimovich, 2008), other work has suggested that experience does not drive improvements in intelligibility scores, but listener attitudes toward non-native speech may (Sheppard, Elliott, & Baese-Berk, 2017). This suggests a complex interplay between exposure and possible social factors. Interestingly, this apparent "exposure-driven" adaptation is subject to some other limits as well.

Laturnus (2018) demonstrated that exposure to five unfamiliar native varieties did not facilitate comprehension of unfamiliar non-native speech. This suggests that the adaptation may be specific to the source of variation. That is, individuals may perceive something fundamentally different about the speech of native speakers and non-native speakers. Interestingly, exposure to multiple native accents does help infants generalize to novel accents (Potter & Saffran, 2017), especially for younger infants who do not show such improvement after exposure to a single accent.

However, it is still unclear what listeners are adapting to. It is possible that there are some "accent-general" properties of non-native accented speech. That is, talkers from a variety of backgrounds may have similar strategies for "solving" the challenges they face when speaking in an L2. Therefore, a listener may be adapting to features that are shared across accented talkers, without reference to language background. An alternate hypothesis, however, is that listeners are simply expanding their categories for a variety of speech sounds (see, e.g., (Schmale, Cristia, & Seidl, 2011). These hypotheses have not directly been compared and should be the target of future research.

Some work has attempted to investigate what factors might influence this sort of adaptation. Xie and Myers (2017) demonstrated that this adaptation only occurs if the generalization talker is similar to the exposure talker or talkers. McCullough and Clopper (2016) examined the phonetic properties of speakers that were perceived to be similar to one another. Their findings suggest that deviations from native talker voice onset time, vowel duration, and spectral qualities of vowels predict the classification of talkers into categories that are perceptually more similar to one another. As in the case of non-native speech sounds, consolidation during sleep improves generalizations to novel talkers (Xie, Earle, & Myers, 2018).

Taken together, the work reviewed here suggests that, although non-native speech is initially more challenging for native listeners to understand, listeners are able to adapt to the speech over time. This adaptation likely requires some combination of the skills seen in the previous two sections, as it requires listeners to both shift existing category boundaries and interpret sounds which do not exist in their native language. However, substantial further work is needed to better understand exactly what listeners are adapting to and what drives the limits on this adaptation.

4. CONCLUSIONS AND SPECULATION

The preceding sections have outline three distinct but related types of perceptual learning for speech: perceptual flexibility for native phonological categories, perceptual learning of non-native phonological categories and perceptual adaptation to non-native speech. Research regarding development of our phonological systems demonstrates that our perception of phonemes becomes language-specific within the first year of life. The studies described here, however, demonstrate a remarkable flexibility in these systems that remains well into adulthood. This flexibility can inform our understanding of the flexibility of our perceptual systems more broadly. However, in addition to this flexibility, the work reviewed here also speaks to the limits of perceptual learning. Below, I outline some important aspects underlying perceptual learning for speech in three conditions.

This balance of flexibility and limitations to learning underlie a critically important issue. The perceptual system must be flexible enough to adapt to new input but must simultaneously be stable enough that each experience does not have the power to fundamentally reorganize the system. Samuel (2011) outlines this issue. One fundamental challenge for the field is to determine under what circumstances the necessity of perceptual flexibility outweighs the costs associated with it. That is, the integrity of the perceptual system must be maintained, therefore change to the system is costly. At the same time, change must occur at some points in order to adapt to an ever-changing environment. Samuel and Kraljic (2009) propose the "Conservative Adjustment/Restructuring Principle" (CA/RP), which proposes that an optimal perceptual system must have a relatively high threshold for change. That is, it should only be modified by very strong evidence that these modifications will aid in future perceptual instances more than they will cause disruption to the integrity of the system. A similar idea was formalized by Kleinschmidt and Jaeger (2015). In their "Ideal Adapter Framework", a listener is modeled as using Bayesian belief updating to adapt to novel stimuli in their input. They describe this process as "Recognize the familiar, generalize to the similar". That is, listeners use their prior knowledge to adapt to similar features they encounter in their input. Their beliefs are constantly updated as a function of their exposure.

However, a fundamental challenge still persists: When will listeners use the flexibility of their perceptual system to adapt to novel instances and when will novel experiences not be integrated into the system. While the work presented here suggests some potential limitations on adaptation in each condition, it is difficult to predict, specifically, when learning will and will not occur in each circumstance. That is, while the balance of flexibility and stability is clearly manifested in the system, and while some researchers have attempted to formally model the cases in which learning has been demonstrated to occur, it is quite challenging to predict how humans will behave in other circumstances.

The aforementioned balance of flexibility and stability is made clearer when one considers how perceptual learning for speech may or may not transfer to changes in speech production. While stability is important for speech perception, it is likely even more critical for speech production. A listener must be able to adapt to each novel speaker with whom they interact, but they are less likely to want to adapt specific phonetic properties of their speech in each circumstance. While it is clear that talkers do shift their productions on a more global scale (e.g., style-shifting), it is less clear that fine-grained phonetic details, such as those addressed in the present paper, shift as a function of interlocuter. Fine-grained phonetic details have been shown to change in response to many features (e.g., lexical properties of a word; Baese-Berk & Goldrick, 2009; Fricke, Baese-Berk, & Goldrick, 2016; Gahl, 2008). However, little evidence has suggested that these factors shift purely as a function of interlocuter. Further, it is clear that production in a non-native language changes very slowly, as most individuals who learn a language later in life speak with a noticeable non-native accent.

In addition to accounting for non-native accents, this balance of flexibility and stability may explain some recent dissociations between perceptual learning and production learning for speech. Many studies have demonstrated mixed results regarding transfer of perceptual learning, especially for non-native categories, to production. For example, Bradlow, Pisoni, Akahane-Yamada, and Tohkura (1997) demonstrated that as a group native Japanese participants improved in production after perception training on English /r/ and /l/. This general improvement was not seen for each individual participant. In fact, some participants demonstrated robust perceptual learning with no production improvement, some demonstrated robust improvement in both domains, and yet other participants improved significantly in production with no significant improvement in perception. The reasons behind these mixed results have been unclear since most theories of acquisition of L2 phonological categories for speech predict that perception and production learning should be very tightly linked. Some intriguing recent evidence has suggested that, under some circumstances, production can actually disrupt perceptual learning (e.g., Baese-Berk & Samuel, 2016; Leach & Samuel, 2007). While the source of this disruption is still being investigated, it is possible that one factor influencing this lack of learning in perception is that in producing tokens, the listener is weighing stability of the perceptual system more strongly than flexibility. Some new evidence has, in fact, suggested that while some individuals demonstrate learning in production without learning in perception, this may actually be limited to cases of repetition, which do not need to tap into phonological categories in the same way as naming or spontaneous production do. That is, while some participants who do not learn in perception do show the ability to improve their productions from pre-to post-test when repeating tokens, only participants who learn in perception make changes when asked to produce tokens spontaneously without repetition (Baese-Berk, 2010). This suggests that perhaps learning differs in perception and production, and one factor modulating this learning is the balance of flexibility and stability described above.

In addition to the fundamental challenge outlined above, it is also important to consider what factors influence an individuals' ability to learn, primary among them the role of variability during learning. In the studies described above variability is typically seen as a factor that improves learning and increases the likelihood of generalization to novel instances. The Ideal Adapter Framework (Kleinschmidt & Jaeger, 2015) predicts that variability should, in fact, improve learning and generalization under many circumstances. However, it is critically important to note that variability does not always improve learning for speech and language. In addition to the work described above, Joe Barcroft and colleagues have directly investigated the circumstances under which vocabulary learning occurs or does not occur as a function of the variability included in the training set. Their work has suggested that, as in the case of learning novel phonetic categories, under some circumstances variability can improve learning and in others it can cause a disruption. Similarly, additional work has suggested that alternating tasks during training on each trial (i.e., perceiving a token and then repeating it) can cause disruptions to perceptual learning (Baese-Berk & Samuel, 2016; Leach & Samuel, 2007), but alternating between active and passive stimulus exposures during training can actually enhance learning (Wright, Baese-Berk, Marrone, & Bradlow, 2015). Clearly, additional work on the impact of variability for perceptual learning of speech is needed to understand the parameters and circumstances under which variability operates as a positive or a negative influence on learning.

One note about the work described here is the importance of including both native and non-native speakers and listeners in our understanding of perceptual learning. In an increasingly global society, communication among speakers who do not share a native language background is frequent. Basing our understanding only on native talkers and listeners excludes many, and perhaps even most, speakers of English and limits our understanding of communication. While the second set of data discussed above, focusing on acquisition of non-native phonological categories, moves away from the perspective of only examining native speakers, it does put a large burden on the non-native listener for conversation. Therefore, the third section, designed to examine how native listeners can improve their ability to understand non-native speakers provides an additional component to the research program laid out here, allowing for all sides of the native and non-native communication chain to be represented in this work.

In some recent work, we have argued that it is critically important to examine all sides of this communication chain in order to truly understand speech perception and perceptual learning (McLaughlin & Baese-Berk, 2018). Traditionally, research programs in psycholinguistics have focused on a monolingual speaker-listener as the ideal. However, this approach has resulted in relatively poor understanding of non-native speakers, in terms of both their own language learning but also how other listeners adapt to their speech. By fully investigating all sides of the communication chain, including both native and non-native speakers and listeners, we can begin to investigate how natural communication occurs, and can use these insights to inform our understanding of perception more broadly.

In conclusion, the work described above demonstrates a simultaneous flexibility and stability for speech perception. These features manifest in three types of perceptual learning: flexibility within phonological categories in a listener's native language, acquisition of novel phonological categories from a listener's non-native language, and adaptation to unfamiliar speech from non-native speakers, which requires adjustment to multiple categories simultaneously. The similarities of these types of learning are clear. Exposure provides listeners with evidence for adaptation. Given sufficient evidence, listeners shift their phonological boundaries and/or acquire novel categories. However, if listeners receive enough evidence that the novel stimuli can be attributed to some external factor (e.g., a pen in a listener's mouth), this adaptation does not occur. Additionally, in most cases described above, listeners benefit from variability. That is, a more variable input allows listeners to generalize to novel instances of a similar stimulus. Taken together, these results demonstrate that an assumption that the phonological system becomes fixed very early in life is incorrect. Instead, the system becomes one which is flexibly tuned to a particular language. Given sufficient exposure, that tuning can be adjusted both within a language and across novel languages.

This work has broad implications for both how we understand speech perception and for our understanding of the perceptual system more broadly speaking. Speech is one of the clearest instances that requires both stability in the perceptual system to tune to language-specific properties of the input and flexibility which allows a listener to interpret new instances from a variety of talkers. The fundamental problem in speech perception is often described as the "lack of invariance" problem. That is, speech input is hugely variable, both within and across talkers, so how listeners generalize to novel instances is not clear. While this is certainly true, the work reviewed here demonstrates that listeners can generalize to novel instances, even when those instances deviate substantially from a listener's typical experience. While the three types of perceptual learning described here may initially seem disparate, taken together they can help us better understand the functionality of the speech perception system and understand it's remarkable flexibility, even when the system is typically perceived as being relatively stable. Finally, the work reviewed here allows us to examine real world linguistic communication between speakers and listeners who may not share linguistic backgrounds, allowing for an ecological validity to the perception problems being examined.

REFERENCES

Antoniou, M., & Wong, P. C. M. (2015). Poor phonetic perceivers are affected by cognitive load when resolving talker variability. *Journal of the Acoustical Society of America, 138*(2), 571–574.

Antoniou, M., & Wong, P. C. M. (2016). Varying irrelevant phonetic features hinders learning of the feature being trained. *Journal of the Acoustical Society of America, 139*(1), 271–278. https://doi.org/10.1121/1.4939736.

Baese-Berk, M. M. (2010). *An examination of the relationship between perception and production* (Unpublished doctoral dissertation). Evanston, IL: Northwestern University.

Baese-Berk, M. M., Bradlow, A. R., & Wright, B. A. (2013). Accent-independent adaptation to foreign accented speech. *Journal of the Acoustical Society of America, 133*(3), EL174–EL180. https://doi.org/10.1121/1.4789864.

Baese-Berk, M., & Goldrick, M. (2009). Mechanisms of interaction in speech production. *Language & Cognitive Processes, 24*(4), 527–554. https://doi.org/10.1080/01690960802299378.

Baese-Berk, M. M., & Samuel, A. G. (2016). Listeners beware: Speech production may be bad for learning speech sounds. *Journal of Memory and Language, 89*, 23–36. https://doi.org/10.1016/j.jml.2015.10.008.

Beddor, P. S., & Strange, W. (1982). Cross-language study of perception of the oral–nasal distinction. *Journal of the Acoustical Society of America, 71*(6), 1551–1561.

Bent, T., Baese-Berk, M., Borrie, S. A., & McKee, M. (2016). Individual differences in the perception of regional, nonnative, and disordered speech varieties. *Journal of the Acoustical Society of America, 140*(5), 3775–3786. https://doi.org/10.1121/1.4966677.

Bent, T., & Bradlow, A. R. (2003). The interlanguage speech intelligibility benefit. *Journal of the Acoustical Society of America, 114*(3), 1600–1610. https://doi.org/10.1121/1.1603234.

Bertleson, P., Vroomen, J., & De Gelder, B. (2003). Visual recalibration of auditory speech identification: A McGurk aftereffect. *Psychological Science, 14*(6), 592–597.

Best, C. T. (1995). A direct realist view of cross-language speech perception. In W. Strange (Ed.), *Speech perception and linguistic experience: Issues in cross-language research* (pp. 171–204). MD: Timonium.

Best, C. T., McRoberts, G. W., & Goodell, E. (2001). Discrimination of non-native consonant contrasts varying in perceptual assimilation to the listener's native phonological system. *Journal of the Acoustical Society of America, 109*(2), 775–794. https://doi.org/10.1121/1.1332378.

Best, C. T., McRoberts, G. W., & Sithole, N. (1988). Examination of perceptual reorganization for nonnative speech contrasts: Zulu click discrimination by English-speaking adults and infants. *Journal of Experimental Psychology: Human Perception and Performance, 14*, 345–360.

Best, C. T., & Tyler, M. D. (2007). Nonnative and second-language speech perception: Commonalities and complementaries. In O. S. Bohn, & M. Munro (Eds.), *Second-language speech learning: The role of language experience in speech perception and production: A Festschrift in Honour of James E. Flege* (pp. 13–34). Amsterdam: John Benjamins.

Bieber, R. E., & Gordon-Salant, S. (2017). Adaptation to novel foreign-accented speech and retention of benefit following training: Influence of aging and hearing loss. *Journal of the Acoustical Society of America, 141*(4), 2800–2811.

Blumstein, S. E., & Stevens, K. N. (1981). Phonetic features and acoustic invariance in speech. *Cognition, 10*(1–3), 25–32.

Bradlow, A. R., Akahane-Yamada, R., Pisoni, D. B., & Tohkura, Y. (1999). Training Japanese listeners to identify English/r/and/l: Long-term retention of learning in perception and production. *Perception & Psychophysics, 61*(5), 977–985.

Bradlow, A. R., & Bent, T. (2008). Perceptual adaptation to non-native speech. *Cognition, 106*(2), 707–729. https://doi.org/10.1016/j.cognition.2007.04.005.

Bradlow, A. R., Pisoni, D., Akahane-Yamada, R., & Tohkura, Y. (1997). Training Japanese listeners to identify English/r/and IV: IV. Some effects of perceptual learning on speech production. *The Journal of the Acoustical Society of America, 101*(4), 2299–2310.

Briere, E. J. (1966). An investigation of phonological interference. *Language, 42*(4), 768–796.

Clarke-Davidson, C., Luce, P. A., & Sawusch, J. R. (2008). Does perceptual learning in speech reflect changes in phonetic category representation or decision bias? *Perception & Psychophysics, 70*(4), 604.

Clarke, C. M., & Garrett, M. F. (2004). Rapid adaptation to foreign-accented English. *Journal of the Acoustical Society of America, 116*(6), 3647–3658. https://doi.org/10.1121/1.1815131.

Drouin, J. R., Theodore, R. M., & Myers, E. B. (2016). Lexically guided perceptual tuning of internal phonetic category structure. *Journal of the Acoustical Society of America, 140*(4), EL307–EL313.

Earle, F. S., Landi, N., & Myers, E. B. (2018). Adults with Specific Language Impairment fail to consolidate speech sounds during sleep. *Neuroscience Letters, 666*, 58–63.

Earle, F. S., & Myers, E. B. (2013). Building phonetic categories: An argument for the role of sleep. *Frontiers in Psychology*.
Earle, F. S., & Myers, E. B. (2015). Overnight consolidation promotes generalization across talkers in the identification of nonnative speech sounds. *Journal of the Acoustical Society of America, 137*(1), EL91–EL97. https://doi.org/10.1121/1.4903918.
Eimas, P. D., & Corbit, J. D. (1973). Selective adaptation of linguistic feature detectors. *Cognitive Psychology, 4*, 99–109.
Eisner, F., & McQueen, J. M. (2006). Perceptual learning in speech: Stability over time. *Journal of the Acoustical Society of America, 119*(4), 1950–1953. https://doi.org/10.1121/1.2178721.
Elman, J. L., & McClelland, J. L. (1988). Cognitive penetration of the mechanisms of perception: Compensation for coarticulation of lexically restored phonemes. *Journal of Memory and Language, 27*(2), 143–165. https://doi.org/10.1016/0749-596X(88)90071-X.
Escudero, P. (2005). *Linguistic perception and second language acquisition: Explaining the attainment of optimal phonological categorization*. Utrecht University LOT.
Escudero, P. (2009). Linguistic perception of "similar" L2 sounds. *Phonology in Perception, 15*, 152–190.
Fenn, K. M., Nusbaum, H. C., & Margoliash, D. (2003). Consolidation during sleep of perceptual learning of spoken language. *Nature, 425*(6958), 614–616.
Flege, J. E. (1987). The production of "new" and "similar" phones in a foreign language: Evidence for the effect of equivalence classification. *Journal of Phonetics, 15*, 47–65.
Flege, J. E. (1995). Second language speech learning: Theory, findings, and problems. In W. Strange (Ed.), *Speech perception and linguistic experience: Issues in cross-language research* (pp. 233–277). MD: Timonium.
Fricke, M., Baese-Berk, M. M., & Goldrick, M. (2016). Dimensions of similarity in the mental lexicon. *Language, Cognition and Neuroscience, 31*(5), 639–645. https://doi.org/10.1080/23273798.2015.1130234.
Fuhrmeister, P., & Myers, E. B. (2017). Non-native phonetic learning is destabilized by exposure to phonological variability before and after training. *Journal of the Acoustical Society of America, 142*(5), EL448–EL454.
Gahl. (2008). *Time* and *Thyme* are not homophones: The effect of lemma frequency on word durations in spontaneous speech. *Language, 84*(3), 474–496. https://doi.org/10.1353/lan.0.0035.
Ganong, W. (1980). Phonetic categorization in auditory word perception. *Journal of Experimental Psychology: Human Perception and Performance, 6*(1), 110–125.
Gass, S., & Varonis, E. M. (1984). The effect of familiarity on the comprehensibility of nonnative speech. *Language Learning, 34*(1), 65–87.
Goto, H. (1971). Auditory perception by normal Japanese adults of the sounds. *Neuropsychologia, 9*(3), 317–323.
Guion, S., Flege, J. E., Akahane-Yamada, R., & Pruitt, J. C. (2000). An investigation of current models of second language speech perception: The case of Japanese adults' perception of English consonants. *Journal of the Acoustical Society of America, 107*(5), 2711–2724.
Hayes-Harb, R., Smith, B. L., Bent, T., & Bradlow, A. R. (2008). The interlanguage speech intelligibility benefit for native speakers of Mandarin: Production and perception of English word-final voicing contrasts. *Journal of Phonetics, 36*(4), 664–679.
Iverson, P., Hazan, V., & Bannister, K. (2005). Phonetic training with acoustic cue manipulations: A comparison of methods for teaching English/r/-/l/to Japanese adults. *Journal of the Acoustical Society of America, 118*(5), 3267.
Iverson, P., & Kuhl, P. K. (1995). Mapping the perceptual magnet effect for speech using signal detection theory and multidimensional scaling. *Journal of the Acoustical Society of America, 97*(1), 553–562.

Iverson, P., & Kuhl, P. K. (1996). Influences of phonetic identification and category goodness on American listeners' perception of/r/and/l/. *Journal of the Acoustical Society of America, 95*(2), 1130–1140.

Jesse, A., & McQueen, J. M. (2011). Positional effects in the lexical retuning of speech perception. *Psychonomic Bulletin & Review, 18*(5), 943–950.

Keating, P., Mikós, M. J., & Ganong, W. (1981). A cross-language study of range of voice onset time in the perception of initial stop voicing. *Journal of the Acoustical Society of America, 70*(5), 1261–1271.

Kennedy, S., & Trofimovich, P. (2008). Intelligibility, comprehensibility, and accentedness of L2 speech: The role of listener experience and semantic context. *Canadian Modern Language Review, 64*(3), 459–489.

Kleinschmidt, D. F., & Jaeger, T. F. (2015). Robust speech perception: Recognize the familiar, generalize to the similar, and adapt to the novel. *Psychological Review*. https://doi.org/10.1037/a0038695.supp.

Kraljic, T., Brennan, S. E., & Samuel, A. G. (2008a). Accommodating variation: Dialects, idiolects, and speech processing. *Cognition, 107*(1), 54–81. https://doi.org/10.1016/j.cognition.2007.07.013.

Kraljic, T., & Samuel, A. G. (2005). Perceptual learning for speech: Is there a return to normal? *Cognitive Psychology, 51*(2), 141–178.

Kraljic, T., & Samuel, A. G. (2006). Generalization in perceptual learning for speech. *Psychonomic Bulletin & Review, 13*(2), 262–268.

Kraljic, T., & Samuel, A. G. (2007). Perceptual adjustments to multiple speakers. *Journal of Memory and Language, 56*(1), 1–15. https://doi.org/10.1016/j.jml.2006.07.010.

Kraljic, T., Samuel, A. G., Brennan, S., & Brennan, S. E. (2008b). First impressions and last resorts: How listeners adjust to speaker variability. *Psychological Science, 19*(4), 332–338.

Kuhl, P. K. (1991). Human adults and human infants show a "perceptual magnet effect" for the prototypes of speech categories, monkeys do not. *Perception & Psychophysics, 50*(2), 93–107.

Kuhl, P. K. (1994). Learning and representation in speech and language. *Current Opinion in Neurobiology, 4*(6), 812–822.

Kuhl, P. K., Conboy, B. T., Conboy, B. T., Coffey-Corina, S., Coffey-Corina, S., Padden, D., et al. (2008). Phonetic learning as a pathway to language: New data and native language magnet theory expanded (NLM-e). *Philosophical Transactions of the Royal Society B: Biological Sciences, 363*(1493), 979–1000. https://doi.org/10.1098/rstb.2007.2154.

Kuhl, P. K., Stevens, E., Hayashi, A., Deguchi, T., Kiritani, S., & Iverson, P. (2006). Infants show a facilitation effect for native language phonetic perception between 6 and 12 months. *Developmental Science, 9*(2), F13–F21. https://doi.org/10.1111/j.1467-7687.2006.00468.x/full.

Kuhl, P. K., Williams, K., Lacerda, F., Stevens, K., & Lindblom, B. (1992). Linguistic experience alters phonetic perception in infants by 6 months of age. *Science, 255*(5044), 606–608.

Laturnus, R. (2018). *Perceptual adaptation to non-native speech: The effects of bias, exposure, and input variation (Unpublished doctoral dissertation)*. New York, NY: New York University.

Leach, L., & Samuel, A. G. (2007). Lexical configuration and lexical engagement: When adults learn new words. *Cognitive Psychology, 55*(4), 306–353. https://doi.org/10.1016/j.cogpsych.2007.01.001.

Lisker, L. I., & Abramson, A. (1964). Cross language study of voicing initial stops. *Journal of the Acoustical Society of America, 35*(11), 384–422.

Logan, J. S., Lively, S. E., & Pisoni, D. B. (1991). Training Japanese listeners to identify English/r/and/l/: A first report. *Journal of the Acoustical Society of America, 89*(2), 874–886. https://doi.org/10.1121/1.1894649.

Mann, V. A., & Repp, B. H. (1981). Influence of preceding fricative on stop consonant perception. *Journal of the Acoustical Society of America, 69*(2), 548–558.
McCullough, E. A., & Clopper, C. G. (2016). Perceptual subcategories within non-native English. *Journal of Phonetics, 55*(C), 19–37. https://doi.org/10.1016/j.wocn.2015.11.002.
McLaughlin, D., & Baese-Berk, M. M. (2018). *The role of the subject: An analysis of experimental methods in nonnative-accented speech perception research* (under review).
McLaughlin, D., Baese-Berk, M. M., Bent, T., Borrie, S. A., & Van Engen, K. J. (2018). Coping with adversity: Individual differences in the perception of noisy and accented speech. *Attention, Perception, & Psychophysics, 1–12.*
McMurray, B., Tanenhaus, M. K., & Aslin, R. N. (2002). Gradient effects of within-category phonetic variation on lexical access. *Cognition, 86*(2), B33–B42.
McMurray, B., Tanenhaus, M. K., & Aslin, R. N. (2009). Within-category VOT affects recovery from "lexical" garden-paths: Evidence against phoneme-level inhibition. *Journal of Memory and Language, 60*(1), 65–91.
McQueen, J. M., Cutler, A., & Norris, D. (2006). Phonological abstraction in the mental lexicon. *Cognitive Science: A Multidisciplinary Journal, 30*(6), 1113–1126.
Mitterer, H., & McQueen, J. M. (2009). Foreign subtitles help but native-language subtitles harm foreign speech perception. *PLoS One, 4*(11), e7785.
Mitterer, H., & Reinisch, E. (2013). No delays in application of perceptual learning in speech recognition: Evidence from eye tracking. *Journal of Memory and Language, 69*(4), 1–19. https://doi.org/10.1016/j.jml.2013.07.002.
Mitterer, H., Scharenborg, O., & McQueen, J. M. (2013). Phonological abstraction without phonemes in speech perception. *Cognition, 129*(2), 356–361. https://doi.org/10.1016/j.cognition.2013.07.011.
Munro, M. J., & Derwing, T. M. (1995). Foreign accent, comprehensibility, and intelligibility in the speech of second language learners. *Language Learning, 45*(1), 73–97.
Myers, E. B., & Mesite, L. M. (2014). Neural systems underlying perceptual adjustment to non-standard speech tokens. *Journal of Memory and Language, 76*, 80–93. https://doi.org/10.1016/j.jml.2014.06.007.
Norris, D., McQueen, J. M., & Cutler, A. (2003). Perceptual learning in speech. *Cognitive Psychology, 47*(2), 204–238.
Perrachione, T. K., Lee, J., Ha, L. Y. Y., & Wong, P. C. M. (2011). Learning a novel phonological contrast depends on interactions between individual differences and training paradigm design. *Journal of the Acoustical Society of America, 130*(1), 461. https://doi.org/10.1121/1.3593366.
Polka, L. (1991). Cross-language speech perception in adults: Phonemic, phonetic, and acoustic contributions. *Journal of the Acoustical Society of America, 89*(6), 2961–2977.
Polka, L. (1992). Characterizing the influence of native language experience on adult speech perception. *Perception & Psychophysics, 52*(1), 37–52.
Potter, C. E., & Saffran, J. R. (2017). Exposure to multiple accents supports infants' understanding of novel accents. *Cognition, 166*, 67–72.
Reinisch, E., Weber, A., & Mitterer, H. (2013). Listeners retune phoneme categories across languages. *Journal of Experimental Psychology: Human Perception and Performance, 39*(1), 75–86. https://doi.org/10.1037/a0027979.
Reinisch, E., Wozny, D. R., Mitterer, H., & Holt, L. (2014). Phonetic category recalibration: What are the categories? *Journal of Phonetics.*
Repp, B. H. (1984). Categorical perception: Issues, methods, and findings. In J. Lass (Ed.), *Speech and language* (Vol. 10, pp. 243–335). New York. (n.d.).
Samuel, A. G. (1986). Red herring detectors and speech perception: In defense of selective adaptation. *Cognitive Psychology, 18*, 452–499.

Samuel, A. G. (2011). The lexicon and phonetic categories: Change is bad, change is necessary. In G. Gaskell, & P. Zwisterlood (Eds.), *Lexical representation: A Multidisciplinary approach* (Vol. 17, pp. 33—50). Mouton de Gruyter.

Samuel, A. G., & Kraljic, T. (2009). Perceptual learning for speech. *Attention, Perception & Psychophysics, 71*(6), 1207—1218. https://doi.org/10.3758/APP.71.6.1207.

Schmale, R., Cristia, A., & Seidl, A. (2011). Contending with foreign accent in early word learning. *Journal of Child Language, 38*(05), 1096—1108. https://doi.org/10.1017/S0305000910000619.

Sheppard, B. E., Elliott, N. C., & Baese-Berk, M. M. (2017). Comprehensibility and intelligibility of international student speech: Comparing perceptions of university EAP instructors and content faculty. *Journal of English for Academic Purposes, 26*, 42—51. https://doi.org/10.1016/j.jeap.2017.01.006.

Sidaras, S. K., Alexander, J. E. D., & Nygaard, L. C. (2009). Perceptual learning of systematic variation in Spanish accented speech. *Journal of the Acoustical Society of America, 125*(5), 3306—3316.

Strange, W., & Dittman, S. (1984). Effects of discrimination training on the perception of/r-l/by Japanese adults learning English. *Perception & Psychophysics, 36*, 131—145.

van Leussen, J.-W., & Escudero, P. (2015). Learning to perceive and recognize a second language: The L2LP model revised. *Frontiers in Psychology, 6*, 1000.

Vroomen, J., van Linden, S., De Gelder, B., & Bertleson, P. (2007). Visual recalibration and selective adaptation in auditory—visual speech perception: Contrasting build-up courses. *Neuropsychologia, 45*(3), 572—577.

Weiand, K. (2007). Implementing Escudero's model for the subset problem. *Rutgers Optimality Archive*, 913.

Werker, J. F., Gilbert, J. H. V., Humphrey, K., & Tees, R. C. (1981). Developmental aspects of cross-language speech perception. *Child Development, 52*, 349—355.

Werker, J. F., & Tees, R. C. (1984). Cross-language speech perception: Evidence for perceptual reorganization during the first year of life. *Infant Behavior and Development, 7*, 49—63. https://doi.org/10.1016/s0163-6383(84)80022-3.

Witteman, M. J., Weber, A., & McQueen, J. M. (2013). Foreign accent strength and listener familiarity with an accent codetermine speed of perceptual adaptation. *Attention, Perception, & Psychophysics, 75*(3), 537—556.

Wright, B. A., Baese-Berk, M. M., Marrone, N., & Bradlow, A. R. (2015). Enhancing speech learning by combining task practice with periods of stimulus exposure without practice. *Journal of the Acoustical Society of America, 138*(2), 928—937. https://doi.org/10.1121/1.4927411.

Xie, X., Earle, F. S., & Myers, E. B. (2018). Sleep facilitates generalisation of accent adaptation to a new talker. *Language, Cognition and Neuroscience, 33*(2), 196—210.

Xie, X., & Fowler, C. A. (2013). Listening with a foreign-accent: The interlanguage speech intelligibility benefit in Mandarin speakers of English. *Journal of Phonetics, 41*(5), 369—378.

Xie, X., & Myers, E. B. (2017). Learning a talker or learning an accent: Acoustic similarity constrains generalization of foreign accent adaptation to new talkers. *Journal of Memory and Language, 97*, 30—46.

Zhang, X., & Samuel, A. G. (2013). Perceptual learning of speech under optimal and adverse conditions. *Journal of Experimental Psychology: Human Perception and Performance, 40*(1), 200—2017.

FURTHER READING

Best, C. T., & Strange, W. (1992). Effects of phonological and phonetic factors on cross-language perception of approximants. *Journal of Phonetics, 20*(3), 305—330.

MacKain, K., Best, C. T., & Strange, W. (1981). Categorical perception of English/r/and/l/ by Japanese bilinguals. *Applied PsychoLinguistics, 2*, 369–390.

Miyawaki, K., Jennings, J. J., Strange, W., Libermann, A. M., Verbrugge, R., & Fujimura, O. (1975). An effect of linguistic experience: The discrimination of [r] and [l] by native speakers of Japanese and English. *Perception & Psychophysics, 18*(5), 331–340.

CHAPTER TWO

A Common Representation of Serial Position in Language and Memory

Simon Fischer-Baum
Department of Psychology, Rice University, Houston, TX, United States
E-mail: simon.j.fischer-baum@rice.edu

Contents

1. Position Representation Schemes 33
2. A General Method for Contrasting Schemes 36
3. Experimental Findings 39
4. Discussion 44
References 52

Abstract

Speaking, spelling, and serial recall require a sequence of items to be produced one at a time in the correct order. The representations that underlie this ability, at a minimum, contain information about the identities and positions of the items in the sequence. This chapter summarizes a series of studies that investigate whether similar principles underlie how position is represented in different cognitive domains. In each domain, a variety of hypotheses have been proposed for how the position of an item is represented. An analysis framework has been developed that allows for controlled cross-domain experimentation that considers a common set of hypotheses to investigate whether there are commonalities in how position is represented. Careful analysis of the patterns of errors produced by neuropsychological case studies and unimpaired adults in a range of tasks support a both-edges representation of position, a scheme in which each item's position is represented both by its distance from the beginning of the sequence and its distance from the end of the sequence. The fact that a similar scheme is used to represent position across a range of cognitive domains suggests that serial order processing may rely on some domain-general representational principles. I discuss implications of this pattern of results for situating language processing in domain-general mechanisms and propose some tentative hypotheses as to why this edge-based representation of position is so prevalent.

Psychology of Learning and Motivation, Volume 68
ISSN 0079-7421
https://doi.org/10.1016/bs.plm.2018.08.002

Our ability to process language relies heavily on our ability to process sequences of linguistic items and their order. Individual words are composed of sequences of segments in spoken language and letters in written language. Sentences are composed of sequences of words. The capacity to track the order of items is clearly critical for language processing. Switching the order of letters or phonemes in a word make *lemons* and *melons* indistinguishable. Sentences with the same words, but in different orders, can have radically different meanings.

Language is not unique in its dependence on serial order processing. We dial 10-digit telephone numbers from memory, and understand the consequences dialing the numbers out of order. When following navigation directions, we need to keep track of the order of left and right turns or else we will end up far from our intended destination. As Karl Lashley pointed out many decades ago, the problem of how we represent and process the serial order of items and events is central to many aspects of cognition (Lashley, 1951).

The fact that serial order is critical in both language and non-language domains naturally leads to the question of whether there are common representations and processes for serial order that cut across domains. This question fits in the context of a larger debate over the role of domain-general and domain-specific processes in language, with extensive debate over the role of domain-general cognitive control (e.g., Nozari & Novick, 2017), working memory (e.g., Tan & Martin, 2018; Tan et al., 2017) or learning mechanisms (e.g., Aslin & Newport, 2012) in language processing. In these debates, domain-generality is taken to mean a variety of things. Language and non-language domains could rely on common capacities with a common neural resource, or the processes in language and non-language domains could rely on similar computational principles, though not necessarily a common neural resource (see Nozari & Novick, 2017 for discussion). The logic of the contribution of serial order processing to these debates is as follows: language relies on the ability to represent not only the items in a sequence — with the items being the letters that form words in written language, the phonemes that form words in spoken language and the words that form sentences — but also the ability to represent the position of those items in a sequence. But the need to represent both the items in a sequence and their positions is not unique to language. It is clear that items for linguistic sequences are unique to language. Phonemes, letters and words are the inventories that make up languages. The way that we represent the position of those items need not be language specific. There may be general schemes for representing position that are observed for sequences of different types; that is, there might be similar computational principles for how serial order is represented across domains. If so, then we will have evidence that our ability to process language depends on domain-specific item representations paired with domain-general position representations.

In the remainder of this chapter, I will argue that these types of general principles of position representation exist. First, I will lay out a series of hypothesis for how position might be represented, which I will refer to as the set of position representation schemes. The strategy of this research has been to consider as a wide of a variety of different position representation schemes as could be imagined. Then, I will describe a general method that can be used to pit each these different position representation schemes against each other, by analyzing corpora of errors produced in different of domains, in order to identify how position is represented for those different types of sequences. Following that, I will describe the results of a series of studies that apply this analysis to sequences that are core to language processing – letters in reading and writing, phonemes in spoken production – and sequences that are thought to be outside of language processing – immediate serial recall experiments with both verbal and nonverbal items. Strikingly, there is a convergence across most of these experiments that indicate a common scheme for representing position, specifically, one in which the order of items is defined by its distance from both the beginning and the end of the sequence. This pattern suggests that this both-edges based representation of position is a cognitive primitive that is critical not only for language but also for other domains (see also Endress, Nespor & Mehler, 2009 for a similar proposal). Finally, I will pose some additional questions for future research based on these findings.

1. POSITION REPRESENTATION SCHEMES

Sequence representations contain information both about the identity of the items, as well as their position. The position of an item in a sequence could be defined using a variety of methods. Consider the position of the E in the word NEST. It could be represented as the second letter from the beginning of the word or as the letter after the N. In the first case, the position of the E is defined relative to a word beginning anchoring-point.

In the second case, the position of the E is defined relative to the letter that precedes it.

Many of the same schemes could be used for difference types of sequences. In a verbal short-term memory experiment, a participant could be asked to recall the list "flour, hitch, zebra, clash" in order. The position of the word "zebra" could be represented as the third word in a sequence, defined relative to the beginning of the sequence. Or it could be the word after the word "hitch", defined by the word that precedes it. Furthermore, because the same schemes could be used to represent position in either domain, a beginning-based scheme or a preceding-item scheme could be a general principle of position representation that applies broadly across cognitive domains.

The research approach has been to consider a large number of possible position representation schemes. These schemes divide broadly into context-independent schemes, context-dependent schemes, and, some domain-specific position representation schemes. The full set of schemes that have been considered in the analyses are included in Table 1. For context-independent schemes, position is defined relative to some anchoring point like the beginning of the sequence, the end of the sequence, the midpoint of the sequence, either the beginning or the end of the sequence, whichever is closer, or with a multidimensional code in which position is defined relative to both the beginning and end of the sequence. For context-dependent schemes, position is defined relative to the other items in the sequence. At its simplest, position is defined relative to the immediately preceding item. More complex versions of context-dependent schemes define position relative to the following item, both the preceding and following item, or the set of items that precede the current item, not just the immediately adjacent one. Versions of context-dependent position representations have been referred to as chaining theories or open bigram theories in different corners of cognitive science. Finally, language-specific representations of position have also been considered, specifically schemes in which position is defined syllabically, have been considered for the sequences of letters and phonemes that make up words.

Proposed position representation schemes differ on other dimensions as well. One large point of contention is the difference between discrete and graded position representations. Classic examples of discrete position representation include the slot-based position representation schemes used

Table 1 Set of Possible Position Representation Schemes Considered With Examples of Theories That Posit These Schemes in Written and Spoken Language and in Short-Term Memory

		Cognitive domain		
		Written language	**Spoken language**	**Short-term memory**
Context-Independent	Beginning-based	Coltheart et al., 2001	Levelt et al., 1999 [a]	Burgess & Hitch, 1999
	End-based			Neath & Crowder, 1990
	Midpoint-based	Caramazza & Hillis, 1990		
	Closest-edge	Jacobs, Rey, Ziegler, & Grainger, 1998		
	Both-edges	Houghton, 2018	Houghton, 1990	Henson, 1998
Context-dependent	Immediately Preceding			
	Multiple Preceding	Grainger, Granier, Farioli, Van Assche, & van Heuven, 2006		Lewandowsky & Murdock, 1989
	Trigram	Seidenberg & McClelland, 1989	Wickelgren, 1969	Wickelgren, 1965
Others	Syllabic	Taft, 1979	Dell, 1986	

[a]Levelt et al. (1999) propose different representations of phoneme position at different levels of phonological representation, with a beginning-based scheme at the immediately post-lexical level.

to code letter position in classic computational models of reading, like the interactive activation model (McClelland & Rumelhart, 1981) and the dual-route connectionist model (Coltheart, Rastle, Perry, Langdon, & Ziegler, 2001). The interactive activation model includes four pools of twenty-six letter units, with each pool representing a single letter position, such that the P in the word POST is represented by a totally different unit than the P in the word SPOT. In discrete schemes, position representations are either the same or different, with no similarity among non-identical positions, such that there is no way of coding the fact that the position of the P in POST is closer to the P in SPOT than it is to the P in STOP. In contrast, more recent computational models of reading have eschewed these slot-based position representations in favor of graded representations of position, in which nearby positions are represented more similarly (e.g. Gomez, Ratcliff, & Perea, 2008; Norris & Kinoshita, 2012). Different answers to the question of "relative to what is position defined?" and "are nearby positions similar to each other?" yield an array of possible position representation schemes.

2. A GENERAL METHOD FOR CONTRASTING SCHEMES

In many cognitive domains — speech production, reading and verbal short-term memory — there has been extensive debate over which of these schemes is used to represent position. Table 1 includes a non-comprehensive sample of papers that propose different position representation schemes for written language, spoken language and short-term memory, focusing only on the "relative to what is position defined?" question. Clearly, methods are needed that can directly contrast these different position representation schemes empirically.

Below I present the outline of a method that can be used to accomplish this goal. A more thorough description of the method can be found in a number of recent papers that have applied it to adjudicate domain-specific debates about position representation (e.g., Spelling: Fischer-Baum, McCloskey, & Rapp, 2010; Short-term memory: Fischer-Baum & McCloskey, 2015; Reading: McCloskey, Schubert & Fischer-Baum, 2013). The method relies on the fact that a clear way to distinguish these position representation schemes is to ask which items are in the same position in two different sequences. Consider the question: which letter in the word CANDY is in the same position as the E in NEST? As discussed above, for different schemes the E in NEST could be described as the second letter or the letter after the N. If the E is the second letter in the word NEST, then the A in CANDY is in the same position as the E in NEST. If the E is the letter after the N in the word NEST, then the D in CANDY is in the same position as the E in NEST.

The analysis focuses on corpora of errors in which participants produce responses in which the sequence contains an intruded item and that intruded item appears in a recent response, what is called in the clinical literature a perseveration error (Cohen & Dehaene, 1998). This approach builds on the psycholinguistics tradition of analyzing speech errors to investigate the nature of the representations and processes that underlie language production (Dell, 1995). The logic is that the mistakes that we make when speaking are not totally random and carefully analyzing them reveals rich linguistic structure. While error analyses are likely most familiar to psycholinguists, careful analysis of errors in other parts of cognitive science similarly provide insights into the nature of cognitive representations (Short-Term Memory: Henson et al., 1996; Visual Cognition: Gregory & McCloskey, 2010). For the current analysis, corpora of errors can be collected from different sources. The papers using this method have investigated patients with acquired language disorders following stroke, neurotypical undergraduates participating in working memory studies in the lab, and deaf high school students producing signed responses. With each of these corpora of errors, the analysis remains exactly the same. Table 2 shows an example of this type of error, in a series of two spelling trials produced by a patient with acquired dysgraphia (Fischer-Baum et al., 2010). The second response contains an error in which the word "edge" is misspelled as ERGE. The R is an intrusion error, it appears in the response but not in the target. As shown in the table, the intruded letter was present in the response immediately preceding the error (FRENCE), raising the possibility that

Table 2 Example of a Two Trial Sequence With Patient CM in Which the Intruded R on Trial 2 is a Possible Perseveration From the Response FRENCE on Trial 1

	Target	**Response**
Trial 1	FRENCH	FRENCE
Trial 2	EDGE	ERGE

the R is a perseveration, produced in ERGE because it was previously produced in FRENCE, what I call the source response.

The analysis tests whether the R appears in the same position in the error as it did in the source response, comparing the rate at which the position matches between the two responses with a rate expected by chance, calculated by a Monte Carlo procedure that randomly selects other responses in the corpora that are distant from the error being analyzed. The answer to that question depends on position representation scheme being considered. According to the beginning-based scheme, the R in ERGE is in the same position as the R in FRENCE as they both are the second letter. But the R in ERGE is not in the same end-based position (third from last vs. fifth from last), the same immediately-preceding item position (after an E vs. after an R) or even the same syllabic position (coda of the first syllable vs. onset of the first syllable).

When looking at the whole corpus of errors, it is frequently the case that the intruded letters appear in the same position in the error as it did in the source response more often than would be expected by chance for all of the position schemes considered. This observation is unsurprising because the same error can count as a position match for multiple schemes. For example, in the error "under" spelled as UNDEL, with the source response MOTEL, the L appears in the same position of the two responses for many schemes (beginning-based, end-based, midpoint-based, preceding-item, syllabic position). An additional set of analyses has been devised to pit these different position schemes against each other, essentially partialling out the errors accounted for by one scheme (e.g. the beginning-based scheme) and then asking whether the remaining perseveration errors match position by a different scheme (e.g. the end-based scheme) more than would be expected by chance. In applying this process iteratively, the analysis method identifies the position representation scheme or combination of schemes that best accounts for the pattern of perseveration errors observed in the corpus.

There are many strengths of this analysis approach for answering the question of whether there are domain-general properties of serial position representation. First, there is a clear way to contrast position representation schemes, including a statistical analysis that can determine which position representation scheme best accounts for the data. Second, the same set of position representations can be compared in different domains, so the conclusion about which position representation scheme best accounts or the data will be derived from the same hypothesis space. Third, the exact same method can be applied to sequences from different domains. As long as a corpus of errors can be collected in which participants produce responses in which intruded items perseveration from recent response more frequently than would be expected by chance, I can analyze whether those perseverations maintain position, and ask position defined how. Taken together, this approach provides an apples-to-apples comparison across domains that can reveal general principles of how position is represented.

3. EXPERIMENTAL FINDINGS

This flexible analysis technique has been applied to a number of different domains already. Fischer-Baum et al. (2010) analyzed letter perseveration errors produced by two individuals with acquired writing problems following stroke, whose writing errors largely consisted of letter perseveration errors, like the one described above, and a third patient who produced similar patterns of spelling errors was reported in Fischer-Baum (2011; Chapter 5). McCloskey, Fischer-Baum, and Schubert (2013), reported a single case study of a brain-damaged individual who made a similar type of error, except in reading, following a stroke. For example, this patient would read the word SAILOR as "sailog" immediately after correctly reading the word "flag." In Fischer-Baum (2011; Chapter 6), I reported a single case study of an individual who makes similar errors, except in spoken production, for example producing the response "hahlo" in response to a picture of an arrow, with many of the immediately preceding responses – house, hat – containing the intruded /h/(a more detailed case report of this patient is available in Olson, Romani, & Halloran, 2007). For each of these case studies, large corpora of perseveration errors were collected by testing the participants extensively over a long period of time. For example, the reading patient was given approximately 30,000 words and nonwords to read aloud, both in the lab and at home. These 30,000 words included approximate 12,000 perseveration errors that were included in the position analyses. While the sample of data from the reading patient was by far the largest, but we collected large corpora from the other patients – between 1500 and 3700 perseveration errors – as well. These large numbers of errors made it possible to have the power to contrast the large set of position representation schemes described above.

For the language domain, we focused on analyzing the errors of patients because neurotypical individuals rarely produce perseveration errors in single word tasks. In short-term memory, however, it is easy to get participants to produce perseveration errors, as participants often intrude items from previous lists into the response to the current list in immediate serial recall tasks (e.g., Conrad, 1960; Estes, 1991; Henson, 1999). Fischer-Baum and McCloskey (2015) reported three verbal immediate serial recall experiments with neurotypical participants that differed in the lengths of the lists presented and the modality of input and output, and Miozzo, Petrova, Fischer-Baum, and Peressotti (2016) presented a similar experiment with deaf participants, recalling sequences of finger-spelled letters. Fischer-Baum (2011; Chapter 9) reported three non-verbal serial recall experiments, two experiments in which participants had to remember sequences of spatial locations and one in which participants were shown sequences of the same object presented at different orientations, and had to recall the sequences of orientations. Each of the seven samples generated a corpus of around 1500 perseveration errors to analyze, ranging from 800 to 3000 depending on the experiment.

The goal of the current chapter is not to provide a detailed analysis of each of these datasets, but to look at the overall pattern of results across experiments. The results are summarized in Table 3. However, I will walk through the analysis of one dataset — the spelling patient LSS reported in Fischer-Baum et al. (2010) — to provide further insight into how conclusions about positions representation scheme are drawn from these analyses. The analysis of LSS included approximately 2300 perseveration errors. Of these errors, 47% matched on beginning-based position and 54% matched on end-based position, both of which exceed what would be expected by chance (~25% chance position matches, *ps* <0.0001). To demonstrate contributions of both beginning- and end-based position, residual analyses was carried out that looked only at the approximately 1000 perseveration errors that did not match on end based position and showed that they matched on beginning based position (36% of the time) more than would be expected by chance (20%, $p < 0.0001$). Similarly, a residual analysis was carried out that looked only at the approximately 1200 perseveration errors that did not match on beginning based position and showed that they matched on end based position (44% of the time) more than would be expected by chance (22%, $p < 0.0001$). Because of these findings, I could conclude that there are contributions of both beginning- and end-based representations of position, or the

Table 3 Summary of the Results of the Position Representation Scheme Analysis Across Data From Sequences of Different Types — Letters in Spelling and in Reading, Phonemes in Speaking, Words and Letters in Verbal Immediate Serial Recall and Locations and Orientations in Nonverbal Immediate Serial Recall

		Begin	End	Graded	Other content-independent	Content-dependent	Syllabic
Spelling	Fischer-Baum et al. (2010); LSS	✓	✓	✓	×	×	×
	Fischer-Baum et al. (2010); CM	✓	✓	✓	×	×	×
	Fischer-Baum (2011; Ch. 5): LHT	✓	✓	✓	×	×	×
Reading	McCloskey et al. (2013): LHD	✓	✓	✓	×	×	×
Speaking	Fischer-Baum (2011; Ch. 6): VS	✓	✓	✓	×	×	×
Verbal Immediate Serial Recall	Fischer-Baum and McCloskey (2015); Exp 0.1	✓	✓	✓	×	(a)	n/a
	Fischer-Baum and McCloskey (2015); Exp 0.2	✓	✓	✓	×	(a)	n/a
	Fischer-Baum and McCloskey (2015); Exp 0.3	✓	✓	✓	×	(a)	n/a
	Miozzo et al. (2016); Deaf Signers	✓	✓	✓	×	×	n/a
Nonverbal Immediate Serial Recall	Fischer-Baum (2011; Ch. 9): Locations 1	✓	×	✓	×	×	n/a
	Fischer-Baum (2011; Ch. 9): Locations 2	✓	×	✓	×	×	n/a
	Fischer-Baum (2011; Ch. 9): Orientations	✓	✓	n/a	×	×	n/a

[a]Small but significant contributions of preceding —item position were observed in all three verbal short-term memory experiments, thought the effects were quite small (~10 errors out of about1500 perseverations).

both-edges scheme. I then considered a graded position representation scheme, and showed that, once exact both-edges position matches were removed, perseveration errors matched nearby positions more than would expected by chance, supporting a graded both-edges representation of position. These results are indicated by checkmarks on the top line of Table 3 under begin, end and graded.

This graded both-edges representation of position was then compared to a variety of other content-independent position representations schemes, to context-dependent position representation schemes and to syllabic position representation schemes. In all cases, there was no additional contribution of these alternative schemes above and beyond the graded both-edges scheme. For example, when graded both-edges position matches were removed from the analysis, the remaining perseveration errors did not match syllabic position more than would be expected by chance. In contrast, when the syllabic position matches were removed from the analysis, the remaining perseverations matched graded both-edges position significantly more often than would be expected by chance. This pattern, showing a significant contribution of the graded both-edges position above and beyond other position representation schemes, but no contribution of the other scheme above and beyond both-edges position is indicated in Table 3 by the xs under the columns for "other content-independent", "content-dependent" and "syllabic."

Each row of Table 3 reflects a unique contribution to the literature, helping to settle domain-specific questions about position representation. As an example, consider the analysis of speech errors in VS (Fischer-Baum, 2011, Chapter 6), which yielded perhaps the most unexpected result. Most theories of speech production assume a syllabic representation of phoneme position, at least at some level of representation, with some of the strongest evidence for this claim coming from the syllabic position constraint – the observation that phoneme swaps in speech errors tend to appear in the same syllabic role (e.g., Boomer & Laver, 1968; Fromkin, 1971; MacKay, 1970; Nootebloom, 1969; Stemberger, 1990). When considered along, there was some evidence for a syllabic position constraint in the patient – that is perseveration errors matched syllabic position more than would be expected by chance – once the graded both-edges scheme was taken in to account, errors maintained graded both-edges position and not syllabic position. This observation of non-syllabic representations of phoneme positions is in line with some theories of speech production. The Levelt, Roelofs, and Meyers (1999) model proposes that during the first stage of phonological encoding, phoneme position is represented by a beginning-based representation of position, with syllabification occurring at a later stage. In a recent paper, Olson, Halloran, and Romani (2015) argue that this same patient's speech errors occur because of an impairment in the selection of phonemes during phonological encoding. Therefore, this analysis provides evidence for the non-syllabic encoding of phoneme position at this level, in line with what was proposed by Levelt and colleagues, and further argues for the graded both-edges representational scheme at this level.

Focusing only on the domain-specific questions can obscure the larger pattern that has become clear from this body of research. As can be seen in Table 3, for nearly all of the data sets that have been analyzed, an identical result is obtained — evidence for a graded, both-edges representation of position, without any contribution of any other position representation schemes. This pattern was seen in all three spelling patients, in the reading patient and in the patient who produced phoneme perseveration errors when speaking. Graded both-edges position representations were observed in the verbal immediate serial recall experiments, though extremely small (about 10 errors out of about 1500 total) contributions of a preceding-item representation of position above and beyond the graded both-edges scheme were also found in three of the verbal immediate serial recall experiments. For the task in which participants had to recall a sequence of orientations for the same objects, there was also support for a both-edges scheme, though the sequences used in the experiment (length 2—4) were too short to test the graded version of that scheme. Looking at all of these experiments together, there appears to be common scheme for representing the position of an item in a sequence, namely one in which the position of an item is defined by its distance from both edges of the sequence, and where the representation of position is graded rather than discrete, such that nearby positions are represented more similarly to each other than more distant positions.

The one place where the results diverge from this overarching pattern of both-edges based position were in the two experiments in which participants had to recall sequences of locations presented on the screen. In one experiment, these locations were scattered across the screen in a Corsi block-like presentation and in the other experiment, these locations were presented on a 4-by-4 grid. In these experiments, perseveration errors maintained position, but only position defined by a graded beginning-based scheme, with no contribution of any other position representation schemes, including the end-based position representation scheme. This surprising result was probed further, and I found that there was no contribution of end-based position even when the analysis was restricted to perseverations in the final position of the sequence. While it is possible that these two studies deviate from the overall pattern for uninteresting reasons, like false negatives that are inherent to null hypothesis testing, this seems unlikely since (1) the two datasets out of twelve that do not show both-edges position representation are very similar to each other — both are sequences of spatial locations and (2) both of these data sets converge on the same position representation scheme — one in which position is defined only by its distance from the beginning of the sequence. Therefore, I generalize the results in the following way: a graded both-edges representation is a general scheme for representing the order of items in a sequence that applies broadly to many of the different types of sequences I have investigated, including linguistic sequences of letters and phonemes that make up words and some sequences in short-term memory, though it does not apply to all types of sequences in short-term memory, with sequences of spatial locations relying on a beginning-based representation of position. The next section provides a framework for making sense of this pattern of results.

4. DISCUSSION

Many cognitive systems require the ability to process sequences of items or events, requiring the system to be able to represent both the items in a sequence and their positions. The fact that the graded both-edges scheme is so widely observed as the mechanism for representing the position of items in a sequence make it a strong candidate for being a cognitive primitive, a part of a general cognitive toolbox that our minds use to make sense of the world around us. The diversity of possible position representation schemes outlined in Table 1 make it clear that there are many ways to solve the problem of representing the order of items in a sequence. The consistency of position representation schemes across disparate domains, shown in Table 3, suggests that our brain relies on a limited number of these possible schemes.

The proposal that both-edges representations of position are a cognitive primitive falls nicely in line with several existing proposals in the literature. Endress et al. (2009) argue that assuming edge-based representation of position as a cognitive primitive can help explain patterns of language learning observed both in laboratory and natural settings. This precise scheme has been proposed by computational modelers of both language (Glasspool & Houghton, 2005; Houghton, 1990, 2018; Houghton, 1990; Houghton, Glasspool, & Shallice, 1994) and working memory (Henson, 1998, 1999) as a common solution to the problem of serial order. The fact that similar conclusions have been reached from these other empirical approaches strengthens the claims made here. The current body of work uses carefully controlled cross-domain experimentation with a common analysis framework and a common set of hypotheses to identify cognitive primitives for position representation. In an effort to make the approach as similar as possible across domains, the kind of data that used to draw conclusions about position representation was limited to perseveration errors. Perhaps there is something unique about these errors that leads to our conclusions about both-edges position encoding. Perhaps the same conclusions would not be reached if other methods were used to probe position representations. This concern is partially alleviated by the fact that other researchers have independently reached the same conclusion through other approaches. Still, one area for future research is to develop other cross-domain experimental approaches that can provide converging evidence for the claims made here. Furthermore, while the current work presents evidence for this cognitive primitive from a number of different experiments, there are still many different types of sequences to examine. In the language domain, the analysis could be applied to the representation of the position of words in a sentence. Here, I would be surprised to find evidence of both-edges representation of position, since there is clear evidence that favors tree structure over linear data structures in syntax (see Perfors, Tenenbaum, & Regier, 2011 for a computational argument favoring this point). There are also other non-linguistics domains that depend on serial order processing — like the planning of actions or the comprehension of complex events that unfold over time — that could be investigated.

The fact that this graded both-edges scheme is used to code position for both linguistic and non-linguistic sequences can help situate language processing in the larger cognitive system. Processing language requires the ability to process a set of items and their serial order. The items themselves are certainly domain specific, as the reading system is specialized to recognize letters, speech production is specialized to produce phonemes. But these results suggest that the ordering mechanisms might not be specialized for a specific domain, instead relying on more general principles of representation that apply broadly across the cognitive system. In this way, our ability to represent and process language can be conceived of as the combination of domain-specific knowledge and domain-general mechanisms for organizing that knowledge. When posed this way, the graded both-edges representation of position might just be one of a series of cognitive primitives that our brains use to organize the input so that we can learn about the world. Endress et al. (2009) argue that another cognitive primitive might be a system for recognizing immediately repeated items or events. Part of the connectionist research program has been to argue that delta-rule learning and back-propagation is another cognitive primitive that we use to learn about regularities in the world, that could explain how human behavior in a wide assortment of domains (Rumelhart, Hinton, & Williams, 1986). Another area for future research would be to more fully investigate the contents of the toolbox of cognitive primitives, identifying the various general systems our minds use to organize both linguistic and non-linguistic information.

If the graded both-edges scheme is a cognitive primitive for representing position for different types of sequences, the natural question is why this scheme as opposed to other possibilities. Both Endress et al. (2009) and Henson (1998) argue that since start and the end of a sequence are its most salient aspects, these edges provide the strongest anchoring points relative to which the items in the sequence can be ordered. This argument makes the most sense for letter sequences in reading, where all of the letters are presented simultaneously and processed in parallel, making it easy to track how far each letter is from both edge-based anchoring points. The argument falters somewhat for a task like immediate serial recall in which words are presented one at a time. When the sequences presented are of variable length participants do not know when the end of the sequence is going to come, making it more difficult to code position of each item relative to the end of the sequence (Farrell & Lelièvre, 2009).

A recent proposal in short-term memory, the "mental whiteboard "hypothesis, might help us make sense of why ends are salient for temporally presented sequences (Abrahamse, Van Dijck, Majerus, & Fias, 2014, 2017). According to this hypothesis, when confronted with a temporal sequence of items or events, the brain generates an internal spatial template and binds each item to specific coordinates in that spatial template in a left to right fashion, the mental equivalent of writing down the sequence on a piece of paper. I will refer to this process as spatialization. With this type of spatial representation of the position of items in a sequence, once the entire sequence has been encoded, both edges of the sequences are salient anchoring points, with the first item being the leftmost and the final item being the rightmost. A variety of sources of evidence support this link between temporal order and space. van Dijck and Fias (2011) reported an interaction between serial position in verbal working memory and hand of response, with judgments about items presented in the first half of the list being faster with the left hand than the right hand, and judgments about items presented in the second half of the list being faster with the right hand than the left hand. Fischer-Baum and Benjamin (2014) report better immediate serial recall of temporal order when the items are presented left to right than when they are presented right to left. Bonato, Saj, and Vuilleumier (2016) report that patients who neglect the left half of space also have more difficulty recalling events that occurred before a central event, relative to controls, than recalling events that occurred afterward.

Assuming spatialization of temporal order information can help to explain why both edges are salience for temporally presented sequences. It can also explain why some, but not all, of the sequence types described in Table 3 show graded both-edges position representation. In the twelve data sets analyzed, only two showed no contribution of end-based representation of position. Both of these data sets involved recalling sequences of locations and both showed clear evidence of beginning-based representation of position. In the framework described above, for these sequences, the beginning of the sequence is a salient edge, but the end of the sequence is not, perhaps because these sequences are not spatialized. Of all of the sequences tested, stimuli which are locations are likely the most difficult stimuli to spatialize.

This point is illustrated in Fig. 1, which contrasts sequences of digits with sequences of locations. For the digits in the digit span task (Fig. 1, top), spatialization provides a straightforward means to recovering serial order information. The identity of each digit is not changed by where it lies in the spatial template and the leftmost digit was the first item presented and the rightmost digit is last. Since maintaining the identity of the items in this task does not rely on a spatial template, space can be repurposed to support memory for serial order. In contrast, the locations in the matrix span task are distinguished from each other solely on the basis of where they are located in space. In Fig. 1, middle, what distinguishes the first and the second item in the sequence is that the first item is in the bottom, right corner and the second item is in the top, left corner. Because space is being used to distinguish the identity of the items, it cannot also be used to maintain their serial order, as illustrated in Fig. 1, middle. If these grids of locations were placed on a mental white board, it would be hard to distinguish which spatial information was being used to represent the items in the sequence from which spatial information is being used to represent the order of the items.

Note that this account predicts that spatialization should only fail when the sequences are of locations defined in two-dimensional space. When different types of spatial information are used to distinguish the items in the sequence, the prediction is that spatialization should still occur. Some of the data reported above already supports this conclusion. When items in a sequence are distinguished by a different type of spatial information — the orientation of the object — the temporal order information can be encoded as spatial coordinates, and the end becomes a salient

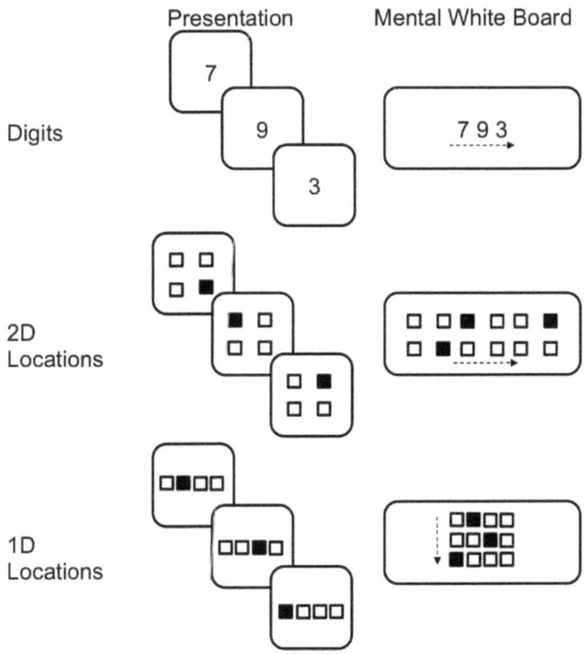

Figure 1 Mental white board hypothesis representations for three sequence types, sequences of digits (top), sequences of locations defined in 2-dimensional space (middle) and sequences of locations defined only along the horizontal dimension (bottom). Note that in the top and the bottom mental white boards, serial order can be easily reconstructed, by reading off items either left to right or top to bottom. In contrast, in the mental white board for the sequences of locations defined in 2-dimensional space, reconstructing serial order is more of a challenge, since both the vertical and horizontal dimensions are used to distinguish the items in the sequence.

edge. Thus, it is unsurprising that both-edges representation of position were identified when analyzing perseveration errors produced when participants recalled sequences of orientations. More strikingly, this account predicts that spatialization should only fail for sequences of locations that are defined in two-dimensional space. For sequences of locations defined in one-dimension, for example along a horizontal line (as shown in Fig. 1, bottom), the mental white board could construct a spatial representation such that the horizontal dimension contains information about the identities of the items in the sequence, while the vertical dimension contains information about their order, with the top-most item being first and the bottom-most item being last. Because spatial information can be used to support serial order information for these types of sequences, the prediction is that both-edges based representations of positions should be observed. In an unpublished experiment, I collected perseveration errors while participants recalled a sequence of locations in which the locations were presented along a single dimension – a horizontal line of 16 possible locations. In line with the prediction, we found that the perseveration errors produced in this experiment matched both beginning- and end-based position, unlike the experiments in which the same number of locations were organized in a 4 by 4 grid. Therefore, one interpretation of the full pattern of results presented in this chapter is that spatialization of temporal order is a cognitive primitive. Graded both-edges representation of position emerge from that spatialization, because with spatialization both edges of the sequence are salient. For all types of sequences for which this spatialization is possible, graded both-edges position is observed. For those sequences in which spatialization is not possible, the only salient edge is the beginning of the sequence, and only beginning-based representations of position are observed.

Note that this proposal goes beyond the mental whiteboard hypothesis which proposes spatialization for temporal order sequences only in the context of working memory. Here, I am arguing that we mentally spatialize sequences of all types, at least including phonemes in speech and letters in writing. There is some evidence for this spatialization in spelling. Caramazza and Hillis (1990) report a patient with right-side neglect following a left-hemisphere stroke who neglects the second half of words in both reading and spelling, linking temporal order with spatial position in writing. To my knowledge, there is no evidence of spatialization in speech, though it could easily be tested, for example seeing whether participants are faster at making judgments with their left hands about phonemes that appear earlier in a word and faster with the right hands about phonemes that appear later in a word. However, spatialization is a prime candidate for being a cognitive primitive as there are many other examples of the brain spatially organizing non-spatial information. For example, the spatial-numerical association of response codes effect has been used to argue that numerical information is organized in a mental number line that is oriented spatially from left-to-right, at least in populations with a left-to-right writing system (Dehaene, Bossini, & Giraux, 1993).

Overall, the claim here is that serial position representation is yet another example of how language processing relies on domain-general mechanisms. But what exactly does it mean to say that serial position representation is domain-general? Nozari and Novick (2017) make a distinction between two notions of domain-generality, the idea that different cognitive systems rely on similar computational principles and that different cognitive systems rely on a common neural resource. The work presented here supports the former notion of domain-generality — that there is a common principle of representation that applies across different sequence types — and says nothing about whether there is a shared neural resource for serial order processing.

As Nozari and Novick (2017) note, common computational principles do not entail a common a neural resource. It is possible, for example, that basic information about neuronal computation place constraints on how neural networks can represent position. This point is beautifully illustrated by Grossberg (1986) who derives a theory of order representation from a few basic assumptions about neuronal computation — self-inhibitory feedback, lateral inhibition and adaptive resonance — and demonstrates how well this theory could explain seemingly disparate findings from different cognitive domains. While Grossberg's theory is not a graded both-edges representation of position, his work illustrates how different sequence types could rely on the same representation scheme not because they rely on a shared neural substrate, but because limitations on the types of computations neurons can carry out means that whenever a neural network is given the responsibility of processing order, the same representational scheme is used.

At the same time, it is also possible that the similarities in the representation of serial position across cognitive domain arise because each of these different cognitive systems rely on a common neural structure for representing and processing order information. It remains an open question, therefore, whether there is a common neural resource that is used for serial order processing across different types of sequences. Cognitive science is filled with methodological tools that could be used to address this question. If there is a common neural substrate for order processing across language and working memory, individuals with deficits in order processing in one domain should also have order processing deficits in the other (e.g., Majerus & Cowan, 2016). Structural equational modeling could be used to ask whether individual differences in serial order processing across different types of sequences can be explained by a common latent variable. Dual-task paradigms could be used to show that demands on order processing in one domain influence order processing performance on a concurrent task in another domain. Neuroimaging studies could identify a common neural locus of order processing across sequence types. Investigations of whether there is a common neural resource for serial order processing could provide additional insights into what it means to have identified graded both-edges as a common representational principle.

As a final note, I want to highlight how this work illustrates the importance of studying language in the context of broader questions about cognitive science. The ability to comprehend and produce language is a remarkable capacity that is arguably unique to humans. Psycholinguistic research has largely focused on developing theories of the representations and processes that enable that capacity. While these theories frequently depend on insights from other aspects of cognitive psychology — perception, memory, cognitive control, statistical learning — the relationship can be characterized as being largely unidirectional. The line of thinking is that language research is specialized. It can draw on more general areas of cognitive psychology, but findings from psycholinguistics research are not relevant for developing theories of how cognitive control works, how statistical learning works or how memory works.

Serial order processing research provides a nice counterpoint to this way of thinking about language research. In his seminal paper on the problem of serial order in behavior, Lashley (1951) marveled at the human capacity to produce complicated sequences of actions rapidly and largely without error. The paper goes through many examples of these sequential behavior, but many of the aspects of serial order behavior that impress him the most have do with language processing. Framed this way, language processing can be seen as a model system for studying a more general aspect of cognition, as a theory of the human capacity for serial order processing needs to be able to explain the complex nature of serial order processing in language (Hartley & Houghton, 1996). Of course understanding language processing is an important goal in its own right. But the role of language research in psychology is not limited to answering this question. Producing and comprehending language are complex human behaviors that engage many different cognitive operations and a full understanding of these cognitive operations requires understanding how they work in the context of language processing.

REFERENCES

Abrahamse, E. E. L., van Dijck, J. J. P., & Fias, W. (2017). Grounding verbal working memory: The case of serial order. *Current Directions in Psychological Science, 26*(5), 429–433.

Abrahamse, E., Van Dijck, J. J. P., Majerus, S., & Fias, W. (2014). Finding the answer in space: The mental whiteboard hypothesis on serial order in working memory. *Frontiers in Human Neuroscience, 8*, 932.

Aslin, R. R. N., & Newport, E. E. L. (2012). Statistical learning: From acquiring specific items to forming general rules. *Current Directions in Psychological Science, 21*(3), 170–176.

Bonato, M., Saj, A., & Vuilleumier, P. (2016). Hemispatial neglect shows that "before" is "left". *Neural Plasticity, 2016.*

Boomer, D. D. S., & Laver, J. J. D. (1968). Slips of the tongue. *British Journal of Disorders of Communication, 3*(1), 2–12.

Burgess, N., & Hitch, G. G. J. (1999). Memory for serial order: A network model of the phonological loop and its timing. *Psychological Review, 106*(3), 551–581.

Caramazza, A., & Hillis, A. A. E. (1990). Spatial representation of words in the brain implied by studies of a unilateral neglect patient. *Nature, 346*(6281), 267–269.

Cohen, L., & Dehaene, S. (1998). Competition between past and present: Assessment and interpretation of verbal perseverations. *Brain, 121*, 1641–1659.

Coltheart, M., Rastle, K., Perry, C., Langdon, R., & Ziegler, J. (2001). DRC: A dual route cascaded model of visual word recognition and reading aloud. *Psychological Review, 108*, 204–256.

Conrad, R. (1960). Serial order intrusions in immediate memory. *British Journal of Psychology, 51*, 45–48.

Dehaene, S., Bossini, S., & Giraux, P. (1993). The mental representation of parity and number magnitude. *Journal of Experimental Psychology. General, 122*(3), 371–396.

Dell, G. G. S. (1986). A spreading-activation theory of retrieval in sentence production. *Psychological Review, 93*(3), 283–321.

Dell, G. S. (1995). Speaking and misspeaking. In L. R. Gleitman, & M. Liberman (Eds.), *An invitation to cognitive science Vol 1* (2nd, pp. 183–208). Cambridge, MA: MIT Press.

Endress, A. A. D., Nespor, M., & Mehler, J. (2009). Perceptual and memory constraints on language acquisition. *Trends in Cognitive Sciences, 13*(8), 348–353.

Estes, W. W. K. (1991). On types of item coding and source of recall in short-term memory. In W. Hockley, & S. Lewandowsky (Eds.), *Relating theory and data: Essays on human memory in honor of Bennet B. Murdock* (pp. 155–173). New York: Psychology Press.

Farrell, S., & Lelièvre, A. (2009). End anchoring in short-term order memory. *Journal of Memory and Language, 60*(1), 209–227.

Fischer-Baum, S. (2011). *Position representation: General principles or domain-specificity?* (Doctoral dissertation). Retrieved from ProQuest Dissertations and Theses. (Accession Order No. AAT 3463614).

Fischer-Baum, S., & Benjamin, A. S. (2014). Time, space and memory for order. *Psychonomic Bulletin & Review, 21*, 1263–1271.

Fischer-Baum, S., & McCloskey, M. (2015). Representation of item position in verbal short-term memory: Evidence from intrusion errors. *Journal of Experimental Psychology Learning Memory and Cognition, 41*, 1426–1446.

Fischer-Baum, S., McCloskey, M., & Rapp, B. (2010). The representation of graphene position: Evidence from acquired dysgraphia. *Cognition, 115*, 466–490.

Fromkin, V. A. (1971). The non-anomalous nature of anomalous utterances. *Language, 47*, 27–52.

Glasspool, D. D. W., & Houghton, G. (2005). Serial order and consonant–vowel structure in a graphemic output buffer model. *Brain and Language, 94*(3), 304–330.

Gomez, P., Ratcliff, R., & Perea, M. (2008). The overlap model: A model of letter position coding. *Psychological Review, 115*(3), 577–600.

Grainger, J., Granier, J. J. P., Farioli, F., Van Assche, E., & van Heuven, W. W. J. (2006). Letter position information and printed word perception: The relative-position priming constraint. *Journal of Experimental Psychology: Human Perception and Performance, 32*(4), 865–884.

Gregory, E., & McCloskey, M. (2010). Mirror-image confusions: Implications for representation and processing of object orientation. *Cognition, 116*(1), 110–129.

Grossberg, S. (1986). The adaptive self-organization of serial order in behavior: Speech, language, and motor control. In E. C. Schwab, & H. C. Nusbaum (Eds.), *Speech perception: Vol. 1. Pattern recognition by humans and machines*. New York: Academic Press.

Hartley, T., & Houghton, G. (1996). A linguistically constrained model of short-term memory. *Journal of Memory and Language, 35*, 1–31.

Henson, R. N., Norris, D. G., Page, M. P., & Baddeley, A. D. (1996). Unchained memory: Error patterns rule out chaining models of immediate serial recall. *The Quarterly Journal of Experimental Psychology Section A, 49*(1), 80–115.

Henson, R. N. A. (1998). Short-term memory for serial order: The start-end model. *Cognitive Psychology, 36*, 73–137.

Henson, R. N. A. (1999). Positional information in short-term memory: Relative or absolute? *Memory & Cognition, 27*, 915–927.

Houghton, G. (1990). The problem of serial order: A neural network model of sequence learning and recall. In R. Dale, C. Mellish, & M. Zock (Eds.), *Current research in natural language generation*. London: Academic Press.

Houghton, G. (2018). Action and perception in literacy: A common-code for spelling and reading. *Psychological Review, 125*(1), 83–116.

Houghton, G., Glasspool, D., & Shallice, T. (1994). Spelling and serial recall: Insights from a competitive queueing model. In G. D. A. Brown, & N. C. Ellis (Eds.), *Handbook of spelling: Theory, process and intervention*. Chichester: John Wiley & Sons.

Jacobs, A., Rey, A., Ziegler, J., & Grainger, J. (1998). MROM-p. An interactive activation, multiple readout model of orthographic and phonological processes in visual word recognition. In J. Grainger, & A. M. Jacobs (Eds.), *Localist connectionist approaches to human cognition*. Hillsdale, N.J.: Lawrence Erlbaum Associates Ltd.

Lashley, K. S. (1951). The problem of serial order in behavior. In L. A. Jeffress (Ed.), *Cerebral mechanisms in behavior*. New York: Wiley.

Levelt, W. J. M., Roelofs, A., & Meyers, A. S. (1999). A theory of lexical access in speech production. *Behavioral and Brain Sciences, 22*, 38–75.

Lewandowsky, S., & Murdock, B. B. B., Jr. (1989). Memory for serial order. *Psychological Review, 96*(1), 25–57.

MacKay, D. G. (1970). Spoonerisms: The structure of errors in the serial order of speech. *Neuropsychologia, 8*, 323–350.

Majerus, S., & Cowan, N. (2016). The nature of verbal short-term impairment in dyslexia: The importance of serial order. *Frontiers in Psychology, 7*, 1522.

McClelland, J. J. L., & Rumelhart, D. D. E. (1981). An interactive activation model of context effects in letter perception: I. An account of basic findings. *Psychological Review, 88*(5), 375–407.

McCloskey, M., Fischer-Baum, S., & Schubert, T. (2013). Representation of letter position in single-word reading: Evidence from acquired dyslexia. *Cognitive Neuropsychology, 30*, 396–438.

Miozzo, M., Petrova, A., Fischer-Baum, S., & Peressotti, F. (2016). Serial position encoding of signs. *Cognition, 154*, 69–80.

Neath, I., & Crowder, R. R. G. (1990). Schedules of presentation and temporal distinctiveness in human memory. *Journal of Experimental Psychology: Learning, Memory, and Cognition, 16*(2), 316–327.

Nootebloom, S. G. (1969). The tongue slips into patterns. In A. A. G. Sciarone, A. J. van Essen, & A. A. van Raad (Eds.), *Nomen: Leyden studies in linguistics and phonetics*. Mouton: The Hauge.

Norris, D., & Kinoshita, S. (2012). Reading through a noisy channel: Why there's nothing special about the perception of orthography. *Psychological Review, 119*(3), 517–545.

Nozari, N., & Novick, J. (2017). Monitoring and control in language production. *Current Directions in Psychological Science, 26*(5), 403–410.

Olson, A., Halloran, E., & Romani, C. (2015). Target/error overlap in jargonaphasia: The case for a one-source model, lexical and non-lexical summation, and the special status of correct responses. *Cortex, 73*, 158–179.

Olson, A. A. C., Romani, C., & Halloran, L. (2007). Localizing the deficit in a case of jargonaphasia. *Cognitive Neuropsychology, 24*(2), 211–238.

Perfors, A., Tenenbaum, J. J. B., & Regier, T. (2011). The learnability of abstract syntactic principles. *Cognition, 118*(3), 306–338.

Rumelhart, D. D. E., Hinton, G. G. E., & Williams, R. R. J. (1986). Learning representations by back-propagating errors. *Nature, 323*(6088), 533–536.

Seidenberg, M. M. S., & McClelland, J. J. L. (1989). A distributed, developmental model of word recognition and naming. *Psychological Review, 96*(4), 523–568.

Stemberger, J. P. (1990). Word shape errors in language production. *Cognition, 35*, 123–157.

Taft, M. (1979). Lexical access-via an orthographic code: The basic orthographic syllabic structure (BOSS). *Journal of Verbal Learning and Verbal Behavior, 18*(1), 21–39.

Tan, Y., & Martin, R. R. C. (2018). Verbal short-term memory capacities and executive function in semantic and syntactic interference resolution during sentence comprehension: Evidence from aphasia. *Neuropsychologia, 113*, 111–125.

Tan, Y., Martin, R. R. C., & Van Dyke, J. J. A. (2017). Semantic and syntactic interference in sentence comprehension: A comparison of working memory models. *Frontiers in Psychology, 8*, 198.

van Dijck, J. J. P., & Fias, W. (2011). A working memory account for spatial–numerical associations. *Cognition, 119*(1), 114–119.

Wickelgren, W. A. (1965). Short-term memory for repeated and non-repeated items. *The Quarterly Journal of Experimental Psychology, 17*, 14–25.

Wickelgren, W. A. (1969). Context-sensitive coding, associative memory and serial order in (speech) behavior. *Psychological Review, 76*, 1–15.

CHAPTER THREE

Neurocomputational Emergentism as a Framework for Language Development

Arturo E. Hernandez[1], Juliana Ronderos and Hannah L. Claussenius-Kalman
Department of Psychology, University of Houston, Houston, TX, United States
[1]Corresponding author. E-mail: aehernandez@uh.edu

Contents

1. Localization of Function 58
2. The Sensorimotor Hypothesis 61
3. Variability of Language Development 67
   3.1 Flexibility in Reading and Cognition 71
   3.2 The Basal Ganglia 73
4. Neuroemergentism: Dynamic Interactions Across Systems 75
5. Conclusion 78
Acknowledgments 78
References 79

Abstract

Neurocomputational Emergentism (Neuroemergentism) is a new framework that provides explanation for the nonlinear dynamics that occur across development. Recent findings in the literature are consistent with the view that the brain involves emergent properties in which processing at a small scale is dynamically aggregated to produce more complex cognition. In this piece, a link is made between Emergentism at the cognitive level with that observed at the neural level. Language development in bilinguals and multilinguals, which shows many nonlinearities, is discussed in light of Neuroemergentism. These results are discussed with regard to reductionism and the importance of sensorimotor processing during early development. The piece ends by considering a model in which reward-based learning systems, declarative memory and sensorimotor processing contribute to language development.

The ways in which language development is affected via dynamic accumulation of change across time has a long tradition within the Emergentist framework (Li, 2015; MacWhinney & O'Grady, 2015). One example

Psychology of Learning and Motivation, Volume 68
ISSN 0079-7421
https://doi.org/10.1016/bs.plm.2018.08.010

from Bates and MacWhinney suggested that language was a new machine built out of old parts (Bates, Benigni, Bretherton, Camaioni, & Volterra, 1979). In this view, language emerges from the interaction of a number of neural and cognitive processes that combine and recombine in order to yield a much richer result, the adult language system. Evidence of this can be seen in seminal work on children with pre- or peri-natal focal lesions showing a very different pattern of symptom-to-lesion mapping relative to adults (Bates et al., 1979; Stiles, Reilly, Levine, Trauner, & Nass, 2012). Most importantly, the areas that predict language impairment change across time depending on the age of the child. The seemingly irregular relationship between age and lesion-symptom mapping across development suggests that the language system itself is built up over time in a highly dynamic and non-linear manner. In this piece, we aim to extend studies in Emergentism to more carefully consider brain-behavior relationships by proposing Neurocomputational Emergentism or Neuroemergentism for short (Hernandez et al., 2018).

Emergentism has a long history in scientific thought beginning with the pioneering work of John Stuart Mill (1974) using examples from chemistry to describe his views. One classic example involves, hydrogen and oxygen, which when combined lead to a completely new compound—water. Stuart Mill also had opinions on Emergence in the field of cognition (Mill, Robson, McRae, & Routledge & Kegan P.University of Toronto, 1974). Following his views on compound interactions he extended his ideas by suggesting a form of "mental chemistry." In this view, mental activity could be the product of other mental events, events that took place in a person's body, or a combination of both. His views on cognition were not only theoretical but also led to direct testing at the empirical level. For example, he suggested that attention could be divided or selective. This view was the subject of follow-up work by a number of investigators including William James and eventually led to the modern area of attention within cognitive psychology. Hence, Emergentism has been at the root of thinking in cognitive and experimental psychology since its beginning.

Emergentist undertones are also present in some recent theoretical work in the neuroimaging literature. In the past ten years, some have suggested that newer cognitive functions (i.e., reading) are a product of phylogenetically old systems that are repurposed for new uses (Anderson, 2010; Dehaene & Cohen, 2007). This has been applied to domains such as reading and arithmetic (Anderson, 2010; Dehaene & Cohen, 2007). The Neuronal Recycling hypothesis (Dehaene & Cohen, 2007) proposes that the brain learns new cultural inventions (i.e., reading and arithmetic) through reconfiguring areas which were initially used for more basic, but related, functions into areas capable of more complex functions through the individual's experience. Similarly, Anderson (2010) argues for the theory of Neural Reuse but expands the concept of recycling beyond cultural constructs such as reading and arithmetic to all novel cognitive functions. The cortical areas which are reused, or borrowed, to accomplish the novel function must have had similar previous functional properties to the novel function being used.

Other frameworks have also considered the dynamic nature of language and cognitive processing. In Language as Shaped by the Brain, Christiansen and Chater (2008) use a similar framework as Neural Reuse to explain language acquisition through the temporal and structural constraints of language processing. In this view, language is acquired as it is continuously processed through individual utterances and through the specific construction combinations learned from the language environment. Christiansen and Chater (2008) argue that there are domain-general constraints to language acquisition and language processing and that change comes through the gradual adaptation of language processing that best fits the existing structural and neural constraints. Neuroconstructivism, as espoused by Karmiloff-Smith (2009), considers not only the importance of ontogenesis but takes it even further through the role of the multidirectional interaction between genes, neural components, cognition and the environment during the lifespan of the individual. The importance of dynamic interactions across multiple levels is one that is at the heart of Emergentist approaches to language.

The following sections seek to flesh out in more detail particular aspects of Neuroemergentism, a newer framework which seeks to extend Emergentism to consider computationally and neurobiologically based aspects of processing First, the focus on localization of cognitive functions within an evolutionary framework will be discussed, specifically within Neuronal Recycling which makes claims about a particular brain region that is dedicated to visual word form processing. In a second section, the sensorimotor hypothesis, a precursor to Neuroemergentism will be discussed as a hypothesis that seeks to consider how the brain systems available for cognitive functions differ depending on age. A subsequent section will focus on specific topics within bilingualism and how Neuroemergentism can be applied to elucidate the nature of these phenomena. Finally, our discussion will conclude by considering Neuroemergentism with regard to three particular brain-based systems: reward-based learning, memory and sensorimotor function. Particular attention will be paid to the ways in which these three systems interact across time in development.

1. LOCALIZATION OF FUNCTION

Localizing cognitive functions is a long-held tradition both in cognitive neuropsychology and cognitive neuroscience (see Miller, Crocker, Spielberg, Infantolino, & Heller, 2013 for an example). Dehaene et al. (2010) take localization one step further by incorporating evolution in what they term Neuronal Recycling. One clear example of this can be seen in the notion of the visual word form area (VWFA), which is in a functional region of the left fusiform gyrus and surrounding cortex. It is posited to involve older neural systems since there has not been enough time on an evolutionary scale for a reading specific adaptation. This results in a straightforward form of reasoning: 1) Humans (actually many primates) had to match an object with a sound, which is enhanced by using the VWFA for matching across visual and auditory modalities; 2) Language appears in humans; 3) Humans begin to write symbols corresponding to spoken language; 4) The systems of reading and writing recruit an area that was formerly involved in visual object recognition and processing of their corresponding sounds. Dehaene then traces out this relationship in more detail. For example, studies using structural interconnectivity of the fusiform gyrus in children who have not yet learned to read can "predict" what portion of this brain structure will show increased activity when reading and thus reveal each individual's own future VWFA (Dehaene et al., 2010). Dehaene et al. have also conducted numerous studies in which they show how the visual word form area is involved in reading wordsFurther studies actually show how particular object categories are reconfigured as children improve their literacy during the first few year of schooling (Dehaene & Dehaene-Lambertz, 2016). In short, they have focused on a particular brain area (the visual word form area) which is taken to be recycled from basic multimodal operations that involve matching visual and auditory footprints of a particular event or object (Dehaene & Cohen, 2007, 2011; Dehaene & Dehaene-Lambertz, 2016).

Despite the consistency of results in favor of the VWFA (Dehaene & Cohen, 2011; Dehaene et al., 2010), there is also considerable research that provides counterevidence to this view (See Dick & Krishnan, 2018 for further discussion). In a seminal counterpoint to Dehaene et al., Price and Devlin (2003) provide considerable evidence that the VWFA was not specialized for visual words. For example, upon reviewing the literature on alexia, a reading deficit due to brain damage, they found no evidence that it was caused by "a one-to-one mapping between visual word form recognition and a specific patch of cortex in the left midfusiform gyrus" (Price & Devlin, 2003, p. 475). The rest of their article reviews evidence from a number of studies that include tasks such as picture naming and repeating auditory words, tasks that do not involve print-to-speech transformations that fueled the VWFA proposal. Price and Devlin (2003) finish by offering three possibilities in this respect: 1) the same area of the brain may have sub-populations involved in different functions; 2) brain activity in the VWFA may involve a single underlying cognitive function not specific to the processing of visual words or even objects in the visual modality; 3) neurons in the midfusiform gyrus show different functions depending on the patterns of co-activation in different parts of the cortex. Price and Devlin (2003) lean slightly toward the last explanation of the three, especially in light of the fact that this region of the brain is highly interconnected with other brain areas (Damasio & Damasio, 1994).

More recently, Vogel, Miezin, Petersen, and Schlaggar (2012), and Vogel, Petersen, and Schlaggar (2014) have followed up on these studies by investigating the nature of brain activity of the VWFA using two separate analyses in monolingual English speakers. First, there was no evidence of greater activity in left occipitotemporal (OT) regions for words or letter strings relative to other stimuli with high—spatial frequency and high-contrast. The clearest example was the Amharic strings that constitute the written system in Ethiopian. In a second analysis, the authors further refined their characterization of this area. Specifically, they investigated an area that was sensitive to the number of strokes forming groups of letters or Amharic strings and also processes learned combinations of characters, such as those in words and pseudowords, as groups. However, this same area did not show activity for learned combinations in consonant and Amharic strings. These results suggest that at least parts of the OT cortex are responsive to letters and hence are not a "word form" area. Although the specific function (and strict localizability) of the VWFA have been the topic of considerable debate, this discussion highlights how areas of the brain may adapt, over time and development, and may adopt fairly specific cognitive functions.

Additional work on an area close to the VWFA that is responsive to faces and termed the fusiform face area (FFA) has also been discussed in a similar light. The original finding of the FFA follows along similar lines to the argument for the VWFA by suggesting that this brain area is specialized for face processing (Kanwisher & Yovel, 2006; Kanwisher, McDermott, & Chun, 1997; Kanwisher, Stanley, & Harris, 1999). Much like discussion with regard to the VWFA there has been considerable controversy with the specificity of brain activity in the FFA. For example, Haist, Lee, and Stiles (2010) were able to replicate and extend studies that had identified results with regard to the FFA. Like other studies they demonstrated, using a passive viewing task, stronger neural activity in the FFA was observed for faces compared to objects such as wristwatches. In another condition, participants were asked to select specific exemplars across both face and non-face categories. In this condition, preference for faces in the FFA showed significant reduction. That is, faces or object stimuli elicited very similar brain activity. Hence, the act of trying to identify a particular item, which is common when viewing faces, also applies to the processing of other objects. A similar finding has emerged in work by Gauthier et al. (2014) with artificially-created stimuli such as greebles (Gauthier, Tarr, Anderson, Skudlarski, & Gore, 1999) or with experts for specific objects such as birds or cars (Gauthier, Skudlarski, Gore, & Anderson, 2000). Taken together these results suggest that it is the underlying process of identifying specific items within a category that is associated with the fusiform face area. Thus, it is not a face area. It just happens to be the case that face processing in humans involves the need to identify particular individuals to a much greater extent. This accumulation of experience can lead to regions seemingly becoming specialized for particular functions with experience.

One particularly interesting phenomenon arises when one considers how the processing of words and faces are related. Studies have found that illiterate adults who learn to read show changes in brain activity in the mid-fusiform gyrus with face "areas" showing greater and greater activity to words with increasing literacy (Dehaene et al., 2010). In a similar vein, children also show a pattern in which areas originally sensitive to faces in the left mid-fusiform gyrus become more and more sensitive to words (Centanni, King, Eddy, Whitfield-Gabrieli, & Gabrieli, 2017). The authors suggest that this increase in left hemisphere sensitivity to words may be the reason that sensitivity to faces shifts to the right hemisphere dominance.

Neuroemergentism, the view that the brain and the concomitant behavior supported by it goes through successive waves of reorganization across time thus contributes to our discussion of the FFA and VWFA. Whereas the discussion has focused on the presence of an FFA or VWFA by one set of researchers and the lack thereof by another set, there has been little discussion of the developmental nature of this dynamic process. Our reading of this literature suggests that it is highly consistent with an Emergentist view in general and with a Neuroemergentist view in particular. Specifically, the addition of reading to a brain that is handling face processing leads to considerable rewiring. Areas previously involved in processing specific exemplars of categories like faces are co-opted for reading leading to a right/left asymmetry with regard to verbal and non-verbal processing. In short, brain areas that might be involved in complex visual processing transform and take on much more specific functions. An understanding of this transformative process would be enhanced by taking a more dynamic view such as that espoused within an Emergentist framework and more specifically within Neuroemergentism.

2. THE SENSORIMOTOR HYPOTHESIS

Understanding the role that sensorimotor processing plays in cognition requires further thinking about how representations at different levels of abstraction interact or compete. Competition was a construct that Bates and MacWhinney (1988) pioneered in their prescient body of work that predated the non-linear dynamical models proposed by connectionists to account for language development. Although the bulk of their work focused on grammatical processing, extensions of the competition model have also sought to look at lexical processing. Bates and Goodman (1997) argued that lexical processing and grammatical processing are highly related. For example, the relationship between vocabulary size and the complexity of grammar in child's speech is very high ($r = 0.84$). In addition, the size of vocabulary also impacts the dominance of irregular and regular verbs in English. With a small vocabulary, very frequent irregular verbs (i.e. go-went, have-had) dominate. As vocabulary size increases, there is an influx of regulars (jump-jumped). Hence, the dominance of irregulars and regulars are due to changes in the numbers and types of words that appear in a child's vocabulary.

Another compelling extension of the Competition Model can be seen in the DevLex model that uses self-organizing phonological and semantic maps (Li, Farkas, & MacWhinney, 2004). In this simulation, the authors chose to model two particular phenomena in the literature using naturalistic parental input. First, Li et al. modeled typically developing children's tendency to produce the wrong word, termed lexical confusion which appears during a vocabulary spurt between 18 and 24 months. During this spurt, DevLex simulations revealed that confusions depended on density—i.e., how many items were produced varied as a function of whether words were in densely populated meaning neighborhoods versus in less dense neighborhoods. For example, young English-speaking children have vocabularies with many more nouns than verbs which may lead to greater word confusions for the former (Gentner, 1982). Results from DevLex suggest that competitive processes such as those suggested by the Competition Model appear in vocabulary acquisition. Furthermore, the model also shows how parental input leads to a dynamic emergent structure in the vocabularies of young children. Li, Farkas and Macwhinney do not explicitly link their results with a neurally-based instantiation of the model. However, they do note that these connectionist models could also be related to brain activity, which is consistent with a Neuroemergentist approach.

The DevLex model also modeled single language age of acquisition effects. One of the most difficult aspects of these phenomena for connectionist models is catastrophic interference, the finding that new learning can completely overwrite older learning. In order to properly model age of acquisition effects, a network must be stable enough to maintain early learned words while at the same time flexible enough to continue making changes. Interestingly, the effects of age of acquisition could also be seen in the density of particular neighborhoods of early and late learned words. Early learned words tended to be alone or have few items that had similar meaning or sounds since at the beginning of learning there were very few items. Younger children simply know fewer words. Late learned words arrived at a time in which there were substantially more words and hence tended to be part of much denser word neighborhoods. Hence, the density of the word neighborhoods at the time of learning distinguishes between early and late learning. Like word confusion, age of acquisition effects arise from increased competition that occurs across different ages.

Discussion about the way in which early and late learning differs across language and non-language domains was proposed by Hernandez and Li (2007), who suggested that early learning requires greater sensorimotor processing compared to late learning, which involves greater dependence on higher-level cognitive processing. This was termed the Sensorimotor

Hypothesis. As we just noted, analysis of this hypothesis involved computational models, including models such as DevLex and other connectionist models at the word level (Li et al., 2004). Specifically, age of acquisition effects appear in the self-organizing maps of DevLex that have a phonological component. Since the processing of speech sounds is by definition reliant on sensory processing, it forms a crucial aspect within the sensorimotor hypothesis. Another source of data that has informed this model comes from a study that investigated the neural activity associated with lexical decisions in a group of German native speakers. In that study, Fiebach, Friederici, Müller, von Cramon, and Hernandez (2003) asked participants to make lexical decisions for early and late learned words that were presented visually. Results revealed increased neural activity for early-learned words in left Heschl's gyrus as well as the mid-precuneus. Heschl's gyrus is involved in the processing of spoken sounds whereas the mid-precuneus is involved in putting together information from multiple senses. Late learned words showed increased activity in the inferior frontal gyrus extending up to BA 47, an area associated with the processing of meaning.

Fiebach et al. (2003) interpreted these differences as arising from the fact that early learning is more impacted by sensory processing than late learning. Other follow-up studies have also supported the sensorimotor hypothesis. For example, Hernandez, Mattarella-Micke, Redding, Woods, and Beilock (2011) tested a group of skilled golfers who learned to play golf before (early learners) or after (late learners) the age of 10. Participants were placed into one of three conditions: putt only (single-task condition), putt while monitoring a stream of words (dual-task condition), or putt while being instructed to pay attention to specific aspects of their golf swing (skill-focused condition). Hernandez et al. (2011) found a significant interaction between age of acquisition and condition. Both the early and late AoA groups showed equal performance in the single-task and the dual-task putting conditions. However, early learners performed worse in the skill-focused condition where they were required to focus on a physical characteristic of their golf swing. The authors suggest that attending to a specific part of a golf swing is more difficult for early learners since it involves breaking an early learned motor chunk into parts. This is consistent with the sensorimotor hypothesis in that it suggests that early learning of a motor skill is more closely tied to a sequence of movements. For late learners, sequences were learned in a more cognitive manner and hence attending to a body part did not disrupt their performance. Taken together, there is evidence that the sensorimotor hypothesis does account for differences between early and late learning.

The sensorimotor hypothesis has its origins in Piagetian Constructivism, which posited the first two or three years as involving learning in a physical sense. However, Piaget never directly gathered data on the brain with regard to his measures. It was Karmiloff-Smith's theorizing that took into account brain development in order to propose Neuroconstructivism (D'Souza & Karmiloff-Smith, 2016; Karmiloff-Smith, 2009). Neuroconstructivism considers development to be the product of interactions at several levels including but not limited to genes, neural components, cognition, and the environment across an individual's lifespan (Karmiloff-Smith, 2009). Neuroconstructivism posits that human abilities are a result of a dynamic nonlinear process across time. The enhanced plasticity of the brain at birth results from a greater number of synapses that are then pruned back during the first few years of life (Huttenlocher, 1999; Matsui et al., 2016). Because the brain is more highly interconnected and not as specialized, Karmiloff-Smith posits that interactions with the environment are crucial in order for the system to become specialized (Karmiloff-Smith, 2006).

One of the main points of the sensorimotor hypothesis is that different brain regions come on line at different points in time and thus differentially influence the performance of very similar tasks in individuals with different learning histories but similar performance profiles. For example, Fiebach et al.'s finding that early learned words led to increased activity in brain areas involved in the processing of speech sounds fits in nicely here. The notion is that early in childhood words are more closely associated with sounds. In that same study late learned words were associated with increased activity in areas involved in semantic processing. Late learned words are learned in relation to other words, as shown in the simulations of word learning. Hence, one can think of late learned words as being connected to other words that share a similar meaning. This all has to do with the particular brain systems that are in place when learning takes place.

In a similar vein, Neuroconstructivists see development as a crucial aspect in understanding of the relationship between the brain and cognition. The common practice in the field of neuropsychology is to create a link between cognitive deficits and particular brain regions. This can also be seen to a certain extent in the work of Dehaene et al. reviewed earlier on the visual word form area. Early work linking disorders to particular genes followed a similar logic. This type of reasoning appears in the study of individuals who show deficits in particular cognitive domains while other domains seem intact, leading to the search for specific genes that map onto performance. The observation of clear deficits in particular domains such as language or spatial processing often overlooks subtle deficits in the intact domains. Neuroconstructivists argue that brain damage is unlikely to lead to selective deficits. Thus, although Neurconstructivism and Neuronal Recycling both reflect some form of Emergentism they differ with regard to the ways in which brain systems develop. Specifically, in Neuroconstructivism small changes that occur early in life affect much larger swaths of the cognitive system, leading to changes across both simple and complex cognitive domains because the underlying basis of the building blocks are altered. The dynamic developmental view taken by Karmiloff-Smith is much less deterministic than Neuronal Recycling as proposed by Dehaene et al. Specifically, it emphasizes how small changes early in development can lead to substantially different outcomes. Finally, Neuroconstructivism is consistent with Price and Devlin's view that there is no VWFA but rather that this portion of the fusiform gyrus is sensitive to many different types of stimuli not just to visually-presented words.

An example of the Neuroconstructivist view can be seen in the characterization of developmental disorders that create very specific deficits. Individuals with Williams Syndrome (WS),show specific deficits in spatial and numerical processing while having relative sparing of language and face-processing. WS is a genetic developmental disorder caused by the deletion of about 28 genes from the long arm of chromosome 7 (Donnai & Karmiloff-Smith, 2000). These patients show proficient use of language and fall within the normal range on norm-referenced face processing tasks. This behavioral profile coexists with a mean IQ of 56 (Mervis et al., 2000) and with very impaired spatial and numerical cognition (Karmiloff-Smith et al., 2004). The brain in individuals with WS is generally smaller and it has overall reduced curvature and increased gyral complexity, in addition to reductions in parietal and limbic regional volumes compared to the typical adult brain (Eisenberg, Jabbi, & Berman, 2010). Furthermore, while the typical adult brain develops gradual specialization and hemispheric localization of function over ontogenetic development (i.e. left hemisphere for language), this does not to occur in the mature brain of individuals with WS, which continues instead to process faces and language bilaterally (Karmiloff-Smith et al., 2004).

Though WS appears on the surface to represent an impairment in specific domains (spatial and numerical cognition) with preservation of others (language and face processing), Neuroconstructivists suggest a possible alternate explanation. Specifically, Karmiloff-Smith and colleagues suggest that the immediate causes of these deficits may be more general than would appear on the surface. In the case of WS they have argued that children may have an underlying deficit in eye movements which over time differentially affect different domains. In some ways, this argument is similar to that provided by Price and Devlin with regard to the visual word form area (Price & Devlin, 2003, 2011). Specifically, the presence of neural activity in the same brain area across multiple domains is consistent with the view that all of these domains share something in common. Neuroconstructivists take this a step farther by suggesting that small differences early in development could cascade differentially into different domains and lead to uneven profiles of cognitive processing.

Another example of how early physical characteristics can have an effect on cognitive development has recently been continued by D'Souza and D'Souza (2018). They argue that certain factors that are on the surface and are not connected directly to language also can also influence the emergence of language functions. One very concrete example offered by D'Souza and D'Souza is the infant's body. The authors argue that by constraining access to real time sensory information, the body serves as a guide for learning and development. The authors also note how infants' sensorimotor limitations may help to restrict the types of information that they manipulate. Short sightedness keeps the realm of visual exploration to the most noticeable parts of any particular object. Physical constraints such as their short arms could limit exploration to one object at a time; their short legs, exploration to only part of a scene.

An example of the ways in which the body constraints development can be seen in a recent study by Pereira, Smith, and Yu (2014). Pereira et al. used head-mounted cameras on infants and mothers which recorded gaze data while infants played with novel objects as the mother spontaneously named them. They found that when the infants grasped an object, the object filled their entire field of view due to the short length of their arms. In addition, infants were much more likely to learn the label of an object if the mother named it while it was in the infant's grasp. Hence, the body of an infant is in essence restricting each label to a single object. These findings reveal the ways in which sensorimotor constraints that serve as the basis of dynamic interactions play a role.

The sensorimotor hypothesis of Hernandez and Li (2007), a precursor to Neuroemergentism, also discusses how the child is constrained by a focus on processing of the world via the senses and their corresponding motor response. Neuroconstructivism sheds light on the sensorimotor hypothesis by signaling the extent to which brain and cognitive development occur as a cascading non-linear process. As we will discuss later, it also serves as a precursor to Neuroemergentism, the framework which we will discuss toward the end of this article.

3. VARIABILITY OF LANGUAGE DEVELOPMENT

As noted earlier, the nature of language acquisition within an emergentist framework such as the competition model involved the interplay between cues to meaning. Although a lot of the early work focused on cross-linguistic differences, Bates and colleagues also considered the effects of individual differences. Seminal work by Nelson (Nelson, 1973) found that children adopted a referential style involving the use of common objects or adopted an expressive style in which many different types of items were used including some canned phrases such as "stop it" and "I love you." Building on these findings, Bates, Bretherton, and Snyder (Bates et al., 1988) looked at language development longitudinally with a group of English speaking children. These results also found that children varied on a continuum between expressive and receptive forms of processing with the vast majority of children lying somewhere in between the two extremes. In addition, these results also show a continuum of development from object names at 13 months, expansion of all other content words (i.e. nouns, verbs and adjectives) at 20 months, and then grammatically complex speech at 28 months. Children that are precocious are early during each one of those stages that involve what might appear on the surface to be different tasks. Thus, language acquisition involves moving from one type of processing to another and we could infer from this that difficulty with object names at 13 months would likely lead to problem with grammatical complexity at 28 months. This is very resonant with the Emergentist approach in which processing early in life transforms across time into a very different type of processing later in life. This refers both to the speed with which children will achieve different milestones but also with respect that expressive and receptive approaches by children to language learning could result in different trajectories through this landscape.

The variability in the nature of language acquisition is also relevant within a multilingual context. For example, Bradley, King, and Hernandez (2013) asked young adult bilinguals and monolinguals to learn a set of vocabulary words in a language that was novel to them (i.e., German). When asked to make semantic judgments of the newly-learned words, monolinguals showed more activation than bilinguals in regions related to executive function and planning, such as the right dorsolateral prefrontal cortex and right supplementary motor area. On the other hand, bilinguals engaging in this task showed more activation in regions involved in language production and articulation, such as the right putamen. This finding demonstrates that encounters with language within one lifetime can influence the brain's methods of representing future languages, and that the neural bases of language seem to build heavily on previous experiences, a theme that is resonant with the sensorimotor hypothesis.

Another group of researchers (Bloch et al., 2009) aimed to understand third language acquisition by studying a group of trilinguals who had all learned their third language after age 9 but differed in the ages at which they had acquired their second language (either at birth, between 1 and 5 years, or after age 9). While in a functional magnetic resonance imaging (fMRI) scanner, the participants completed a silent narration task in each of their three languages. Then researchers compared brain activation associated with third language production across participants. They found that earlier L2 acquisition was associated with a more homogeneous set of activation patterns, and that as age of L2 acquisition increased, so did individual variability in activation. Earlier L2 acquisition was more closely related to regions traditionally implicated in language production such as left Broca's and bilateral Wernicke's area. As age of L2 acquisition increased, there was more involvement of a wider range of brain areas that was more subject to individual variability. Additionally, Dutch/French/English multilinguals engage a broader range of brain areas the later they acquired each language (Vingerhoets et al., 2003). It seems to be the case that learning two languages simultaneously in childhood affects the representation and localization of future (i.e., third) languages, even when the third language is learned more than a few years later.

The idea that there are multiple avenues to language acquisition is further supported by research examining brain structure. One (Archila-Suerte, Woods, Chiarello, & Hernandez, 2018) found that the neuroanatomical profiles of bilingual children with balanced proficiency in English and Spanish were different from those of bilingual children with unbalanced proficiency (i.e., those who had higher proficiency in Spanish, the native language). Specifically, balanced bilingualism, compared to unbalanced bilingualism, was related to thinner cortices of the left superior temporal gyrus, inferior frontal gyrus, and middle frontal gyrus, areas in which thicker cortex has been related to more difficulty processing a second language (Mårtensson et al., 2012), and a larger bilateral putamen, thought to be involved in phonological skills and word production (Cherodath, Rao, Midha, T A, & Singh, 2016). Such differences in brain morphology highlight the notion that different experiences in language lead to corresponding adaptations in the brain.

Multilingual acquisition also sheds light on the fact that age of acquisition plays a role in the learning of language as posited by the sensorimotor hypothesis. The reason for this phenomenon may be that first language learning occurs early in development, when sensorimotor structures such as the basal ganglia are in the midst of development. Given that Bloch et al. (2009) noted brain activation during the L1 task that was not just more homogenous than L2 or L3 tasks, but more likely to be localized to the basal ganglia and classic language regions (Broca's and Wernicke's areas), it is likely that there is a shift from sensorimotor learning during early development to learning that is dependent on higher-level cognition in later development (Hernandez & Li, 2007). When a second language is introduced in childhood, the basal ganglia adapts to permit accurate and efficient switching between languages (Stocco & Prat, 2014). This would explain why bilinguals have been found to recruit motor structures when using a new language, but use of a new language for monolinguals seems to depend on the caudate and prefrontal cortex (Bradley et al., 2013). Abutalebi and Green (2007) propose that the prefrontal cortex is recruited when language processing is effortful, requires cognitive control, and cannot be achieved through automatic processing. Together, the above-mentioned results provide support for the idea that the neural substrates recruited during language learning depend heavily on previous experience, which in turn affects future language learning.

Given that new languages increasingly build on previous knowledge, De Bot, Jaensch, and García Mayo (2015) stress the importance of considering all of a multilingual's languages in tandem with one another, rather than treating them as separate entities. For instance, some argue that multilingualism is facilitated through interacting processes, rather than fixed building blocks where a second language is laid on top of the first language and then the third language is laid on top of the second (Herdina & Jessner, 2002, 2013). This idea fits nicely with the scalpel model of third language acquisition (Slabakova, 2017), which argues that L3 grammar pulls with scalpel-like precision from useful grammar properties of the L1 and L2.

Rather than the first language being the most dominant, both languages interact with the third language and provide multiple jumping-off points for successful third language acquisition. Thus, multilingual learning may provide a very interesting alternative view for the development of language even in monolinguals and of the learning of skills in general. Previous knowledge at different points in time provide different types of "jumping off" points in a nonlinear, dynamic and cascading fashion.

The dynamic nature of multilingual acquisition also sheds light on a different discussion within the language learning literature related to language attrition in learners of more than one language. Language attrition refers to the loss of language skills across time when people are exposed to another language (Schmid, 2011). It is especially apparent in immersion contexts where use of a language is discontinued or reduced. Researchers discuss examples of what appears to be complete loss in some individuals who experience extreme foreign language immersion in childhood. In other cases, there is a loss of sensitivity to the phonological contrasts in a native language. This is an important and fascinating topic not just because of its relevance to bilingualism but also because of what it tells us about plasticity in the child and adult language systems (for a review see Schmid, 2011).

There is no doubt that individuals exposed to a new language show changes in the other language(s) they speak. The multilingual literature uses this as a given. However, it is unclear exactly why researchers use the word attrition. The use of this word implies that something is lost. We think a better term would be language absorption and/or diffusion. Our memories always involve a mixing of old and new memories. Language is similar in that sense. Rather than a language being lost, which is what attrition implies, we would argue that it is absorbed or diffused which fits in with the notion of being repurposed. This is very resonant with the views of neuronal recycling, neural reuse and Neuroconstructivism discussed earlier. It is also consistent with a Neuroemergentist approach.

For an adult, absorption may be too simple a term and perhaps we could consider language diffusion. In short, the representations and processes acquired for a particular language are absorbed, repurposed, and reconfigured for a different use. This view can be seen in the earlier discussion which highlights how a first language script affects reading in a second language. Thus, a child who is exposed to Spanish at home but learns in English at a young age may show a reduction in Spanish knowledge over time. Later we will discuss how learning of two languages can affect awareness of speech sounds and meaning.

The dynamic reconfiguration process that occurs in the learning of two languages, or the transition from one dominant language to another one, shows us that the relationship between two languages is not straightforward. Hernandez (2013) described the notion of two languages as being like two species in one ecosystem both cooperating and competing for resources at any given time, which is consistent with data on language and cognition. One interesting question that remains is attempting to better integrate work in bilingual and multilingual language development with the general language development literature. For example, Brian MacWhinney proposed a Unified Competition Model (MacWhinney, 2004) which conceptualizes first and second language learning as involving both risk-generating and protective processes. Risk-generating processes allow for the system to learn new things, whereas protective processes allow for systems to maintain existing knowledge. MacWhinney suggests that the difference between L1 and L2 acquisition (and beyond!) hinges on the particular configuration of these processes. The ability to learn new things, flexibility, while maintaining the old, stability, has also been considered in the cognitive neuroscience literature. In the following sections we will consider stability and flexibility in reading before proceeding to the neural areas that might be involved in this process.

3.1 Flexibility in Reading and Cognition

Another way to look at Neuroemergent processes is to consider how literacy might alter language representation. For example, reading in English by children who speak another language is colored by the nature of the other language. Take the case of children in Singapore who have Chinese, Tamil or Malay as a first language (L1) and learn to read in English. Results reveal that reading in English may vary depending on the type of script in L1. Chinese involves a logographic alphabetic system in which one-to-one correspondence between sound and print is not clear. Malay, on the other hand, is alphabetic which leads to a much more direct connection between the sound(s) of a written letter or set of letters. Children who learn Chinese as their L1 are particularly adept at reading sight words in English. However, because their L1 is not alphabetic, they have difficulty with decoding words relative to those who have Malay as their L1. This has to do with the fact that the native language is biasing either a semantic or phonological based approach to learning to read in the second language. In a similar vein, studies have found enhanced phonological awareness in Spanish-English bilinguals relative to both English monolinguals as well as Chinese-English bilinguals. This is due in part to the transparency of the sound-to-print mapping in Spanish. Other researchers have suggested similar beneficial transfer effects that depend on language typology. For example, work with children in Spanish-English dual-immersion programs finds benefit for both Spanish and English native speakers with regard to derivational morphology (i.e. -able, -ment, -city) in English (Kuo, Ramirez, de Marin, Kim, & Unal-Gezer, 2017). That is, learning that *plastic* and *plástico* can become *plasticity* and *plasticidad* leads children to start to notice this much more than English speaking children in English-only schooling. This shows the ways in which one language can add to the other in line with a Neuroemergentist view.

Finally, more recent work in our laboratory has looked at the influence of individual differences and language learning history on proficiency in a bilingual's two languages. This study built on previous work that identified two genetic variants related to dopamine functioning (e.g., the ANKK1/TaqIa polymorphism within the *DRD2* gene and the Val158Met polymorphism within the *COMT* gene) which had been shown to predict cognitive flexibility and learning (i.e., Colzato et al., 2010; Stelzel et al., 2010). These genetic variants have also been found to predict second language learning (Mamiya et al., 2016), although this relationship may not be stable across development (Sugiura et al., 2017). Results with a group of Spanish-English bilinguals revealed a three-way interaction between the genetic variants and AoA when looking at balanced proficiency level as the outcome variable. At earlier AoAs, individuals with the genetic variant associated with higher levels of subcortical dopamine (A1+) showed the highest levels of bilingual proficiency. At later AoAs, individuals with the genetic variant associated with cortical dopamine levels that are balanced between stability and flexibility (Val/Met) showed the highest levels of bilingual proficiency. These results are consistent with the sensorimotor hypothesis. Specifically, development of language involves subcortical areas of the brain early in life and involves cortical processes later in life (Hernandez & Li, 2007). It also shows the importance of both stability and flexibility in bilingual language development (Green & Abutalebi, 2013).

3.2 The Basal Ganglia

In the previous section, we discussed how bilingualism and multilingualism naturally elicit a dynamic and fluid model of cognitive and language development that is highly resonant with Neuroemergentism. In this section, we elaborate on neural models of bilingualism and how they relate to general models of cognition. The relationship between the frontal-striatal tract and the need to adapt to tasks which have competing information is one that has received considerable attention recently in the bilingual literature. Stocco, Yamasaki, Natalenko, and Prat (2014) were one of the first sets of researchers to suggest that bilingualism, in fact, provides brain training by strengthening the way subcortical structures, more specifically the basal ganglia, control and re-route signals to the pre-frontal cortex. Neuroimaging studies support the view that the neural representation of both languages in bilingual individuals overlaps greatly and that dynamic processes are involved between cortical and subcortical structures to produce the intended language (Abutalebi & Green, 2007). Furthermore, these studies suggest that the differences in terms of the regions recruited to produce language result from recruitment of additional regions required to produce the less automatic control over the second language (L2).

This overlap of a common neural substrate implies that as languages compete for access to these shared neural regions in a bilingual individual, a mechanism must exist to control which language to use. Stocco, Lebiere, O'Reilly, and Anderson (2012) argue that the same neural mechanisms used in executive function are linked to the neural substrates controlling language switching in bilingual individuals. Stocco et al. (2012) point to studies that report involvement of the basal ganglia during language switching and translating to support their hypothesis (Lehtonen et al., 2005; Price, Green, & von Studnitz, 1999). Although the basal ganglia are largely associated with their role in learning and skill acquisition, Stocco et al. (2012) propose a link between the bilingual language experience and improved executive functions through "training" of the fronto-striatal loops which are involved in controlling both. Recently, more neuroimaging studies have looked at the role of the basal ganglia in language processing supporting this proposal and the importance of this region in language control (Prat & Just, 2011).

In order to explain the role of the basal ganglia in the brain of bilingual individuals, Stocco et al. (2012) adopt the conditional routing model

(Stocco, Lebiere, & Anderson, 2010) as a computational model to examine the neural circuit involved. The conditional routing model assumes that the basal ganglia interferes in the exchange of signals between cortical regions (Stocco et al., 2012). During the learning process or in situations where automatized cognitive routines cannot accomplish a task (for example, in task-switching), the basal ganglia can amplify signals from a selected source region to enhance the likelihood that this signal will influence behavior despite weaker cortico-cortical network connections. This is important to understand how the brain in bilingual individuals selects and switches between possible language outputs and grammatical processes to produce the appropriate ones for a specific situation. Through the learning process, complex rules in grammar become automated rules in the basal ganglia permanently encoded in the form of patterns that determine signal routing (Stocco et al., 2010). Damage to the basal ganglia supports this, since damage would impair the automated rule-based behaviors in the more proficient language and spare, to some extent, a second language where rules are less explicitly encoded in the cortex. Stocco et al. (2012) link language and executive function through the lens of the conditional routing model and the common set of rules and computations that depend on the basal ganglia circuitry. They specifically point to the capacity to shift between mental tasks and to apply complex rules that depend on the capacity of the basal ganglia to properly route and amplify the signals from the appropriate regions in the network for the specific task. This is the basis of their theory that practice in one cognitive function (for example, bilingualism) can have positive impacts on other cognitive functions (for example, executive functions) that utilize the same communication circuitry via the basal ganglia.

Stocco and Prat (2014) propose that being raised bilingual "trains the brain", or rather the basal ganglia circuitry, to explain improved executive function in bilingual individuals (if there is indeed a bilingual executive advantage; see Paap et al., 2015 for discussion). Bilingualism imposes two competing forces challenging the basal ganglia circuitry. On the one hand the two languages must be kept distinct (each have unique morphosyntactical rules), however, they will share overlapping context, information, and brain resources. The basal ganglia (and specifically, the striatum) have an important role in maintaining these distinct representations of languages while selecting the relevant rules and the capacity to override habitual responses, in the appropriate circumstances, depending on the intended language. Stocco and Prat (2014) also argue that efficient practice in two or more languages has the unexpected result of increased flexibility to override signals to the prefrontal cortex controlling existing cortico-cortical connections. This flexibility "trains" the brain to do this for other executive functions that are required in response to changing task demands.

Although the notion of a bilingual advantage is one that has met with considerable skepticism, the notion of flexibility and stability is one that is resonant with a Neuroemergentist view of language development. One interesting point of debate is exactly how to conceptualize this process. For example, Dehaene propose that the VWFA involves the co-opting of areas that were originally involved in different functions (Dehaene & Cohen, 2007, 2011; Dehaene & Dehaene-Lambertz, 2016). Price and Devlin argue that the VWFA is involved in many different functions (Price & Devlin, 2003, 2011). Despite their differences both views are consistent with a Neuroemergentist view in which new functions emerge from the combination of simple building blocks. A similar argument could be made with regard to bilingualism and whether it involves growing a separate brain area for each language. A simpler alternative for the bilingual and multilingual brain is to adapt to the new problem space. This approach to problem solving will necessarily involve neural structures that have not traditionally been considered in the cognitive neuroscience of language development. Moving forward, research with models involving the basal ganglia will help to further uncover the flexibility and stability that are involved in the maintenance of old knowledge or skills in the face of the learning of new knowledge or skills, a hallmark of language development.

4. NEUROEMERGENTISM: DYNAMIC INTERACTIONS ACROSS SYSTEMS

The previous sections summarized the emergent nature of language processing including the sensorimotor hypothesis, and newer work with neuroplasticity in bilingual and multilingual language learning. In previous work, Hernandez et al. (2018) proposed Neurocomputational Emergentism, Neuroemergentism for short, to describe the ways in which behavior and the concomitant neural systems organize and reorganize themselves across time. This conceptual model is one that lends itself nicely to a neurocognitive account of development given its nonlinear dynamical nature. However, the particular neural substrates with which we could more clearly instantiate such an approach have not been as well articulated.

In the previous section, Stocco et al. propose a model to investigate the ways in which the brain handles competing rules. This approach to modeling flexibility and adaptability involves reinforcement learning (Stocco, 2018). An extension of this is to consider how reinforcement learning interacts with other brain systems. One approach, suggested by Stocco (2018), is to look at one system devoted to reinforcement learning and a second system devoted to declarative memory. The work of Stocco et al. is part of a broader movement entitled computational cognitive neuroscience. Computations at the neural level often show Emergentist properties. For example, researchers discuss network dynamics which contain interactions between populations of relatively simple neurons that coalesce into functional neighborhoods or networks that can encompass many different brain areas. Hence, the neurocomputations underlying cognition have Emergentist properties as complex cognition arises from relatively simple building blocks.

In suggesting a Neuroemergentist view of cognition one moves beyond a modular/interactive dichotomy by suggesting a more componential view of language involving memory and reward-based learning. One crucial aspect of language development could be thought of as consisting of reward-based learning and memory. Hence, some aspects of language such as syntax might rely to a greater extent on the basal ganglia which has been associated with reward-based learning whereas the learning of semantics relies to a greater extent on memory, which is driven by the hippocampus. In this sense, Stocco's view resembles the Declarative/Procedural model of Ullman et al. (Ullman, 2001, 2006).

The Declarative/Procedural model (Ullman, 2005, 2016) provides evidence that developmental changes support different types of learning across development. In this view, Ullman argues that grammatical and lexical processing rely on different memory systems, which in turn have their roots in different neural mechanisms (Squire & Wixted, 2011). Ullman's model proposes that learning that happens early in development is facilitated by the procedural memory system. As the brain develops, learning shifts to rely on a more dominant declarative memory system (Ullman, 2016). The procedural memory system underlies perceptual motor and cognitive skills that permit acquisition of habits and rule-learning. This means that routine and amount of experience have a strong influence on the development of procedural memory (Ashby, Tuerner, & Horvitz, 2010). Procedural memory is rooted in basal ganglia circuitry and has connections to the supplementary motor area and inferior frontal gyrus (BA 44/45), and possibly the cerebellum (Mochikuzi-Kawai, 2008). The declarative memory system, on the other hand, is rooted in the medial temporal lobe and connecting structures. This system is involved in knowledge for facts and events and has been implicated in word learning (Kutas & Federmeier, 2011; Raboyeau, Marcotte, Adrover-Roig, & Ansaldo, 2010).

The D/P model therefore predicts that the cognitive processes that are learned early in life will rely more on procedural memory (Ullman, 2005, 2016). In terms of language, syntactic processing should be easier for early learners than late learners of a language due to high activation of the procedural memory system (Friederici, 2006). This is one reason native speakers of a language (whether they are monolinguals or multilinguals speaking in their first language) often have a natural "feel" for what "sounds right." By the same token, because late learners have less access to the procedural memory system and more access to the declarative system, they should find syntactic processing to be harder than other types of processing (e.g., lexical processing for vocabulary) (Ullman, 2016). In this sense, cognition occurs through a transformative process, where new input is organized, represented, and produced through multiple systems.

One interesting point of further debate is to reconcile this view of procedural memory with that of the sensorimotor processing. The use of a reward-based learning approach is particularly interesting in light of the fact that it has been associated with neurotransmitters such as dopamine as well as with particular influences on the frontostriatal tract (Westbrook & Frank, 2018). Two separate neurobiologically based systems not only contribute to the evaluation of models of cognition but also makes the case for the underlying importance of a neurocomputational approach. The consideration of the hippocampus and its relationship to vocabulary also adds an additional dimension to language processing. However, the introduction of reward-based learning suggests that the cognitive system dedicated to language may be composed of other non-linguistic systems much as proposed in neuronal recycling, reuse and in Bates and MacWhinney's Emergentist approaches to language.

If reward-based learning involves the procedural part of knowledge and the hippocampus contributes declarative aspects, we are still left with the ways in which the system is built across time. Here is where sensorimotor processing comes into play. During early development competition could exist at more sensory levels, for example, between phonemes or between the motor programs associated with speech sounds. Krizman, Skoe, Marian, and Kraus (2014) discuss the ways in which the auditory system is modulated by the cognitive control system. Thus, early development would be characterized as involving competition across different levels of the language hierarchy. As Bates et al. noted it could begin at the level of object names around one year of age and progress to the level of grammatical complexity sometime around a child's third birthday. For the bilingual child this competition, could occur at lower levels in the very sounds of language and up to words, clauses and sentences. This shows the ways in which the cognitive control system may be connected to more and more sophisticated and complex informational types across development.

Identifying what underlying systems exist at birth and how they transform across development remains a significant challenge regardless of our theoretical approach. Considering different theoretical viewpoints clearly leads to all kinds of new questions. Stocco's and Ullman's models show us that a deeper conversation from two approaches that share some features but differ in other ways is likely to sharpen our view of language, cognition and development to a great extent.

5. CONCLUSION

Neurocomputational Emergentism considers change within a developmental framework and in a dynamical view in that small simple things can transform into something much more complex later. Whereas the main impetus has been to account for development with diverse language populations, the proposed systems are not specific to learning of multiple languages but are likely present in the learning of a single language. While the present piece proposes three systems that dynamically interact, reward-based learning, memory and sensorimotor function, this is likely a simplification and over time additional systems may be needed in order to further develop this model. Thus, Neuroemergentism is still mostly a conceptual framework. By extending work in Emergentism, Dynamical Systems, and Neurocognitive Development to the neural and computational level, it is our sincere hope to derive a model that can more accurately capture the dynamic nature of language and cognitive development.

ACKNOWLEDGMENTS

This research was supported in part by grant # 5R03HD079873-02, Effects of genetic differences and bilingual status on cognitive control, to AEH and P50 HD052117, Texas Center for Learning Disabilities, from the Eunice Kennedy Shriver National Institute of Child Health and Human Development to the University of Houston.

REFERENCES

Abutalebi, J., & Green, D. (2007). Bilingual language production: The neurocognition of language representation and control. *Journal of Neurolinguistics, 20*(3), 242–275. https://doi.org/10.1016/j.jneuroling.2006.10.003.

Anderson, M. L. (2010). Neural reuse: A fundamental organizational principle of the brain. *Behavioral and Brain Sciences, 33*, 245–266. https://doi.org/10.1017/S0140525X10000853. discussion 266–313.

Archila-Suerte, P., Woods, E. A., Chiarello, C., & Hernandez, A. E. (2018). Neuroanatomical profiles of bilingual children. *Developmental Science*. https://doi.org/10.1111/desc.12654 [Epub ahead of print].

Ashby, F. G., Tuerner, B. O., & Horvitz, J. C. (2010). Cortical and basal ganglia contributions to habit learning and automaticity. *Trends in Cognitive Sciences, 14*(5), 208–215.

Bates, E. A., Benigni, L., Bretherton, I., Camaioni, L., & Volterra, V. (1979). *The emergence of symbols: Cognition and communication in infancy*. New York: Academic Press.

Bates, E. A., Bretherton, I., Snyder, L., Beeghly, M., Shore, C., McNew, S., … O'Connell, B. (1988). *From first words to grammar: Individual differences and dissociable mechanisms*. New York, NY, US: Cambridge University Press.

Bates, E. A., & Goodman, J. C. (1997). On the inseparability of grammar and the lexicon: Evidence from acquisition, aphasia and real-time processing. *Language & Cognitive Processes, 12*(5–6), 507–584. https://doi.org/10.1080/016909697386628.

Bates, E. A., & MacWhinney, B. (1988). *What is functionalism?*. Retrieved from: http://ezproxy.lib.uh.edu/login?url=http://search.ebscohost.com/login.aspx?direct=true&db=eric&AN=ED302075&site=ehost-live.

Bloch, C., Kaiser, A., Kuenzli, E., Zappatore, D., Haller, S., Franceschini, R., … Nitsch, C. (2009). The age of second language acquisition determines the variability in activation elicited by narration in three languages in Broca's and Wernicke's area. *Neuropsychologia, 47*(3), 625–633.

Bradley, K. A. L., King, K. E., & Hernandez, A. E. (2013). Language experience differentiates prefrontal and subcortical activation of the cognitive control network in novel word learning. *Neuroimage, 67*, 101–110. https://doi.org/10.1016/j.neuroimage.2012.11.018.

Centanni, T. M., King, L. W., Eddy, M. D., Whitfield-Gabrieli, S., & Gabrieli, J. D. E. (2017). Development of sensitivity versus specificity for print in the visual word form area. *Brain and Language, 170*, 62–70. https://doi.org/10.1016/j.bandl.2017.03.009.

Cherodath, S., Rao, C., Midha, R., T A, S., & Singh, N. C. (2016). A role for putamen in phonological processing in children. *Bilingualism: Language and Cognition, 20*(2), 318–326. https://doi.org/10.1017/S1366728916000614.

Christiansen, M. H., & Chater, N. (2008). Language as shaped by the brain. *Behavioral and Brain Sciences, 31*(5). https://doi.org/10.1017/S0140525x08004998, 489–+.

Colzato, L. S., Waszak, F., Nieuwenhuis, S. T., Posthuma, D., & Hommel, B. (2010). The flexible mind is associated with the Catechol-O-methyltransferase (COMT) Val158Met polymorphism: Evidence for a role of dopamine in the control of task switching. *Neuropsychologia, 48*, 2764–2768.

D'Souza, D., & Karmiloff-Smith, A. (2016). Why a developmental perspective is critical for understanding human cognition. *Behavioral and Brain Sciences, 39*, e122. https://doi.org/10.1017/S0140525X15001569.

Damasio, A. R., & Damasio, H. (1994). Cortical systems for retrieval of concrete knowledge: The convergence zone framework. In C. Koch, J. L. Davis, C. Koch, & J. L. Davis (Eds.), *Large-scale neuronal theories of the brain* (pp. 61–74). Cambridge, MA, USA: The MIT Press.

De Bot, K., Jaensch, C., & García Mayo, M. d. P. (2015). What is special about L3 processing? *Bilingualism: Language and Cognition, 18*(2), 130–144. https://doi.org/10.1017/S1366728913000448.

Dehaene, S., & Cohen, L. (2007). Cultural recycling of cortical maps. *Neuron, 56*(2), 384–398. https://doi.org/10.1016/j.neuron.2007.10.004.
Dehaene, S., & Cohen, L. (2011). The unique role of the visual word form area in reading. *Trends in Cognitive Sciences, 15*(6), 254–262. https://doi.org/10.1016/j.tics.2011.04.003.
Dehaene, S., & Dehaene-Lambertz, G. (2016). Is the brain prewired for letters? *Nature Neuroscience, 19*(9), 1192–1193. https://doi.org/10.1038/nn.4369.
Dehaene, S., Pegado, F., Braga, L. W., Ventura, P., Nunes Filho, G., Jobert, A., ... Cohen, L. (2010). How learning to read changes the cortical networks for vision and language. *Science, 330*(6009), 1359–1364. https://doi.org/10.1126/science.1194140.
Dick, F., & Krishnan, S. (2018). What underlies the emergence of stimulus- and domain-specific neural responses? *Journal of Neurolinguistics* (this issue).
Donnai, D., & Karmiloff-Smith, A. (2000). Williams syndrome: From genotype through to the cognitive phenotype. *American Journal of Medical Genetics, 97*(2), 164–171.
D'Souza, D., & D'Souza, H. (2018). Emergent and constrained: Understanding brain and cognitive development. *Journal of Neurolinguistics* (this issue).
Eisenberg, D. P., Jabbi, M., & Berman, K. F. (2010). Bridging the gene-behavior divide through neuroimaging deletion syndromes: Velocardiofacial (22q11.2 Deletion) and Williams (7q11.23 Deletion) syndromes. *Neuroimage, 53*(3), 857–869. https://doi.org/10.1016/j.neuroimage.2010.02.070.
Fiebach, C., Friederici, A. D., Müller, K., von Cramon, D. Y., & Hernandez, A. E. (2003). Distinct brain representations for early and late learned words. *Neuroimage, 19*(4), 1627–1637. https://doi.org/10.1016/S1053-8119(03)00227-1.
Friederici, A. D. (2006). Broca's area and the ventral premotor cortex in language: Functional differentiation and specificity. *Cortex, 42*(4), 472–475.
Gauthier, I., McGugin, R. W., Richler, J. J., Herzmann, G., Speegle, M., & Van Gulick, A. E. (2014). Experience moderates overlap between object and face recognition, suggesting a common ability. *Journal of Vision, 14*(8), 7. https://doi.org/10.1167/14.8.7.
Gauthier, I., Skudlarski, P., Gore, J. C., & Anderson, A. W. (2000). Expertise for cars and birds recruits brain areas involved in face recognition. *Nature Neuroscience, 3*(2), 191–197. https://doi.org/10.1038/72140.
Gauthier, I., Tarr, M. J., Anderson, A. W., Skudlarski, P., & Gore, J. C. (1999). Activation of the middle fusiform 'face area' increases with expertise in recognizing novel objects. *Nature Neuroscience, 2*(6), 568–573. https://doi.org/10.1038/9224.
Gentner, D. (1982). Why nouns are learned before verbs: Linguistic relativity versus natural partitioning. In S. A. Kuczaj (Ed.), *Language, thought and culture: Vol. 2. Language development* (pp. 301–334). Hillsdale, NJ: Erlbaum.
Green, D. W., & Abutalebi, J. (2013). Language control in bilinguals: The adaptive control hypothesis. *Journal of Cognitive Psychology, 25*(5), 515–530. https://doi.org/10.1080/20445911.2013.796377.
Haist, F., Lee, K., & Stiles, J. (2010). Individuating faces and common objects produces equal responses in putative face-processing areas in the ventral occipitotemporal cortex. *Frontiers in Human Neuroscience, 4*, 181. https://doi.org/10.3389/fnhum.2010.00181.
Herdina, P., & Jessner, U. (2002). A dynamic model of multilingualism: Perspectives of change in psycholinguistics. *Multilingual Matters, 121*.
Herdina, P., & Jessner, U. (2013). The implications of language attrition for dynamic systems theory: Next steps and consequences. *International Journal of Bilingualism, 17*(6), 752–756. https://doi.org/10.1177/1367006912454625.
Hernandez, A. E. (2013). *The bilingual brain*. New York: Oxford University Press.
Hernandez, A. E., Claussenius-Kalman, H. L., Ronderos, J., Castilla-Earls, A. P., Sun, L., Weiss, S. D., & Young, D. R. (2018). Neuroemergentism: A framework for studying cognition and the brain. *Journal of Neurolinguistics* (in press).

Hernandez, A. E., & Li, P. (2007). Age of acquisition: Its neural and computational mechanisms. *Psychological Bulletin, 133*(4), 638–650.
Hernandez, A. E., Mattarella-Micke, A., Redding, R. W. T., Woods, E. A., & Beilock, S. (2011). Age of acquisition in sport: Starting early matters. *American Journal of Psychology, 124*(3), 253–260.
Huttenlocher, P. R. (1999). Synaptogenesis in human cerebral cortex and the concept of critical periods. In N. Fox, L. A. Leavitt, & J. G. Warhol (Eds.), *The role of early experience in infant development*. Johnson & Johnson Pediatric Institute.
Kanwisher, N., McDermott, J., & Chun, M. M. (1997). The fusiform face area: A module in human extrastriate cortex specialized for face perception. *Journal of Neuroscience, 17*(11), 4302–4311.
Kanwisher, N., Stanley, D., & Harris, A. (1999). The fusiform face area is selective for faces not animals. *Neuroreport, 10*(1), 183–187.
Kanwisher, N., & Yovel, G. (2006). The fusiform face area: A cortical region specialized for the perception of faces. *Philosophical Transactions of the Royal Society B: Biological Sciences, 361*(1476), 2109–2128. https://doi.org/10.1098/rstb.2006.1934.
Karmiloff-Smith, A. (2006). The tortuous route from genes to behavior: A neuroconstructivist approach. *Cognitive, Affective, & Behavioral Neuroscience, 6*(1), 9–17.
Karmiloff-Smith, A. (2009). Nativism versus neuroconstructivism: Rethinking the study of developmental disorders. *Developmental Psychology, 45*(1), 56–63. https://doi.org/10.1037/a0014506.
Karmiloff-Smith, A., Thomas, M., Annaz, D., Humphreys, K., Ewing, S., Brace, N., ... Campbell, R. (2004). Exploring the Williams syndrome face-processing debate: The importance of building developmental trajectories. *Journal of Child Psychological Psychiatry, 45*(7), 1258–1274. https://doi.org/10.1111/j.1469-7610.2004.00322.x.
Krizman, J., Skoe, E., Marian, V., & Kraus, N. (2014). Bilingualism increases neural response consistency and attentional control: Evidence for sensory and cognitive coupling. *Brain and Language, 128*(1), 34–40. https://doi.org/10.1016/j.bandl.2013.11.006.
Kuo, L.-J., Ramirez, G., de Marin, S., Kim, T.-J., & Unal-Gezer, M. (2017). Bilingualism and morphological awareness: A study with children from general education and Spanish-English dual language programs. *Educational Psychology, 37*(2), 94–111.
Kutas, M., & Federmeier, K. D. (2011). Thirty years and counting: Finding meaning in the N400 component of the event-related brain potential (ERP). *Annual Review of Psychology, 62*, 621–647.
Lehtonen, M. H., Laine, M., Niemi, J., Thomsen, T., Vorobyev, V. A., & Hugdahl, K. (2005). Brain correlates of sentence translation in Finnish-Norwegian bilinguals. *Neuroreport, 16*(6), 607–610.
Li, P. (2015). Bilingualism as a dynamic process. In B. MacWhinney, & W. O'Grady (Eds.), *The handbook of language emergence*. Malden, Mass: Wiley Blackwell.
Li, P., Farkas, I., & MacWhinney, B. (2004). Early lexical development in a self-organizing neural network. *Neural Networks, 17*, 1345–1362. https://doi.org/10.1016/j.neunet.2004.07.004.
MacWhinney, B. (2004). A unified model of language acquisition. In J. K. A. D. Groot (Ed.), *Handbook of bilingualism: Psycholinguistic approaches*. New York: Oxford University Press.
MacWhinney, B., & O'Grady, W. (2015). *The handbook of language emergence*. Malden, Mass: Wiley Blackwell.
Mamiya, P. C., Richards, T. L., Coe, B. P., Eichler, E. E., & Kuhl, P. K. (2016). Brain white matter structure and COMT gene are linked to second-language learning in adults. *Proceedings of the National Academy of Sciences, 113*(26), 7249.
Mårtensson, J., Eriksson, J., Bodammer, N. C., Lindgren, M., Johansson, M., Nyberg, L., & Lövdén, M. (2012). Growth of language-related brain areas after foreign language learning. *Neuroimage, 63*(1), 240–244. https://doi.org/10.1016/j.neuroimage.2012.06.043.

Matsui, M., Tanaka, C., Niu, L., Noguchi, K., Bilker, W., Wierzbicki, M., & Gur, R. (2016). Age-related volumetric changes of prefrontal gray and white matter from healthy infancy to adulthood. *International Journal of Clinical and Experimental Neurology, 4*(1), 1–8. https://doi.org/10.12691/ijcen-4-1-1.

Mervis, C. B., Robinson, B. F., Bertrand, J., Morris, C. A., Klein-Tasman, B. P., & Armstrong, S. C. (2000). The Williams syndrome cognitive profile. *Brain and Cognition, 44*(3), 604–628. https://doi.org/10.1006/brcg.2000.1232.

Miller, G., Crocker, L., Spielberg, J., Infantolino, Z., & Heller, W. (2013). Issues in localization of brain function: The case of lateralized frontal cortex in cognition, emotion, and psychopathology. *Frontiers in Integrative Neuroscience,* 7(2). https://doi.org/10.3389/fnint.2013.00002.

Mill, J. S., Robson, J. M., McRae, R. F., Routledge & Kegan, P., & University of Toronto. (1974). *Collected works of John Stuart Mill* (Vol. 7, p. 7). Toronto: University of Toronto Press; London: Routledge & Kegan Paul.

Mochikuzi-Kawai. (2008). The neural basis of procedural memory. *Brain and Nerve, 60*(7), 825–832.

Nelson, K. (1973). Structure and strategy in learning to talk. *Monographs of the Society for Research in Child Development, 38*(1/2), 1–135. https://doi.org/10.2307/1165788.

Paap, K. R., Johnson, H. A., & Sawi, O. (2015). Bilingual advantages in executive functioning either do not exist or are restricted to very specific and undetermined circumstances. *Cortex, 69*, 265–278. https://doi.org/10.1016/j.cortex.2015.04.014.

Pereira, A. F., Smith, L. B., & Yu, C. (2014). A bottom-up view of toddler word learning. *Psychonomic Bulletin & Review, 21*(1), 178–185. https://doi.org/10.3758/s13423-013-0466-4.

Prat, C. S., & Just, M. A. (2011). Exploring the neural dynamics underpinning individual differences in sentence comprehension. *Cerebral Cortex, 21*(8), 1747–1760. https://doi.org/10.1093/cercor/bhq241.

Price, C. J., & Devlin, J. T. (2003). The myth of the visual word form area. *Neuroimage, 19*(3), 473–481.

Price, C. J., & Devlin, J. T. (2011). The interactive account of ventral occipitotemporal contributions to reading. *Trends in Cognitive Sciences, 15*(6), 246–253. https://doi.org/10.1016/j.tics.2011.04.001.

Price, C. J., Green, D. W., & von Studnitz, R. (1999). A functional imaging study of translation and language switching. *Brain: a Journal of Neurology, 122*(Pt 12), 2221–2235.

Raboyeau, G., Marcotte, K., Adrover-Roig, D., & Ansaldo, A. I. (2010). Brain activation and lexical learning: The impact of learning phase and word type. *Neuroimage, 49*(3), 2850–2861. https://doi.org/10.1016/j.neuroimage.2009.10.007.

Schmid, M. S. (2011). *Language attrition*. Cambridge: Cambridge University Press.

Slabakova, R. (2017). The scalpel model of third language acquisition. *International Journal of Bilingualism, 21*(6), 651–665. https://doi.org/10.1177/1367006916655413.

Squire, L. R., & Wixted, J. T. (2011). The cognitive neuroscience of human memory since H.M. *Annual Review of Neuroscience, 34*, 259–288.

Stelzel, C., Basten, U., Montag, C., Reuter, M., & Fiebach, C. J. (2010). Frontostriatal involvement in task switching depends on genetic differences in d2 receptor density. *J Neurosci, 30*(42), 14205–14212. https://doi.org/10.1523/JNEUROSCI.1062-10.2010.

Stiles, J., Reilly, J. S., Levine, S. C., Trauner, D. A., & Nass, R. (2012). *Neural plasticity and cognitive development : Insights from children with perinatal brain injury*. New York: Oxford University Press.

Stocco, A. (2018). Commentary on neuroemergentism: A framework for studying cognition and the brain. The neurocomputations of neuroemergentism: Long-term memory + reinforcement learning = language? *Journal of Neurolinguistics* (this issue).

Stocco, A., Lebiere, C., & Anderson, J. R. (2010). Conditional routing of information to the cortex: A model of the basal ganglia's role in cognitive coordination. *Psychological Review, 117*(2), 541–574. https://doi.org/10.1037/a0019077.

Stocco, A., Lebiere, C., O'Reilly, R. C., & Anderson, J. R. (2012). Distinct contributions of the caudate nucleus, rostral prefrontal cortex, and parietal cortex to the execution of instructed tasks. *Cognitive, Affective, & Behavioral Neuroscience, 12*(4), 611–628. https://doi.org/10.3758/s13415-012-0117-7.

Stocco, A., & Prat, C. S. (2014). Bilingualism trains specific brain circuits involved in flexible rule selection and application. *Brain and Language, 137*, 50–61. https://doi.org/10.1016/j.bandl.2014.07.005.

Stocco, A., Yamasaki, B., Natalenko, R., & Prat, C. S. (2014). Bilingual brain training: A neurobiological framework of how bilingual experience improves executive function. *International Journal of Bilingualism, 18*(1), 67–92. https://doi.org/10.1177/1367006912456617.

Sugiura, L., Toyota, T., Matsuba-Kurita, H., Iwayama, Y., Mazuka, R., Yoshikawa, T., & Hagiwara, H. (2017). Age-Dependent Effects of Catechol-O-Methyltransferase (COMT) Gene Val158Met Polymorphism on Language Function in Developing Children. *Cereb Cortex, 27*(1), 104–116. https://doi.org/10.1093/cercor/bhw371.

Ullman, M. T. (2001). The neural basis of lexicon and grammar in first and second language: The declarative/procedural model. *Bilingualism: Language and Cognition, 4*, 105–122.

Ullman, M. T. (2005). A cognitive neuroscience perspective on second language acquisition: The declarative/procedural model. In C. Sanz (Ed.), *Mind and context in adult second language acquisition: Methods, theory, and practice* (pp. 141–178). Washington, DC: Georgetown University Press.

Ullman, M. T. (2006). The declarative/procedural model and the shallow structure hypothesis. *Applied PsychoLinguistics, 27*(1), 97–105.

Ullman, M. T. (2016). The declarative/procedural model: A neurobiological model of language learning, knowledge, and use A2 - Hickok, Gregory. In S. L. Small (Ed.), *Neurobiology of language* (pp. 953–968). San Diego: Academic Press (Chapter 76).

Vingerhoets, G., Van Borsel, J., Tesink, C., van den Noort, M., Deblaere, K., Seurinck, R., … Achten, E. (2003). Multilingualism: An fMRI study. *Neuroimage, 20*(4), 2181–2196.

Vogel, A. C., Miezin, F. M., Petersen, S. E., & Schlaggar, B. L. (2012). The putative visual word form area is functionally connected to the dorsal attention network. *Cerebral Cortex, 22*(3), 537–549. https://doi.org/10.1093/cercor/bhr100.

Vogel, A. C., Petersen, S. E., & Schlaggar, B. L. (2014). The VWFA: it's not just for words anymore. *Frontiers in Human Neuroscience, 8*, 88. https://doi.org/10.3389/fnhum.2014.00088.

Westbrook, A., & Frank, M. (2018). Dopamine and proximity in motivation and cognitive control. *Current Opinion In Behavioral Sciences, 22*, 28–34. https://doi.org/10.1016/j.cobeha.2017.12.011.

CHAPTER FOUR

Syntactic Adaptation

Edith Kaan[1] and Eunjin Chun[1]
Department of Linguistics, University of Florida, Gainesville, FL, United States
[1]Corresponding authors: E-mail: kaan@ufl.edu; chuneunjin97@gmail.com

Contents

1. Introduction 86
2. Evidence for Syntactic Adaptation in Comprehension 88
   2.1 Fine et al. (2013) 88
   2.2 Hallmarks of Syntactic Adaptation 89
3. Mechanisms Underlying Syntactic Adaptation 90
   3.1 Activation of Syntactic Structures 90
   3.2 Error-Based Learning Models 90
   3.3 Belief-Updating Models 91
4. Evidence for Syntactic Adaptation? 93
   4.1 Other Structures Tested and Techniques Used 93
   4.2 Adaptation Is Not Always Observed 95
   4.3 Decreases in Ambiguity Effects 96
      4.3.1 Ambiguity Effects do Not Always Decrease Over the Course of the Study 96
      4.3.2 Methodological Issues 97
   4.4 Differences Between Exposure Groups 98
   4.5 Reversal Effects 98
   4.6 Syntactic Effects? 99
      4.6.1 Adaptation May Not Be to the Syntactic Structure Intended 99
      4.6.2 Adaptation May Not Be Syntactic 100
   4.7 Adaptation as Linking Contextual Cues and Syntactic Properties 101
5. What Is the Relation Between Adaptation and Learning? 102
   5.1 Differences in Adaptation inside and outside the Lab 102
   5.2 Learning From Exposure May Not Be Driven by a Unitary Mechanism 103
   5.3 The Way Forward: Factors Driving Adaptation 104
   5.4 What Determines Syntactic Adaptation? 105
      5.4.1 What Is Deviant? 105
      5.4.2 Goals and Social Function 106
   5.5 What Determines Generalization? 107
      5.5.1 Variation 107
      5.5.2 Similarity 107
   5.6 What Memory Systems Are Involved? 108
6. Concluding Remarks 110
Acknowledgments 110
References 111

Psychology of Learning and Motivation, Volume 68
ISSN 0079-7421
https://doi.org/10.1016/bs.plm.2018.08.003

Abstract

Listening to other people talk, we very quickly get used to their accent, word choice, or specific meaning of phrases. It is less clear to what extent listeners and readers adapt to abstract syntactic structures: do sentences with a particular structure get easier to process as more sentences of the same type have been seen or heard? Recent models of sentence processing propose that syntactic adaptation, language change, and language learning are driven by the same mechanisms. It is therefore no surprise that syntactic adaptation has drawn a lot of recent interest in the field of language processing research. In this chapter we review and discuss the evidence put forward for syntactic adaptation, mainly in comprehension. We then discuss the relation between syntactic adaptation as studied in the laboratory, and language adaptation and learning in daily life. Finally, we discuss and list some factors that may underlie various forms of adaptation, and sketch directions for future research.

1. INTRODUCTION

Listening to other people talk, we very quickly get used to their accent, word choice, or specific meaning of phrases, even sometimes up to the point that we change our own way of speaking. Such adaptation of behavior in response to changes in the environment, be it linguistic or non-linguistic, is a very common observation and occurs in many species (Clark, 2013).

From a linguistic point of view, investigating how and when language users adapt to the language surrounding them is important since this may shed light on how speakers communicate, how language changes from generation to generation or through contact between speakers of different languages, and how language can be learned by exposure by children as well as adults. Recent theories and models attempt to unify language production, adaptation, and learning (Chang, Dell, & Bock, 2006). It is therefore not surprising that linguistic adaptation has drawn a lot of recent attention in language processing research.

There is ample evidence that adaptation occurs to unusual or accented speech sounds (Bradlow & Bent, 2008; Kraljic & Samuel, 2007; Xie, Weatherholtz, et al., 2018), intonation (Kurumada, Brown, & Tanenhaus, 2017), how objects are referred to (Metzing & Brennan, 2003), and other aspects of meaning (Yildirim, Degen, Tanenhaus, & Jaeger, 2016). Data are less clear, however, as to the extent to which adaptation takes place to abstract syntactic structures, such as the use of a passive construction, or of a certain word order, regardless of the specific words used. By "syntactic adaptation" with respect to language production, we mean that speakers are more likely to produce a sentence with a particular syntactic structure themselves after repeatedly being exposed to sentences with this structure; with respect to comprehension we mean that a listener or reader has an easier time processing sentences with a particular syntactic structure as more sentences of this particular structure have been encountered. This can also be reflected in a change in how readers or listeners process sentences that are syntactically ambiguous. An example is *the sister of the actor who was on the balcony*. In this case the relative clause *who was on the balcony* can modify either *sister* or *actor*. Language users often have a preference for one reading over another, but their preference can be changed after repeated exposure to a particular resolution of the ambiguity (Chun, 2018). "Syntactic adaptation" has also been referred to as "structural persistence" or "cumulative priming". In this chapter, we distinguish syntactic adaptation from (trial-to-trial) syntactic priming, although priming can be seen as an instance of adaptation (Chang, Janciauskas, & Fitz, 2012; Fine, Jaeger, Farmer, & Qian, 2013; Jaeger & Snider, 2008). Syntactic priming is the well-established and well-replicated phenomenon that a speaker is likely to use the structure they just heard (Bock, 1986; Mahowald, James, Futrell, & Gibson, 2016), or that a reader needs less time to process a complex syntactic structure when the sentence is preceded by a sentence of a similar type (Tooley & Traxler, 2010). We use the term "syntactic adaptation" to describe a more general situation in which comprehension or production is not or not only affected by the most recently encountered structure, but by the cumulative prior exposure to structures of the same type. Although syntactic adaptation was first studied mainly in production (e.g. Kaschak & Borreggine, 2008; Kaschak, 2007; Kaschak, Loney, & Borreggine, 2006), most recent studies on syntactic adaptation have involved reading and listening. We will therefore mainly focus on comprehension rather than production.

Below we will first discuss what kind of experimental data has been considered evidence of syntactic adaptation, and how adaptation can be accounted for in terms of cognitive mechanisms. Next, we will show that adaptation effects are not always observed or replicated, and that it is not always clear that the adaptation is to a syntactic structure. We then discuss the relation between syntactic adaptation as studied in the laboratory, and language adaptation and learning in daily life. Finally, we discuss and list some factors that may underlie various forms of adaptation, and sketch directions for future research.

2. EVIDENCE FOR SYNTACTIC ADAPTATION IN COMPREHENSION

2.1 Fine et al. (2013)

A study that has been often cited as evidence for syntactic adaptation in language comprehension is (Fine et al., 2013). In this study native English readers were presented with sentences such as in (1).

(1) The experienced soldiers...

a. ... warned about the dangers before the midnight raid.

b. ... spoke about the dangers before the midnight raid.

c. ... warned about the dangers conducted the midnight raid.

d. ... who were told about the dangers conducted the midnight raid

In (1a) and (1c) the verb (*warned*) is ambiguous between a simple past and a passive participle. In (1a) the sentence continues in a way that is compatible with the simple past (main clause) interpretation of the verb. This analysis is typically preferred by English speakers. In (1c) the verb is the start of a reduced relative clause, modifying *soldiers*. This interpretation only becomes clear at the verb *conducted*. The sentences in (1b) and (1d) are unambiguous control sentences: in (1b) the verb (*spoke*) can only be a main clause verb; in (1d) *who were* ... unambiguously marks the start of a relative clause. Participants in the Fine et al. study read sentences like these word by word in a self-paced reading experiment. Replicating other studies (MacDonald, Just, & Carpenter, 1992), reading times were longer for the disambiguating region (*before/conducted the midnight*) when the verb was ambiguous and the information disambiguated towards the infrequent reduced relative clause reading of the verb (1c). Critically, this difference between the ambiguous (1c) and unambiguous (1d) relative clause conditions became smaller over the course of the study. This was mainly because readers started to read the disambiguating region in the (c) condition faster; the reading times for the non-ambiguous and main clause versions (1a, b) were not as much affected by the number of prior sentences of those types. To check whether exposure could indeed change readers' expectations for a particular structure, Fine et al. conducted another experiment in which they first presented one group of readers with sentences containing a relative clause such as (1c) and (1d), and another group of readers with a set of sentences of a completely different type (block 1). Next, all participants saw a block of sentences containing relative clauses (block 2). This was followed by another block of sentences such as (1a) and (1b) which contained main clauses, but no relative clauses (block 3). As expected, the group that had seen relative clauses in the first block, and hence had seen more reduced relative clauses overall, showed smaller ambiguity effects in block 2 than the group that had not been exposed to reduced relative clauses in block 1. However, this difference between the groups was only marginally significant. Strikingly, however, the group that had been exposed to more reduced relative clauses showed longer reading times in block 3 when the ambiguous verb was disambiguated to a main clause interpretation (1a) compared to an unambiguous main clause control (1b). This latter pattern goes against the general bias for main clause continuations over reduced relative continuations. Taken together, data from the Fine et al. (2013) study suggest that readers can adapt to the syntactic structure of the sentences they have read: sentences with reduced relatives became easier to read as a function of exposure to other sentences with this structure.

2.2 Hallmarks of Syntactic Adaptation

The Fine et al. (2013) study illustrates what can be considered the hallmarks of syntactic adaptation. First, difficulty with an initially infrequent and non-preferred structure, e.g. a reduced relative clause, decreases as a function of exposure to items with this structure: as more structures of the same type have been seen, reading times for the initially difficult condition become shorter. Recent exposure affects the processing of infrequent or atypical structures more than that of frequent or typical structures. This is referred to as the "inverse frequency effect", and has also been observed in trial-to-trial priming (e.g. Bernolet & Hartsuiker, 2010; Hartsuiker & Westenberg, 2000; Jaeger & Snider, 2013; Scheepers, 2003). Second, the decrease in difficulty between the infrequent and frequent structure over the experiment is less strong for a control group with no or less recent exposure to the structure. This rules out that the decrease in difficulty is due to participants getting used to the experimental paradigm in general. Instead, the decrease can be considered specific to having been exposed to sentences with a particular structure. Third, a reversal of the initial preference can be seen: after exposure to structures that are normally infrequent structures or non-preferred continuations, processing times become longer for structures or continuations that are normally preferred. Finally, the Fine et al. data support the view that adaptation is to an abstract syntactic structure. Verbs were not repeated in experiment 1 in Fine et al. (2013). It is therefore not likely that the adaptation was specific to the particular verbs used. Rather, exposure to sentences of a particular type made it easier down the line to read other sentences with different words, but with the same structure (see also Fine & Jaeger, 2016; Tooley & Traxler, 2018). The Fine et al. (2013) study and results meet these four criteria, and therefore can be considered evidence of syntactic adaptation (but see below).

3. MECHANISMS UNDERLYING SYNTACTIC ADAPTATION

3.1 Activation of Syntactic Structures

What underlies syntactic adaptation? Models proposed thus far have been extensions of models proposed for trial-to-trial syntactic priming. One of such models is the ACT-R model by Reitter, Keller, and Moore (2011). This model was designed as a model of production, but can easily be applied to comprehension. In this model, syntactic frames indicate how syntactic categories, such as nouns, prepositions and verbs, can be combined. This information is stored in declarative memory. As a syntactic structure is encountered more often, the base-level activation increases of the corresponding frame. The activation level of the syntactic frame is very high after it has very recently been accessed (as in trial-to-trial priming). The activation decays rapidly afterward, but the base-level activation increases each time the structure is used. The higher the base-level activation, the easier it is to access and process the structure. Changes in base-level activation are logarithmic, and are therefore largest for structures that are initially less frequent than for structures that are frequent and already have a high base-level activation. As such, this model can account for the inverse frequency effect, that is, the observation that initially non-preferred or infrequent structures become easier to process as more of these structures have been encountered, whereas structures that already are frequent in the language are not much affected by increased exposure.

3.2 Error-Based Learning Models

Models such as the ACT-R (Reitter et al., 2011) explain adaptation in terms of changes in the activation of the syntactic representations in declarative memory (cf. residual activation accounts of priming, Pickering & Branigan, 1998). Another approach is to view adaptation as a change in procedural knowledge, that is, the ways in which sentences are assembled during processing. One of such models is the Dual Path model, an error-based implicit learning model based on a connectionist network (Chang et al., 2006). In this approach, the same mechanisms underlie both language learning and priming, which by extension includes cumulative adaptation (Chang et al., 2012). In this framework, language users use their prior knowledge and their production system to generate predictions of what is to come in the linguistic input. If these predictions are not borne out, the difference between what was expected and what is observed in the input is used to adjust their future processing. This adjustment is considered a form of implicit learning since this happens mainly without the language user's awareness or explicit feedback, and since it may have long-lasting effects (Bock & Griffin, 2000). To illustrate how this model works, let's consider American-English speakers in a hypothetical study investigating syntactic priming. While listening to or reading sentences, participants in this study may expect a sentence to continue in a certain way. For instance, based on previous experience with American English, participants may have a high expectation for a "giving" event to be expressed as a double object construction (*the man gives the dog a bone*), in which a verb is followed by a noun that is the recipient (*dog*) of the action. If the sentence they hear or read instead has a theme following the verb and contains a prepositional dative (as in *the man gives a bone to the dog*), the participants experience a prediction error. This deviation between what is expected and what is perceived leads to an adjustment of the strength of connections in the processing system such that the participants become more likely to produce and predict a prepositional dative in subsequent trials. Like the ACT-R model, this model can account for the observation that adaptation is largest for structures that are initially less frequent: since infrequent structures are least expected, the prediction error is largest for these structures, and the system will change more to accommodate this unexpected input than is the case with more frequent structures. The idea put forward by Chang et al. (2006) and others is that language users are life-time language learners: this continuous testing of predictions and updating of knowledge happens during language use in general, and eventually leads to the convergence of an individual's knowledge and own utterances with those of the surrounding language community.

3.3 Belief-Updating Models

A related approach are models based on Bayesian belief-updating (Jaeger & Snider, 2013; Kleinschmidt & Jaeger, 2015; Qian, Jaeger, & Aslin, 2012). According to these models, language users keep track of the probability distributions of syntactic and other features they encounter. On the basis of their experience with the language, language users will have certain expectations as to whether or how often certain (syntactic) features occur in a particular context. If language users encounter something that is unexpected to some degree, they update their probability representations to more closely align with those of the environment, thus optimizing information transmission (Levy, 2008). This is equivalent to minimizing future prediction errors in the error-based learning models discussed above. Belief-updating models account for inverse frequency effects in a way similar to error-based learning models. Infrequent structures are less expected on the basis of prior knowledge. The language user's knowledge of the probability distribution will therefore shift more when an infrequent structure is encountered than a frequent structure. Infrequent structures become easier to process (more expected) with increased exposure, whereas structures that were frequent to start with are affected to a smaller extent. With sufficient exposure to initially infrequent structures, these structures may eventually become more expected than the structures that were initially preferred, accounting for the reversal of the preference as observed in Fine et al. (2013).

The difference between the belief-updating and implicit learning models on the one hand, and the representation-based models like the ACT-R on the other, is that the latter explain adaptation by assuming that the activation of a syntactic representation in memory is modulated by exposure, whereas the other two models explain adaptation as a change in processing procedures or expected probabilities. Problematic for representation-based approaches is that listeners and readers can adapt to ungrammatical structures (Caffarra & Martin, 2018; Coulson, King, & Kutas, 1998; Hahne & Friederici, 1999; Hanulikova, van Alphen, van Goch, & Weber, 2012), or structures unfamiliar to them (Fraundorf & Jaeger, 2016; Kaschak & Glenberg, 2004). Since these structures normally do not occur in the individual's language, these may not have a representation in declarative memory, and hence adaptation cannot be explained by a change in base-level activation. Another potential problem for a representational approach is that listeners can simultaneously adapt to two talkers who each have systematically different structural preferences (Kamide, 2012; Kroczek & Gunter, 2017). For instance, Kroczek and Gunter (2017) exposed German listeners to two speakers: one speaker often used the Subject-Object-Verb order in German whereas the other often used the Object-Subject-Verb order. After exposure, the listeners expected a certain word order when they heard one speaker, and the other word order when they heard the other speaker. Representation-based models cannot readily account for these observations since talker identity is typically not part of the linguistic representation in these models. Unlike verbs, which are part of the linguistic information stored in memory, talker identity is not connected to a particular syntactic structure in declarative memory. This means that a structure is activated in abstraction of the talker, and that the system cannot keep track of which speaker used which construction. A potential solution would be to encode talker-specific information as part of an event-semantics module through which associations can be formed with syntactic structures. Talker-specific syntactic adaptation then is a form of associative learning. In the Reitter et al. (2011) ACT-R model, however, associative learning is short-lived. This approach can therefore not explain the long-lasting talker-specific syntactic adaptation reported by Kroczek and Gunter (2017).

In belief-updating models, on the other hand, language users keep track of different sets of probability distributions of syntactic and other features for each talker or context that they consider distinct (Kleinschmidt & Jaeger, 2015; Qian et al., 2012). In other words, language users keep track of what is typical for a specific situation. Also in error-based learning models, the language user takes aspects of the context, which may include talker-identity, into consideration when forming predictions. What is "typical" in one context can be considered something atypical (unsystematic, ungrammatical or inappropriate) in another situation, and vice versa. Both the belief-updating and error-based implicit learning models can therefore account for talker-specific adaptation as well as the observation that ungrammaticalities can be adapted to.

4. EVIDENCE FOR SYNTACTIC ADAPTATION?

4.1 Other Structures Tested and Techniques Used

The Fine et al. (2013) study is one among many recent studies investigating adaptation. These studies cover a variety of syntactic structures. A non-exhaustive overview of the syntactic structures tested is given in Table 1. Experimental techniques used in these studies include self-paced reading, as used by Fine et al. (2013), measuring eye-movements during reading (Atkinson, 2016; Farmer, Fine, Yan, Cheimariou, & Jaeger, 2014; Tooley & Traxler, 2018) or listening (Chun, 2018; Henry, Hopp, & Jackson, 2017; Kamide, 2012; Ryskin, Qi, Duff, & Brown-Schmidt, 2016b), electrophysiology (Coulson et al., 1998; Hahne & Friederici, 1999), functional Magnetic Resonance Imaging (Noppeney & Price, 2004), as well as sentence completion, grammaticality judgment or question answering (Chun, 2018; Kroczek & Gunter, 2017; Luka & Barsalou, 2005). The latter technique measures changes in interpretational preferences. For instance, Chun (2018) exposed English listeners to sentences such as (2a), in which the relative clause (*that will wear the shoes*) is intended to modify the last noun phrase heard, *the woman* in this case. Subsequently, listeners were presented with a block of sentences such as (2b) in which the relative clause modified the first noun mentioned (*cat* in this case).

(2) **a.** I know the cat of the woman that will wear the shoes.
b. I know the cat of the woman that will wear the collar.

Before, between, and after the exposure blocks participants were tested on their interpretation of sentences containing ambiguous relative clauses such as (3a). Sentences were followed by questions such as in (3b). In this example, the answer *hairdresser* indicates that the readers preferred the relative clause to modify the last mentioned noun phrase, the answer *client* suggests that the reader preferred the relative clause to modify the first mentioned noun phrase.

(3) **a.** I met the client of the hairdresser who is talking loudly.
b. Who is talking loudly? The client or the hairdresser?

Chun (2018) observed that after exposure to sentences such as (2a), participants gave more answers corresponding to the last-mentioned noun phrase than before exposure. After exposure to a block of sentences such as (2b) readers gave more answers corresponding to the first-mentioned noun phrase after than they did before exposure. This suggests that exposure can influence the reader's interpretational preferences.

Table 1 Overview of Structures Tested in Adaptation Studies

Structure tested	Example
Reduced relative clauses[1,2,3,4,5,6]	*The experienced soldiers warned about the dangers <u>conducted</u> the midnight raid.*
Relative clauses modifying the last-mentioned or earlier mentioned noun[7,8]	*The cat of the woman that will wear the <u>collar</u>/<u>shoes</u>.*
Object relative clause[9]	*The clerk that the typist trained told the truth.*
Word order variations[10, 11, 12]	Subject-Object-Verb or Object-Subject-Verb in German
Verbs followed by a clause rather than a direct object noun phrase[13,14]	*Monroe checked his wallet <u>was</u> still in his pocket*

Table 1 Overview of Structures Tested in Adaptation Studies—cont'd

Structure tested	Example
Questions or relative clauses in which *what/who* is initially interpreted as the object of the verb rather than the following preposition ("filled gap" effects)[15, 16]	*I wonder what the plumber fixed the sink with.*
And coordinating two clauses rather than two direct objects[16]	*The chef stirred the soup and the sauce was stirred by the assistant.*
Verb bias for an instrument or a modifier reading of a prepositional phrase[17]	*You should rub the duck with the sponge.*
Direct object/subject ambiguities[12,18]	*After the headmaster had left the school deteriorated rapidly.*
Overt or null pronoun referring to objects or subjects (in Portuguese)[19]	*The athlete*[Subj] *consulted the orthopedist*[Obj] *at the hospital when he/Ø returned from the journey to Italy.*
Need Verb-*ed* construction[20,21]	*The car needs cleaned.*

[1]Farmer et al. (2014); [2]Fine et al. (2013); [3]Long and Prat (2008); [4]Malone and Mauner (2018); [5]Harrington Stack, James, and Watson (2018); [6]Tooley and Traxler (2018); [7]Chun (2018); [8]Kamide (2012); [9]Wells, Christiansen, Race, Acheson, and MacDonald (2009); [10]Henry et al. (2017); [11]Kroczek and Gunter (2017); [12]Noppeney and Price (2004); [13]Fine, Qian, Jaeger, and Jacobs (2010); [14]Myslín and Levy (2016); [15]Atkinson (2016); [16]Kaan, Futch, Fernández Fuertes, Mujcinovic, and Álvarez de la Fuente (2018); [17]Ryskin et al. (2016b); [18]McCann and Kaan (2017); [19]Fernandes, Luegi, Correa Soares, de la Fuente, and Hemforth (2018); [20]Fraundorf and Jaeger (2015); [21]Kaschak and Glenberg (2004). Note: underscore in the example indicates the point of disambiguation.

4.2 Adaptation Is Not Always Observed

As can be seen in Table 1, quite a few studies have investigated syntactic adaptation in comprehension. However, evidence for syntactic adaptation has been rather inconsistent. We mentioned that the following observations are typically considered the hallmarks of syntactic adaptation: First, a decrease of processing difficulty (e.g., decrease in reading times) is seen for the complex or non-preferred continuation versus the simple or preferred controls condition as a function of exposure. We will refer to this as a decreased ambiguity or complexity effect. In question answering studies, a change in interpretational preferences is seen as a result of exposure. Second, this change in processing difficulty or interpretational preference is less strong for a control group with no or less recent exposure to the structure investigated. Third, a reverse ambiguity effect may be seen: after exposure to normally infrequent structures or non-preferred continuations, these structures now become the preferred structures, and processing times become longer for structures or continuations that were preferred at the start of the study. Fourth, adaptation is to an abstract syntactic structure, not specific to lexical items. When taking these four criteria into consideration, evidence of syntactic adaptation has not always been found or replicated. We will consider each of these four criteria in turn.

4.3 Decreases in Ambiguity Effects

4.3.1 Ambiguity Effects do Not Always Decrease Over the Course of the Study

First, many studies have not observed a decrease in ambiguity or complexity effects, or changes in preferences as a function of exposure over the course of the study. For instance, Stack, James, and Watson (2018) Experiment 1 failed to replicate the decrease in the ambiguity effect for reduced relatives as observed in Fine et al. (2013). McCann and Kaan (2017) reported no effect of exposure on the processing of object/subject ambiguities. An example is *As the swimmer trained the kids were yelling in the other pool*, in which *the kids* is initially interpreted as the object of *trained*, but needs to be the subject of *were yelling*. Kaan, Futch, Fernández Fuertes, Mujcinovic, and Álvarez de la Fuente (2018) tested so-called filled-gaps. These occur in sentences such as *I wonder what the plumber fixed the sink with*. Initially *what* is taken to be the direct object of the following verb (*fixed*). However, the verb is already followed by a direct object noun phrase (*the sink*). This noun phrase is not expected, leading to an increase in reading times on the direct object in *what* sentences versus control sentences (the "filled-gap effect"). Recent exposure to filled-gap sentences did not robustly decrease reading times to the direct object in *what* sentences (but see Atkinson, 2016). Also, talker-specific changes of preference (Kamide, 2012; Kroczek & Gunter, 2017) have not been consistently replicated (Liu, Burchill, Tanenhaus, & Jaeger, 2017; Ryskin, Qi, Duff, & Brown-Schmidt, 2016a). Furthermore, some populations and structures may be more prone to showing or eliciting a decrease in difficulty than others. Sikos, Martin, Fitzgerald, and Grodner (2016) report a decrease in the ambiguity effect for reduced relative clauses when these were embedded in simple, frequent constructions, as in *The dean knew that the professor evaluated by the observer should talk to a counselor*, but not when these were embedded in complex, infrequent structures such as *The dean knew what the professor evaluated by the observer should present during the lecture*. In our lab we exposed readers to *and* ambiguities such as *The chef stirred the soup and the sauce was stirred by the assistant*. Readers initially take *the soup and the sauce* as the direct object of *the chef stirred*. However, this analysis can no longer be correct at *was*; the interpretation needs to be changed to one in which *and* connects two clauses, and *the sauce* is the subject of *was stirred by the assistant*. We found a decrease in the ambiguity effect at *was* as readers were more exposed to these sentences, but only when the readers were native English speakers. Spanish second-language learners of English showed no change in the size of the ambiguity effect (Kaan et al., 2018). Adaptation, in the sense of seeing a change in parsing preference as a function of exposure, or decrease in processing difficulty for infrequent structures, is therefore not ubiquitously found.

4.3.2 Methodological Issues

Failure to find or replicate adaptation effects may of course be due to methodological issues. First, adaptation effects are rather small, and failure to find adaptation effects may therefore be due to the lack of power, that is, the number of items and participants may not have been sufficient to detect effects of exposure. However, even with sufficient power, Stack et al. (2018) failed to replicate the reversal effects and group differences reported by Fine et al. (2013). On the other hand, increasing the number of items may obscure adaptation effects: adaptation may be so quick that it occurs during the first few items only. This change may hard to detect if a large number of items is used.

Another methodological issue is that adaptation effects may be driven by only a few data points. As mentioned above, Kaan et al. (2018) conducted a self-paced reading study investigating reading times to filled-gap constructions, testing whether reading times to the filled-gap decreased as more such structures had been seen. A statistically significant decrease in filled-gap effects over time was found after omitting reading times longer than 2000 ms from the analysis (which was the method employed by Fine et al. 2013). However, adaptation effects were no longer observed after applying a more traditional and stringent procedure for removing outlier data points (longer than 2.5 standard deviations from the subject mean). This suggests that some adaptation effects may be driven by just a few long reading times.

Given these potential problems with quantitative measures, changes in syntactic preferences may best be tested by means of forced choice question answering tasks before, during and after exposure (Chun, 2018; Kroczek & Gunter, 2017), however not all syntactic constructions are suitable for this.

4.4 Differences Between Exposure Groups

Even if a decrease in reading times or differences in other measures are found between two conditions as a function of exposure, this is no conclusive evidence that syntactic adaptation has taken place. For instance, as participants get used to the task, they may start reading faster overall. Shorter reading times may be associated with smaller differences between conditions, and hence with smaller ambiguity effects. A decrease in the size of the ambiguity effect therefore does not necessarily imply adaptation to the syntactic structure in question (see also Fine, Qian, Jaeger, & Jacobs, 2010; Harrington Stack et al., 2018). Relatedly, one needs to be sure that a decrease in observable processing difficulty over the course of the study is not due to a decrease in attention, or to readers starting to read more shallowly. A solution to both issues is to insert comprehension questions and/or include a control group that has less exposure to the critical items but has seen as many trials in the task as the experimental group. However, only very few studies compare groups with different prior exposure to the structure in question. One example is Experiment 2 of Fine et al. (2013) reported in the first section of this chapter. In this study, the group with prior exposure to the critical structure showed a smaller ambiguity effect than the group without. However, this effect was not very strong, and was not replicated by Harrington Stack et al. (2018). Note that between-group comparisons are not without problems either, since groups may differ in their language proficiency, cognitive control, or other factors that may affect their language behavior. Groups would therefore need to be matched on these factors, or tested on a baseline task before exposure (Wells, Christiansen, Race, Acheson, & MacDonald, 2009) in order to rule out pre-existing differences.

4.5 Reversal Effects

Third, not many studies have probed changes in processing of the structure that is initially preferred as a function of exposure to the non-preferred structure. Fine et al. (2013) report more difficulty for main clause continuations for the participant group who had had more exposure to reduced relative clauses. This reversal effect was replicated by Sikos et al. (2016) but not by Harrington Stack et al. (2018). Atkinson (2016) failed to find reversal effects for filled-gap effects. Chun (2018) reports that the preference for interpreting the relative clause as belonging to the last mentioned noun had weakened after exposure to sentences such as (2b), but never turned into a preference for the first-mentioned noun. Finally, Kaschak and

Glenberg (2004) even report a facilitation of the originally preferred structure after exposure to the infrequent or unusual structure in their study.

A reversal of the initial processing preference has therefore hardly been observed, if tested at all. An observable reversal of preference may be taken as strong evidence for adaptation; the absence of a reversal effect, however, does not mean there is no adaptation. If the initial preference for a particular structure is very strong, it may take a very long exposure to the alternate structure before this preference, based on life-long experience, will reverse, if at all. In addition, as hypothesized by Atkinson (2016), the activation of the two alternate structures may be unrelated: a boost in the activation of one (or change of probability of one), does not necessarily imply an inhibition or change in the probability of the other.

4.6 Syntactic Effects?

4.6.1 Adaptation May Not Be to the Syntactic Structure Intended

If a participant shows a change in parsing preferences or a decrease in ambiguity of complexity effects as a function of exposure, this does not necessarily imply that the participants adapted to the abstract syntactic structure intended by the researcher, or to a syntactic structure at all. For instance, in a replication of Fine (2013), Malone and Mauner (2018) showed that readers not only adapted to the reduced relative structure, but also to the oddity of having a transitive verb without a following direct object. For instance in *The drunk tourists heard on their balcony*..., the phrase *on the balcony* is unexpected given that *heard* is typically followed by a direct object noun phrase or a clause, as in *The drunk tourists heard the manager*. If the sentence is completed as in *The drunk tourists heard on their balcony were asked to leave*, there are two potential problems for the reader: the non-typical use of a transitive verb without a direct object, and the verb (*were*) signaling that the preceding phrase is a relative clause. In Malone and Mauner's replication of Fine (2013), readers initially took longer to read *heard on the balcony* in the ambiguous conditions (*The drunk tourists heard on their balcony were asked to leave*) versus unambiguous relative clauses (*The drunk tourists who were heard on the balcony*...). This difference became smaller over the course of the study, and more so if the verb (*heard*) normally had a strong bias to occur with an object noun phrase. Note that the sentence is disambiguated towards a reduced relative clause only later, at the main clause verb *were*. Malone and Mauner's results therefore suggest that readers adapted to the unusual use of the transitive verbs (namely that there was no direct object following) in addition to, or instead of, adapting to the relative clause structure.

Similarly, Wells et al. (2009) reported that object relatives, such as *the clerk that the typist trained told the truth*, became easier to read after repeated exposure to this structure. However, as the authors observe themselves, the object relatives used in the study all contained animate noun phrases. In natural language use, sentences with object relatives typically contain inanimate noun phrases and pronouns. The adaptation observed in the study may therefore be due to readers getting used to the a-typical occurrence of animate noun phrases in object relatives, rather than adapting to the object relative structure itself.

4.6.2 Adaptation May Not Be Syntactic

What may appear as adaptation to a syntactic structure may also, or instead, be an adaptation to particular aspects of processing the structure, especially if the structure is unusual. Kaschak and Glenberg (2004) propose that unusual continuations of sentences can leave a trace in episodic memory of the processing that goes into the comprehension. Adaptation effects can be accounted for by assuming that these processing episodes are easier to retrieve every time a similar structure is encountered. In support of this, Kaschak and Glenberg (2004) report that adaptation to an unusual structure also facilitates the processing of the preferred structure that was originally considered. In particular, if one hears *This meal needs cooked* ... one may expect that the next word is a noun (*This meal needs cooked vegetables*). This interpretation may still be selected and maintained in episodic memory even if the sentence continues in a less usual way such as in the non-standard *need* Verb-*ed* construction (*This meal needs cooked given that dinner is in an hour*). Kaschak and Glenberg (2004) indeed report facilitation for the initially frequent construction (*This meal needs cooked vegetables*) after exposure to the unusual structures, supporting the view that what is adapted to are aspects of processing, not syntactic structures.

This approach can explain another finding. Fraundorf and Jaeger (2016) also used the *need* Verb-*ed* construction in their study. One group of participants came from a region of the US in which this structure is commonly used and was therefore familiar with this non-standard variation. Another group came from a different region of the US and was not familiar with this dialectal structure. The latter group, who was initially not familiar with the structure, adapted to it rather quickly. This group then was presented with another, similar, unfamiliar structure, which they also adapted to quickly. Interestingly, the group of participants who were familiar with the dialectal structure in their daily lives did not show adaptation to this other unusual structure. This suggests that readers who were initially unfamiliar with the dialectal structure did not so much adapt to this specific syntactic structure, but rather to the unusualness of the items or to the application of uncommon processing strategies, which then generalized to similar atypical constructions (Kaschak & Glenberg, 2004). Apparent syntactic adaptation effects may therefore not be adaptation to a syntactic structure.

4.7 Adaptation as Linking Contextual Cues and Syntactic Properties

Considering the four criteria for adaptation, the evidence for syntactic adaptation is rather inconsistent. However, a number of studies do report a decrease of ambiguity effects or change in preference, meeting the first criterion. Assuming that methodological concerns can be ruled out in these cases, it is clear that something changes in the way listeners and readers process particular sentences as they get more exposed to sentences of that type. The question then is, what changes? And why does this change occur in some cases but not in others?

The view of adaptation that we endorse here is that language users establish a link in memory between particular linguistic characteristics and characteristics of the situation, for instance, of the talker or the experimental context. This is also the view incorporated by belief-updating and error-based implicit learning models discussed above. This approach can also account for why adaptation effects often do not easily generalize to other experimental paradigms (Atkinson, 2016; Kaschak, Kutta, & Coyle, 2014) or the participants' own language use outside the lab. A laboratory situation is very different from daily life, and hence participants may associate the particular syntactic structures used with being in a study, and will not carry over these new frequency distributions over to other situations if these are not considered sufficiently similar. This account can also explain why cue(talker)-specific adaptation has not been systematically observed in the lab (Kamide, 2012; Liu et al., 2017; Ryskin, Qi, Duff, & Brown-Schmidt, 2017; Ryskin et al., 2016a). In cases where it is not obvious that a particular structure is systematically linked to a particular situation, talker or/and verb (for instance, because the structure is not unusual, the alternation has no different communicative or semantic function, or the mapping is too complex), the association may not be picked up by the listener or reader, and hence may not be stored in memory. In this case, no adaptation occurs that is specific to the talker or other characteristic of the context. On the other hand, when a rather unusual structure is systematically produced by a particular talker (Kroczek & Gunter, 2017), the link between the talker and the particularities of the structure can be detected and is tracked. Although this approach of course raises the question of what is "typical", and how the connections between the situation and the linguistic characteristics are made, among other things, this approach opens the way to explaining not **that** adaptation occurs, but **when** it occurs and when it does not.

5. WHAT IS THE RELATION BETWEEN ADAPTATION AND LEARNING?

5.1 Differences in Adaptation inside and outside the Lab

The initial excitement regarding syntactic adaptation was driven by the idea that syntactic adaptation as observed in the lab may give insight in language learning. Given that lab experiments on syntactic adaptation are necessarily rather limited, what does adaptation as studied in the lab have to do with adaptation observed in common language use and language learning through exposure? After all, children learn their native language through exposure, and there is ample evidence that adult language users are extremely sensitive to differences in frequencies in the language they are surrounded by (e.g. MacDonald, 2013), and can adapt to changes in distributional frequencies even at an abstract level. To give an example, Dussias and Sagarra (2007) looked at the interpretation of relative clause attachment ambiguities, such as *the sister of the actor who was on the balcony*. Monolingual English speakers prefer the relative clause (underscored in the example) to be about the last-mentioned noun (*actor*). Native Spanish speakers, reading these sentences in Spanish, have a preference for the relative clause to modify the first-mentioned noun (*sister*). However, after having lived in an English speaking country for an average of 7 years, and having been naturally exposed to English, native Spanish speakers had changed their preferred interpretation and preferred the relative clause to modify the last-mentioned noun, even in their native Spanish. This suggests that in daily life people change their abstract, syntactic preferences based on exposure.

Given the abundant evidence that people adjust their language use on the basis of exposure, it may at first sight be rather surprising that syntactic adaptation effects in the lab have been so hard to obtain, especially in comprehension. However, the effects of natural exposure on syntactic preferences are typically very slow, with some learners taking months or even years to change aspects of their language processing. For instance, the Dussias and Sagarra (2007) study cited above also included Spanish speakers who had been immersed in an English language environment for an average of 8.5 months. Their parsing preferences were not different from Spanish speakers who had not been much exposed to English. It is therefore more surprising that syntactic adaptation effects **can** be observed in the lab, and that these can occur so quickly (after eight trials in Fine et al., 2013). This is a clear difference between syntactic adaptation in the lab and in natural exposure.

Another difference is generalizability. As discussed in the above, if observed in the lab, adaptation effects are not very likely to carry over between tasks, including one's own spontaneous production. This is different from language learning, in which generalizing to other situations, and incorporating properties of the language exposed to into one's own language use are considered evidence of learning.

5.2 Learning From Exposure May Not Be Driven by a Unitary Mechanism

These apparent differences between adaptation as observed in the lab and through natural exposure may for a great deal be attributed to differences in how the linguistic input is presented. In exposure tasks in the laboratory, a limited number of rather homogeneous sentences with an unusual structure is presented within a short amount of time. As we will further discuss below, this can explain fast, context-specific adaptation in the lab. In natural language exposure, on the other hand, many different sentences of a particular structure are encountered, produced by different speakers or writers in different communicative contexts, intermixed with many other sentences, and repetitions are typically spread out over a long time period. This may explain why learning is slow and less context-specific. Adaptation may therefore work differently, and even may recruit different cognitive systems, depending on the nature of the task, input, modality, timing, goals, and other elements. These are all aspects in which laboratory tasks differ from each other and from exposure to language in daily life.

Learning from exposure may not be the same in all situations. This ties in with recent insights from statistical learning paradigms suggesting that learning regularities from the input is not a unitary phenomenon. In statistical learning paradigms, participants are exposed to sequences of syllables, sounds or shapes that, unbeknownst to them, have statistical regularities. For instance in the classic studies by Saffran and colleagues (Saffran, Aslin,

& Newport, 1996; Saffran, Newport, Aslin, Tunick, & Barrueco, 1997), a continuous stream of nonsense syllables was presented (*bupadapatubitutibupatubibupada …*), in which the sequence of some syllables was fixed. For instance, *bu-pa-da, pa-tu-bi* and *tu-bi-bu* were always presented as a sequence, but the order of these three-syllable sequences was varied. When tested after exposure, adults as well as children were above chance at distinguishing sequences that had been fixed sequences in the stream (e.g. *bupada*) from sequences that were not fixed (e.g. *batipa*), indicating they had extracted the regularities from the speech stream. Performance on one statistical learning task typically does not correlate with performance on other statistical learning tasks. This suggests that different systems are recruited from task to task. At an abstract level, these systems are similar in that they extract distributional properties from the input, but how this is implemented depends on the modality, stimulus properties, and the type of regularity to be extracted (Erickson, Kaschak, Thiessen, & Berry, 2016; Frost, Armstrong, Siegelman, & Christiansen, 2015; Siegelman, Bogaerts, & Frost, 2017). In the same vein, adaptation as a result of language exposure may work differently and involve different systems in natural situations than in the lab, and even differently from study to study.

5.3 The Way Forward: Factors Driving Adaptation

Future research needs to identify factors and systems that are important for language learning through natural exposure, yet can be tested in the lab. By making language exposure in the lab more like language exposure in the real world, one can identify which systems and factors are important for language learning through exposure.

Above we endorsed the view that syntactic adaptation involves establishing a link between particular syntactic characteristics and characteristics of the situation. This can be modeled in belief-updating models as setting up and maintaining different probability distributions of syntactic features given specific talker- or other context-related cues (Jaeger & Snider, 2013; Kleinschmidt & Jaeger, 2015; Qian et al., 2012).

In our view, belief-updating does not necessarily result from the workings of one system. The way prior information factors in, how information is encoded, regularities are detected and cue-based associations are made, how and to what extent information is updated, and how it is stored and retrieved may be different depending on the circumstances. Some proponents of belief-updating models do not invoke different cognitive systems, but attribute apparent differences in adaptation to differences in the distribution and type of information (Fine & Jaeger, 2016; Jaeger & Snider, 2013). Much more experimental and computational work therefore needs to be carried out to provide more details as to what components can be distinguished, and how these map onto cognitive and neural systems. However, the following interrelated issues are particularly important and deserve further study, especially if one wants to relate adaptation in the laboratory to adaptation in natural exposure situations: what determines adaptation, what is generalized, and which memory systems are involved?

5.4 What Determines Syntactic Adaptation?

5.4.1 What Is Deviant?

Adaptation is, in essence, a change in processing as a result of exposure. This implies that there need to be aspects in the input that are different from what the language user expects before exposure. People differ in their prior experience. What is considered typical may therefore vary by individuals and groups. Something can be very usual for some people or speakers of some language varieties, but rather unusual or unknown to others (Fernandes, Luegi, Correa Soares, de la Fuente, & Hemforth, 2018; Kaan & Chun, 2017; Kutta, Kaschak, Porcellini, & Jones, 2017). People therefore differ as to what is deviant to them, and would differ as to what they show larger adaptation towards, or can adapt to in the first place. Adaptation may be hard if the structure is not already in the listener's or reader's inventory, or not easily derivable from existing knowledge. This may account for the lack of observable adaptation in the lab in second-language learners (Kaan et al., 2018), or in native speakers when the structure is very uncommon and complex (McCann & Kaan, 2017; Sikos et al., 2016).

Encountering an unusual syntactic structure or other characteristic however does not imply adaptation to this structure. An unusual word order may just be regarded as a speech error or dysfluency, and will not lead to new associations or an update of beliefs. In other situations, the input may be only slightly deviant from what is the norm in the current context (Qian et al., 2012). Some regularities may even go unnoticed, for instance when two talkers use the same verbs and verb frames, but differ in their mapping between the verbs and frames (Ryskin et al., 2017). Other aspects of the input may indeed be considered deviant, and will lead to adaptation, either as a shift in the probability distribution of the current set maintained, or as the set-up or retrieval of a different set probability distributions that are specific to that talker or context (Qian et al., 2012).

5.4.2 Goals and Social Function

Adaptation also depends on the listener's or reader's goals and attitudes (cf. the utility function in belief-updating models, Kuperberg & Jaeger, 2016). If the current way of processing works well enough for the purpose of the listener or reader, and the current cues are deemed reliable, there is no need to change processing or update knowledge. This may be related to blocking effects in language learning, in which initially learned associations impede the learning of other cues (Arnon & Ramscar, 2012; Ellis & Sagarra, 2010). For instance if learners initially rely on temporal adverbials, such as *yesterday*, to express tense, they have difficulties learning verb inflections expressing tense (Ellis & Sagarra, 2010). On the other hand, comprehenders may change their processing if they no longer deem a talker or a cue reliable (Henry et al., 2017; Hopp, 2016).

An important difference between lab exposure tasks and natural language exposure is that the latter is embedded in a social and communicative context, whereas adaptation studies typically involve isolated sentences lacking a communicative goal. A communicative context may encourage language users to change their current processing or representations to align with those of the interlocutors so as to make communication more efficient. Social and other cues that may steer adaptation, such as talker identity or the use of particular verb, may therefore be picked up in natural learning situations but may go unnoticed in a lab context. Syntactic structures often do not much affect meaning or discourse function when presented in isolation. The absence of communicative information in the lab may then also explain why adaptation to syntactic structures is less robust than adaptation to meaning (Yildirim et al., 2016), intonation (Kurumada et al., 2017), or accents (Bradlow & Bent, 2008), cf. Liu et al. (2017); and why adaptation in production appears to be more robust than in comprehension (cf. Atkinson, 2016; Kutta & Kaschak, 2012). Evidence that communicative goals are important for adaptation comes from production studies using trial-to-trial priming. Typically, participants are much more likely to re-use the structure they just heard when they are in a dialog situation rather than when they are producing sentences in a monologue situation (Branigan, 2007), especially if the dialog partners are working toward the same goal (Reitter & Moore, 2014). Furthermore, speakers are more likely to repeat the structure that they just heard from a person who is addressing them rather than when this person is talking to somebody else (Branigan, Pickering, McLean, & Cleland, 2007). Syntactic adaptation in the lab may therefore be more robust when the exposure is part of an interactive context, as in natural exposure situations.

In addition, social context may be important for changes in the learner's own language production. If the learner identifies certain input as coming from an "in group", or group they want to belong to, the set of probability distributions derived by adaptation to this group may form the learner's default, own-use model of language (for social effects on priming and adaptation, see e.g. Chun, Barrow, & Kaan, 2016a, 2016b; Lev-Ari, 2016, 2018; Weatherholtz, Campbell-Kibler, & Jaeger, 2014).

5.5 What Determines Generalization?

5.5.1 Variation

Another difference observed between adaptation in the lab and language learning is that adaptation in the lab seems to be restricted to the specifics of the context, whereas a language learner is able to generalize over situations. How are generalizations made, and how can this be encouraged in the lab? We mentioned before that a lab situation is very different from daily life and is clearly demarcated. Participants may therefore adapt to the specifics of the language restricted to a laboratory context or even the specific task (Atkinson, 2016). In natural language exposure, different contexts are not as clearly outlined. In addition the same syntactic structures or features are encountered in many different contexts and uttered by many different speakers. This is different from exposure phases in the lab, which are typically rather uniform. Future research could test to what extent variation in the exposure phase affects the generalizability of the adaptation (along the lines of studies on learning phonetic distinctions and vocabulary, e.g. Barcroft & Sommers, 2005; Clopper & Pisoni, 2004; Lively, Logan, & Pisoni, 1993).

5.5.2 Similarity

Generalization can also be seen as transferring a specific set of expectations from a prior speaker or situation to new but similar speaker or situation. For instance, when a non-native speaker with a particular accent makes certain types of syntactic error, one regards these errors as typical for non-native speakers with that particular accent (Caffarra & Martin, 2018; Hanulikova et al., 2012), which in turn steers adaptation to non-native speakers that sound similar. This can account for why native Spanish speakers expect English-accented Spanish speakers to make errors with grammatical gender, but not with number agreement (Caffarra & Martin, 2018). Similarly, learners of a third or fourth language typically transfer properties from languages they learned previously to the new language they are learning, but are more likely to do so for the language they deem more similar to the language currently being learned (Rothman, 2011). Also this can be accounted for by generalizing what is typical in one situation to a similar context (Pajak, Fine, Kleinschmidt, & Jaeger, 2016).

Generalization is thus a function of perceived similarity between situations or speakers. By manipulating perceived similarity in the lab, one can study what type of features are relevant and how similar the speaker or situation needs to be in order for adaptation to be carried over and be generalized.

5.6 What Memory Systems Are Involved?

Trial-to-trial syntactic priming and, by extension, syntactic adaptation have been considered a form of implicit learning, involving changes in long-term memory (Bock & Griffin, 2000; Chang et al., 2006). Arguments for this view were that (a) participants are typically not aware of the experimental manipulation and the change in their language processing as a result of exposure; (b) the priming is to an abstract structure, independent of the lexical items; (c) priming effects are long lasting (over ten intervening sentences in Bock & Griffin, 2000); and (d) priming does not require intact short-term memory (Ferreira, Bock, Wilson, & Cohen, 2008). However, several researchers have argued for the involvement of a short-term memory system in trial-to-trial priming as well (Chang et al., 2006, 2012; Hartsuiker & Bernolet, 2017; Hartsuiker, Bernolet, Schoonbaert, Speybroeck, & Vanderelst, 2008; Reitter et al., 2011; Tooley & Traxler, 2018). Involvement of a short-term memory system can account for the lexical boost effect, that is, the observation that trial-to-trial priming effects are larger when the prime and target share lexical items. Lexical boost effects are typically short lived (Hartsuiker et al., 2008), but have been found to last across two or three sentences in comprehension (Tooley & Bock, 2014). The lexical boost effect can be accounted for, for instance, by assuming that the lexical items of the prime sentence and their links to the abstract syntactic representation can remain active in short-term memory when the target is processed (Reitter et al., 2011). This increases the likelihood that the same structure is used when the lexical items are repeated in the target production, and makes it easier to access the structure in comprehension (for a different explanation see Fine & Jaeger, 2016; Jaeger & Snider, 2013). Second, low-proficient second-language learners can show syntactic trial-to-trial production priming effects in the absence of lexical overlap between the prime and target, even though they do not yet have an abstract structural representation in long term memory (Bernolet, Hartsuiker, & Pickering, 2013; Jackson & Ruf, 2016). This can be accounted for by assuming that the participant keeps the prime sentence in short-term memory and edits parts of it to make the target sentence (Bernolet et al., 2013; Hartsuiker & Bernolet, 2017).

Similarly, cumulative syntactic adaptation may involve a combination of memory systems, with different systems being recruited to different extents depending on the task and specifics of the stimuli. The critical sentences in adaptation studies are rather densely spaced in time, and are typically rather uncommon structures, especially in comprehension studies. These structures, or other peculiarities of (processing) the stimulus items, may therefore draw the participants' attention and may remain active in short-term memory or episodic memory, linked to other specifics of the context (Coyle & Kaschak, 2008; Kaschak & Glenberg, 2004). Assuming that changes in short-term memory can occur fast, whereas changes in long-term memory are rather slow, a difference in reliance on short-term or episodic memory, likely in addition to other (long-term) systems, may explain why adaptation can occur within a few trials in the lab, and why this time-course is different from adaptation in natural exposure to syntactic structures. Future research needs to be conducted to further identify which cognitive systems are involved in adaptation, and under which conditions (for preliminary studies in this direction see Fine & Jaeger, 2016; Kutta et al., 2017; Tooley & Traxler, 2018).

Another important question is how context-specific frequency distributions are stored, consolidated and reactivated in memory. In natural language exposure, specific structures are repeated over long stretches of time, which involves regular re-access and re-consolidation, making the memory representations stronger, and, in some cases, more general (Simon, Gómez, Nadel, & Scalf, 2017). Adaptation studies that have used multiple sessions, hence allowing for re-access and re-consolidation, have reported effects of exposure lasting up to at least several days after the last exposure session (Wells et al., 2009), even up to nine months (Kroczek & Gunter, 2017). Systematically and orthogonally varying the number of critical items in the exposure phase and their spread over time may provide more insight in how context-specific distributional information is encoded, re-accessed, re-consolidated, and changed in the process (see Xie, Earle, & Myers, 2018, for research along these lines on foreign accent adaptation). This in turn will give us more insight in long-term changes and in language learning.

6. CONCLUDING REMARKS

Even though there is abundant evidence that language users adapt to the features of the language surrounding them, the evidence for adaptation to abstract syntactic structures after exposure in the lab is equivocal. Syntactic adaptation effects have not been systematically observed, and hardly generalize to different contexts. We endorse the view that adaptation as seen in the lab is the formation of an association between properties of the language stimuli and features of the context, resulting in changes in expectations of what is linguistically typical for contexts with those features. Although adaptation in the lab appears to be rather different from language learning through natural exposure to language, studying adaptation in the lab gives us insight into mechanisms and systems underlying changes in processing as a function of exposure to the linguistic environment. Based on recent insights in statistical learning, it is not likely that distribution-based learning, including syntactic adaptation, is a general, unitary mechanism. Different systems may factor in depending on the nature of the exposure, prior knowledge, specifics of the context, goals, and temporal spacing of the critical input.

Future research on syntactic adaptation should be aimed at identifying factors that facilitate adaptation and more clearly connect lab exposure with naturalistic exposure. Above we mentioned the importance of communicative contexts and goals. This may encourage adaptation and make it easier for the listener to pick up on information that co-occurs with a particular syntactic structure (e.g. talker or verb type). Furthermore, manipulating both the amount of exposure and the distribution over time and situations to more resemble natural learning contexts may make it easier to study in the lab which factors underlie language learning through naturalistic exposure. These kinds of studies can best be carried out using artificial languages, so as to reduce the effect of prior knowledge. This line of research may give more insight in how children as well as adults learn a language, and how their language use or language in general changes through interaction and exposure.

ACKNOWLEDGMENTS

This work was supported in part by a Southeastern Conference Visiting Faculty Travel Grant awarded to EK, and a *Language Learning* dissertation grant awarded to EC.

REFERENCES

Arnon, I., & Ramscar, M. (2012). Granularity and the acquisition of grammatical gender: How order-of-acquisition affects what gets learned. *Cognition, 122*(3), 292–305. https://doi.org/10.1016/j.cognition.2011.10.009.

Atkinson, E. (2016). *Active dependency completion in adults and children: Representations and adaptation* (Ph.D). Johns Hopkins University.

Barcroft, J., & Sommers, M. S. (2005). Effects of acoustic variability on second language vocabulary learning. *Studies in Second Language Acquisition, 27*(3), 387–414. https://doi.org/10.1017/S0272263105050175.

Bernolet, S., & Hartsuiker, R. J. (2010). Does verb bias modulate syntactic priming? *Cognition, 114*(3), 455–461. https://doi.org/10.1016/j.cognition.2009.11.005.

Bernolet, S., Hartsuiker, R. J., & Pickering, M. J. (2013). From language-specific to shared syntactic representations: The influence of second language proficiency on syntactic sharing in bilinguals. *Cognition, 127*(3), 287–306. https://doi.org/10.1016/j.cognition.2013.02.005.

Bock, K. (1986). Syntactic persistence in language production. *Cognitive Psychology, 18*(3), 355–387. https://doi.org/10.1016/0010-0285(86)90004-6.

Bock, K., & Griffin, Z. M. (2000). The persistence of structural priming: Transient activation or implicit learning? *Journal of Experimental Psychology: General, 129*(2), 177–192. https://doi.org/10.1037/0096-3445.129.2.177.

Bradlow, A. R., & Bent, T. (2008). Perceptual adaptation to non-native speech. *Cognition, 106*(2), 707–729. https://doi.org/10.1016/j.cognition.2007.04.005.

Branigan, H. P. (2007). Syntactic priming. *Language and Linguistics Compass, 1*(1–2), 1–16. https://doi.org/10.1111/j.1749-818X.2006.00001.x.

Branigan, H. P., Pickering, M. J., McLean, J. F., & Cleland, A. A. (2007). Syntactic alignment and participant role in dialogue. *Cognition, 104*(2), 163–197. https://doi.org/10.1016/j.cognition.2006.05.006.

Caffarra, S., & Martin, C. D. (2018). Not all errors are the same: ERP sensitivity to error typicality in foreign accented speech perception. *Cortex*. https://doi.org/10.1016/j.cortex.2018.03.007.

Chang, F., Dell, G. S., & Bock, K. (2006). Becoming syntactic. *Psychological Review, 113*(2), 234–272. https://doi.org/10.1037/0033-295x.113.2.234.

Chang, F., Janciauskas, M., & Fitz, H. (2012). Language adaptation and learning: Getting explicit about implicit learning. *Language & Linguistics Compass, 6*(5), 259–278. https://doi.org/10.1002/lnc3.337.

Chun, E. (2018). *The role of prediction in adaptation: An evaluation of error-based learning accounts* (Ph.D.). Gainesville, FL: University of Florida.

Chun, E., Barrow, J., & Kaan, E. (2016a). Korean L2 learners' structural priming mediated by speakers with different English accents. In *Paper presented at the 29th annual CUNY conference on human sentence processing*. Gainesville, FL: University of Florida.

Chun, E., Barrow, J., & Kaan, E. (2016b). Native English speakers' structural alignment mediated by foreign-accented speech. *Linguistics Vanguard, 2*.

Clark, A. (2013). Whatever next? Predictive brains, situated agents, and the future of cognitive science. *Behavioral and Brain Sciences, 36*(3), 181–204. https://doi.org/10.1017/S0140525X12000477.

Clopper, C. G., & Pisoni, D. B. (2004). Effects of talker variability on perceptual learning of dialects. *Language and Speech, 47*(3), 207–238. https://doi.org/10.1177/00238309040470030101.

Coulson, S., King, J. W., & Kutas, M. (1998). Expect the unexpected: Event-related brain response to morphosyntactic violations. *Language & Cognitive Processes, 13*(1), 21–58.

Coyle, J. M., & Kaschak, M. P. (2008). Patterns of experience with verbs affect long-term cumulative structural priming. *Psychonomic Bulletin & Review, 15*, 967–970.

Dussias, P. E., & Sagarra, N. (2007). The effect of exposure on syntactic parsing in Spanish–English bilinguals. *Bilingualism: Language and Cognition, 10*(1), 101–116.

Ellis, N. C., & Sagarra, N. (2010). The bounds of adult language acquisition. Blocking and learned attention. *Studies in Second Language Acquisition, 32*(04), 553–580. https://doi.org/10.1017/S0272263110000264.

Erickson, L. C., Kaschak, M. P., Thiessen, E. D., & Berry, C. A. S. (2016). Individual differences in statistical learning: Conceptual and measurement issues. *Collabra, 2*(1), 1–17. https://doi.org/10.1525/collabra.41.

Farmer, T. A., Fine, A. B., Yan, S., Cheimariou, S., & Jaeger, T. F. (2014). Error-driven learning of higher-level expectancies during reading. In P. Bello, M. Guarini, M. McShane, & B. Scassellati (Eds.), *Proceedings of the 36th annual meeting of the cognitive science society* (pp. 2181–2186). Austin, TX: Cognitive Science Society.

Fernandes, E. G., Luegi, P., Correa Soares, E., de la Fuente, I., & Hemforth, B. (2018). Adaptation in pronoun resolution: Evidence from Brazilian and European Portuguese. *Journal of Experimental Psychology: Learning, Memory, and Cognition.* https://doi.org/10.1037/xlm0000569. No Pagination Specified-No Pagination Specified.

Ferreira, V. S., Bock, K., Wilson, M. P., & Cohen, N. J. (2008). Memory for syntax despite amnesia. *Psychological Science, 19*(9), 940–946. https://doi.org/10.1111/j.1467-9280.2008.02180.x.

Fine, A. B. (2013). *Prediction, error, and adaptation during online sentence comprehension* (Ph.D.). University of Rochester.

Fine, A. B., & Jaeger, T. F. (2016). The role of verb repetition in cumulative structural priming in comprehension. *Journal of Experimental Psychology: Learning, Memory, and Cognition, 42*(9), 1362–1376. https://doi.org/10.1037/xlm0000236.

Fine, A. B., Jaeger, T. F., Farmer, T. A., & Qian, T. (2013). Rapid expectation adaptation during syntactic comprehension. *PLoS ONE, 8*(10), 1–18. https://doi.org/10.1371/journal.pone.0077661.

Fine, A. B., Qian, T., Jaeger, T. F., & Jacobs, R. A. (2010). Is there syntactic adaptation in language comprehension?. In *Proceedings of the 2010 workshop on cognitive modeling and computational linguistics* (pp. 18–26). Uppsala University.

Fraundorf, S. H., & Jaeger, T. F. (2015). The influence of experience on processing of dialectal and conventional structures. In *Paper presented at the poster presented at the 28th annual CUNY conference on human sentence processing, Los Angeles, CA.*

Fraundorf, S. H., & Jaeger, T. F. (2016). Readers generalize adaptation to newly-encountered dialectal structures to other unfamiliar structures. *Journal of Memory and Language, 91*, 28–58. https://doi.org/10.1016/j.jml.2016.05.006.

Frost, R., Armstrong, B. C., Siegelman, N., & Christiansen, M. H. (2015). Domain generality versus modality specificity: The paradox of statistical learning. *Trends in Cognitive Sciences, 19*(3), 117–125. https://doi.org/10.1016/j.tics.2014.12.010.

Hahne, A., & Friederici, A. D. (1999). Electrophysiological evidence for two steps in syntactic analysis: Early automatic and late controlled processes. *Journal of Cognitive Neuroscience, 11*, 194.

Hanulikova, A., van Alphen, P. M., van Goch, M. M., & Weber, A. (2012/04). When one person's mistake is another's standard usage: The effect of foreign accent on syntactic processing. *Journal of Cognitive Neuroscience, 24*, 878–887.

Harrington Stack, C., James, A., & Watson, D. G. (2018). A failure to replicate rapid syntactic adaptation in comprehension. *Memory and Cognition.* https://doi.org/10.3758/s13421-018-0808-6.

Hartsuiker, R. J., & Bernolet, S. (2017). The development of shared syntax in second language learning. *Bilingualism: Language and Cognition, 20*(2), 219–234. https://doi.org/10.1017/S1366728915000164.

Hartsuiker, R. J., Bernolet, S., Schoonbaert, S., Speybroeck, S., & Vanderelst, D. (2008). Syntactic priming persists while the lexical boost decays: Evidence from written and spoken dialogue. *Journal of Memory and Language, 58*(2), 214–238. https://doi.org/10.1016/j.jml.2007.07.003.

Hartsuiker, R. J., & Westenberg, C. (2000). Word order priming in written and spoken sentence production. *Cognition, 75*(2), B27–B39. https://doi.org/10.1016/S0010-0277(99)00080-3.

Henry, N., Hopp, H., & Jackson, C. N. (2017). Cue additivity and adaptivity in predictive processing. *Language, Cognition and Neuroscience, 32*(10), 1229–1249. https://doi.org/10.1080/23273798.2017.1327080.

Hopp, H. (2016). Learning (not) to predict: Grammatical gender processing in second language acquisition. *Second Language Research, 32*(2), 277–307. https://doi.org/10.1177/0267658315624960.

Jackson, C. N., & Ruf, H. T. (2016). The priming of word order in second language German. *Applied PsychoLinguistics, 38*(2), 315–345. https://doi.org/10.1017/S0142716416000205.

Jaeger, T. F., & Snider, N. E. (2008). Implicit learning and syntactic persistence: Surprisal and cumulativity. In *Proceedings of the 29th annual cognitive science society* (pp. 1061–1066). Austin,TX: Cognitive Science Society.

Jaeger, T. F., & Snider, N. E. (2013). Alignment as a consequence of expectation adaptation: Syntactic priming is affected by the prime's prediction error given both prior and recent experience. *Cognition, 127*(1), 57–83. https://doi.org/10.1016/j.cognition.2012.10.013.

Kaan, E., & Chun, E. (2017). Priming and adaptation in native speakers and second-language learners. *Bilingualism: Language and Cognition, First View*, 1–15. https://doi.org/10.1017/S1366728916001231.

Kaan, E., Futch, C., Fernández Fuertes, R., Mujcinovic, S., & Álvarez de la Fuente, E. (2018). Adaptation to syntactic structures in native and non-native sentence comprehension. *Applied PsychoLinguistics* (accepted for publication).

Kamide, Y. (2012). Learning individual talkers' structural preferences. *Cognition, 124*(1), 66–71. https://doi.org/10.1016/j.cognition.2012.03.001.

Kaschak, M. P. (2007). Long-term structural priming affects subsequent patterns of language production. *Memory & Cognition, 35*(5), 925–937.

Kaschak, M. P., & Borreggine, K. L. (2008). Is long-term structural priming affected by patterns of experience with individual verbs? *Journal of Memory and Language, 58*, 862–878. https://doi.org/10.1016/j.jml.2006.12.002.

Kaschak, M. P., & Glenberg, A. M. (2004). This construction needs learned. *Journal of Experimental Psychology: General, 133*(3), 450–467. https://doi.org/10.1037/0096-3445.133.3.450.

Kaschak, M. P., Kutta, T. J., & Coyle, J. M. (2014). Long and short term cumulative structural priming effects. *Language, Cognition and Neuroscience, 29*(6), 728–743. https://doi.org/10.1080/01690965.2011.641387.

Kaschak, M. P., Loney, R. A., & Borreggine, K. L. (2006). Recent experience affects the strength of structural priming. *Cognition, 99*(3), B73–B82. https://doi.org/10.1016/j.cognition.2005.07.002.

Kleinschmidt, D. F., & Jaeger, T. F. (2015). Robust speech perception: Recognize the familiar, generalize to the similar, and adapt to the novel. *Psychological Review, 122*(2), 148–203. https://doi.org/10.1037/a0038695.

Kraljic, T., & Samuel, A. G. (2007). Perceptual adjustments to multiple speakers. *Journal of Memory and Language, 56*(1), 1–15. https://doi.org/10.1016/j.jml.2006.07.010.

Kroczek, L. O. H., & Gunter, T. C. (2017). Communicative predictions can overrule linguistic priors. *Scientific Reports, 7*(1), 17581. https://doi.org/10.1038/s41598-017-17907-9.
Kuperberg, G. R., & Jaeger, T. F. (2016). What do we mean by prediction in language comprehension? *Language, Cognition and Neuroscience, 31*(1), 32–59. https://doi.org/10.1080/23273798.2015.1102299.
Kurumada, C., Brown, M., & Tanenhaus, M. K. (2017). Effects of distributional information on categorization of prosodic contours. *Psychonomic Bulletin & Review*. https://doi.org/10.3758/s13423-017-1332-6.
Kutta, T. J., & Kaschak, M. P. (2012). Changes in task-extrinsic context do not affect the persistence of long-term cumulative structural priming. *Acta Psychologica, 141*(3), 408–414. https://doi.org/10.1016/j.actpsy.2012.09.007.
Kutta, T. J., Kaschak, M. P., Porcellini, A., & Jones, J. L. (2017). Implicit and explicit memory factors in cumulative structural priming. *Collabra: Psychology, 3*(1), 13. https://doi.org/10.1525/collabra.59.
Lev-Ari, S. (2016). Selective grammatical convergence: Learning from desirable speakers. *Discourse Processes, 53*(8), 657–674. https://doi.org/10.1080/0163853X.2015.1094716.
Lev-Ari, S. (2018). Social network size can influence linguistic malleability and the propagation of linguistic change. *Cognition, 176*, 31–39. https://doi.org/10.1016/j.cognition.2018.03.003.
Levy, R. (2008). Expectation-based syntactic comprehension. *Cognition, 106*(3), 1126–1177. https://doi.org/10.1016/j.cognition.2007.05.006.
Liu, L., Burchill, Z., Tanenhaus, M. K., & Jaeger, T. F. (2017). Failure to replicate talker-specific syntactic adaptation. In *Proceedings of the 39th annual meeting of the cognitive science society* (pp. 2616–2621).
Lively, S. E., Logan, J. S., & Pisoni, D. B. (1993). Training Japanese listeners to identify English/r/and/l/. II: The role of phonetic environment and talker variability in learning new perceptual categories. *Journal of the Acoustical Society of America, 94*, 1242–1255.
Long, D. L., & Prat, C. S. (2008). Individual differences in syntactic ambiguity resolution: Readers vary in their use of plausibility information. *Memory & Cognition, 36*(2), 375–391.
Luka, B. J., & Barsalou, L. W. (2005). Structural facilitation: Mere exposure effects for grammatical acceptability as evidence for syntactic priming in comprehension. *Journal of Memory and Language, 52*(3), 436–459. https://doi.org/10.1016/j.jml.2005.01.013.
MacDonald, M. C. (2013). How language production shapes language form and comprehension. *Frontiers in Psychology, 4*(226). https://doi.org/10.3389/fpsyg.2013.00226.
MacDonald, M. C., Just, M. A., & Carpenter, P. A. (1992). Working memory constraints on the processing of syntactic ambiguity. *Cognitive Psychology, 24*(1), 56–98. https://doi.org/10.1016/0010-0285(92)90003-K.
Mahowald, K., James, A., Futrell, R., & Gibson, E. (2016). A meta-analysis of syntactic priming in language production. *Journal of Memory and Language, 91*, 5–27. https://doi.org/10.1016/j.jml.2016.03.009.
Malone, A., & Mauner, G. (2018). What do readers adapt to in syntactic adaptation?. In *Paper presented at the CUNY conference on human sentence processing*. Davis, CA: UC Davis.
McCann, D., & Kaan, E. (2017). Non-adaptation to garden paths. In *Paper presented at the 30th annual CUNY conference on human sentence processing*. Cambridge, MA: MIT.
Metzing, C., & Brennan, S. E. (2003). When conceptual pacts are broken: Partner-specific effects on the comprehension of referring expressions. *Journal of Memory and Language, 49*(2), 201–213. https://doi.org/10.1016/S0749-596X(03)00028-7.
Myslín, M., & Levy, R. (2016). Comprehension priming as rational expectation for repetition: Evidence from syntactic processing. *Cognition, 147*, 29–56. https://doi.org/10.1016/j.cognition.2015.10.021.
Noppeney, U., & Price, C. J. (2004). An fMRI study of syntactic adaptation. *Journal of Cognitive Neuroscience, 16*(4), 702–713. https://doi.org/10.1162/089892904323057399.

Pajak, B., Fine, A. B., Kleinschmidt, D. F., & Jaeger, T. F. (2016). Learning additional languages as hierarchical probabilistic inference: Insights from L1 processing. *Language Learning, 66*(4), 900–944. https://doi.org/10.1111/lang.12168.

Pickering, M. J., & Branigan, H. P. (1998). The representation of verbs: Evidence from syntactic priming in language production. *Journal of Memory and Language, 39*(4), 633–651. https://doi.org/10.1006/jmla.1998.2592.

Qian, T., Jaeger, T. F., & Aslin, R. N. (2012). Learning to represent a multi-context environment: More than detecting changes. *Frontiers in Psychology, 3*. https://doi.org/10.3389/fpsyg.2012.00228.

Reitter, D., Keller, F., & Moore, J. D. (2011). A computational cognitive model of syntactic priming. *Cognitive Science, 35*(4), 587–637. https://doi.org/10.1111/j.1551-6709.2010.01165.x.

Reitter, D., & Moore, J. D. (2014). Alignment and task success in spoken dialogue. *Journal of Memory and Language, 76*, 29–46. https://doi.org/10.1016/j.jml.2014.05.008.

Rothman, J. (2011). L3 syntactic transfer selectivity and typological determinacy: The typological primacy model. *Second Language Research, 27*(1), 107–127. https://doi.org/10.1177/0267658310386439.

Ryskin, R. A., Qi, Z., Duff, M. C., & Brown-Schmidt, S. (2016a). Constraints on adaptation to syntactic variability between and within speakers. In *Paper presented at the 29th annual CUNY conference on human sentence processing*. Gainesville, FL: University of Florida.

Ryskin, R. A., Qi, Z., Duff, M. C., & Brown-Schmidt, S. (2016b). Verb biases are shaped through lifelong learning. *Journal of Experimental Psychology: Learning, Memory, and Cognition*. https://doi.org/10.1037/xlm0000341. No Pagination Specified.

Ryskin, R. A., Qi, Z., Duff, M. C., & Brown-Schmidt, S. (2017). Syntactic variability between and within speakers: When to adapt, when to generalize?. In *Paper presented at the 30th annual CUNY conference on human sentence processing*. Cambridge, MA: MIT.

Saffran, J. R., Aslin, R. N., & Newport, E. L. (1996). Statistical learning by 8-month-old infants. *Science, 274*(5294), 1926.

Saffran, J. R., Newport, E. L., Aslin, R. N., Tunick, R. A., & Barrueco, S. (1997). Incidental language learning: Listening (and learning) out of the corner of your ear. *Psychological Science, 8*(2), 101–105.

Scheepers, C. (2003). Syntactic priming of relative clause attachments: Persistence of structural configuration in sentence production. *Cognition, 89*(3), 179–205. https://doi.org/10.1016/S0010-0277(03)00119-7.

Siegelman, N., Bogaerts, L., & Frost, R. (2017). Measuring individual differences in statistical learning: Current pitfalls and possible solutions. *Behavior Research Methods, 49*(2), 418–432. https://doi.org/10.3758/s13428-016-0719-z.

Sikos, L., Martin, H., Fitzgerald, L., & Grodner, D. (2016). Memory-based limits on surprisal-based syntactic adaptation. In *Paper presented at the 29th annual CUNY conference on human sentence processing, Gainesville, FL*.

Simon, K. C. N. S., Gómez, R. L., Nadel, L., & Scalf, P. E. (2017). Brain correlates of memory reconsolidation: A role for the TPJ. *Neurobiology of Learning and Memory, 142*, 154–161. https://doi.org/10.1016/j.nlm.2017.03.003.

Tooley, K. M., & Bock, K. (2014). On the parity of structural persistence in language production and comprehension. *Cognition, 132*(2), 101–136. https://doi.org/10.1016/j.cognition.2014.04.002.

Tooley, K. M., & Traxler, M. J. (2010). Syntactic priming effects in comprehension: A critical review. *Language and Linguistics Compass, 4*(10), 925–937. https://doi.org/10.1111/j.1749-818X.2010.00249.x.

Tooley, K. M., & Traxler, M. J. (2018). Implicit learning of structure occurs in parallel with lexically-mediated syntactic priming effects in sentence comprehension. *Journal of Memory and Language, 98*(Supplement C), 59–76. https://doi.org/10.1016/j.jml.2017.09.004.

Weatherholtz, K., Campbell-Kibler, K., & Jaeger, T. F. (2014). Socially-mediated syntactic alignment. *Language Variation and Change, 26*(3), 387–420. https://doi.org/10.1017/S0954394514000155.

Wells, J. B., Christiansen, M. H., Race, D. S., Acheson, D. J., & MacDonald, M. C. (2009). Experience and sentence processing: Statistical learning and relative clause comprehension. *Cognitive Psychology, 58*(2), 250–271. https://doi.org/10.1016/j.cogpsych.2008.08.002.

Xie, X., Earle, F. S., & Myers, E. B. (2018). Sleep facilitates generalisation of accent adaptation to a new talker. *Language, Cognition and Neuroscience, 33*(2), 196–210. https://doi.org/10.1080/23273798.2017.1369551.

Xie, X., Weatherholtz, K., Bainton, L., Rowe, E., Burchill, Z., Liu, L., & Jaeger, T. F. (2018). Rapid adaptation to foreign-accented speech and its transfer to an unfamiliar talker. *Journal of the Acoustical Society of America, 143*(4), 2013. https://doi.org/10.1121/1.5027410.

Yildirim, I., Degen, J., Tanenhaus, M. K., & Jaeger, T. F. (2016). Talker-specificity and adaptation in quantifier interpretation. *Journal of Memory and Language, 87*, 128–143. https://doi.org/10.1016/j.jml.2015.08.003.

CHAPTER FIVE

Neural Indices of Structured Sentence Representation: State of the Art

Ellen Lau
Department of Linguistics, University of Maryland, College Park, MD, United States
E-mail: ellenlau@umd.edu

Contents

1. Linguistic and Non-linguistic Relations 119
2. Neural Correlates of Linguistic Relations: Three Candidates 121
   2.1 Instantiating Local Linguistic Relations 122
   2.2 Classic Conceptions of Syntactic Working Memory 129
   2.3 Broader Conceptions of Syntactic Memory 134
3. Conclusion 138
References 139

Abstract

Natural language is characterized by structured, hierarchical relationships between morphemes, some of which can span an unbounded amount of intervening material. Determining how such relationships are neurally encoded in on-the-fly language comprehension is a fascinating challenge for cognitive neuroscience, and depends on foundational assumptions about both human parsing and the memory architecture. This chapter reviews current approaches towards discovering neurophysiological correlates to the encoding of structured sentence representations.

Consider the difference being given a list of things to pick up at the store which reads:

chocolate

eggs

Now consider being asked *Please go to the store and pick up some chocolate eggs*. In the former case, the two items *chocolate* and *eggs* become related in our mental representation at some level, by virtue of sharing membership in the temporary category 'things to pick up at the store right now'. But

Psychology of Learning and Motivation, Volume 68
ISSN 0079-7421
https://doi.org/10.1016/bs.plm.2018.08.004

in the latter case, the same two words become related in a different way. Here the listener understands this sequence of two words to correspond to a single phrase, which refers to a single referent in the discourse (a set of chocolate eggs), and where the conceptual properties of that referent follow from applying some of the conceptual properties labelled by *chocolate* to those labelled by *eggs*. This difference in how the same sequence is analyzed derives from the context and from the listener's knowledge of the grammar of English, which specifies that the head of a noun phrase follows noun or adjectival modifiers—in other words, *chocolate egg* means a type of egg that is made of chocolate, not a type of chocolate that is made out of eggs.

Natural language is characterized by structured, hierarchical relationships between words, including non-local relationships which can span across a potentially unbounded amount of intervening material. Although we've begun with a simple case, much more complex sets of relations between elements are routinely observed in natural utterances. For example, in the sentence *I like cold and creamy brownies with ice cream*, listeners relate the two descriptors of *cold* and *creamy* to *brownies* (even though they might be more plausible descriptors of *ice cream*), they also relate the locative phrase *with ice cream* to *brownies*, and furthermore the entire unit *cold and creamy brownies with ice cream* is related to the verb *like*. The hierarchical, non-linear phrasal relationships that form sentence structure are also famously illustrated by agreement phenomena; in subject-verb agreement in English, the form of the verb depends not on the number of the noun linearly preceding it, but on the number of the noun that 'heads' the phrase, hence *brownies with ice cream are delicious* and not *brownies with ice cream is delicious*. These are some of the many facts that motivate the descriptors 'structured' and 'hierarchical' for natural language. In other sentences a noun phrase may be linearly distant from the verb that it is related to, as in questions like *What kind of eggs did you ask your sister to buy?* In principle, the linear distance between the argument and the verb is unlimited, e.g. *What kind of eggs did you ask your sister to tell her daughter to buy?* This is the kind of fact that motivates the descriptor 'non-local'. The kind of relationships evidenced by these facts are often graphically depicted by linguists using 'tree diagrams' as in Figure 1, but it is important to recognize that while these diagrams can be used to illustrate different subtheories of linguistic relations, the format of the diagrams themselves is not a theory of mental representations (as is wrongly implied when newcomers to the field say that they want to use neuroimaging to 'find out whether syntactic trees are really in the brain'). Rather, tree diagrams are just a convenient shorthand for illustrating the structured relations between words and phrases that *must* be represented by the brain in order to explain the sentences that people produce and understand.

This chapter reviews the current state-of-the-art in cognitive neuroscience towards discovering how the brain encodes this relational information in real time to allow successful language comprehension. This question is a challenging and important one, as there is little consensus about how short-term or working memory operations are neurally implemented in general, and relatively little is known about how relational information is encoded in short-term memory in particular. Therefore, much of the work reviewed here does not yet take the form of testing an explicit proposal for neural encoding of structured representations in language, but rather begins by asking a somewhat simpler question: what, if any, current neurophysiological measures track properties of these representations?

1. LINGUISTIC AND NON-LINGUISTIC RELATIONS

An important distinction to keep in mind throughout the following is between *linguistic* and *non-linguistic* representations. Even without language, humans and many other organisms seem able to represent and reason about concepts—for example, the ability of preverbal infants to associate different sensory and non-sensory properties with the same caregiver representation. It similarly seems that even in the absence of language, we can rapidly encode novel relations between existing conceptual representations—such as the fact that my mother possesses an apple, which I can immediately encode when I see her pick one up and put it in her pocket. We also know that in many cases humans can relatively easily retain these new conceptual relations in long-term memory, at least for the duration of a few minutes or hours.

In learning a language, we acquire 'words' (more technically, *morphemes*) that label some subset of these concepts. These linguistic objects come with their own set of language-specific restrictions about what kind of relations can be created between them. For example, even without knowing the words *chocolate* and *egg*, you could likely learn these conceptual categories from experience, and you could also learn that there is a prototypical way in which these concepts are combined in the world (e.g., that chocolate eggs *can* be any size but are most often smaller than ostrich eggs). But there is nothing about the concepts of *chocolate* and *egg* in themselves that causes the English phrase *chocolate egg* to refer to a kind of egg and *egg chocolate* to refer to a kind of chocolate—this is a fact about language, and not about the concepts. Therefore, sentence comprehension can be roughly viewed as a process of encoding *linguistic* relations between morphemes and phrases, which can then in turn drive the encoding of new relations between the concepts or discourse referents that the words label in long-term memory. 'Parsing' is the term that is usually used to refer to the process of determining the linguistic relations. The current chapter focuses on work whose aim was primarily to better understand the neural encoding of the linguistic relations, even though it is difficult in practice to clearly dissociate neural activity associated with encoding linguistic and non-linguistic relations.

Note that one reason to suppose that the encoding of linguistic relations might be qualitatively and mechanistically distinct from the conceptual relations is that there is no need to retain the linguistic relations after the sentence is over—we don't usually need to remember how a sentence was framed, we just need to remember the non-linguistic message that it conveyed. Indeed, classic behavioral work suggested that information about the form of the sentence became difficult to retrieve once the sentence was over, while the meaning persists over the long term (Sachs, 1967; Jarvella, 1971; although see; Murphy & Shapiro, 1994). As I will discuss in more detail below, this observation is very relevant for evaluating the likelihood that non-invasive measures of neural activity can track currently encoded linguistic relations at all. Retention of information in long-term memory should not rely on ongoing neural activity, if it is to be metabolically plausible, and therefore if linguistic relational information were carried forward in time through rapid encoding in long-term memory, we would have no hope of observing indices of those relations in subsequent neural activity. However, it is not so implausible to suppose that retention of information in short-term or working memory depends on ongoing neural activity—this is a popular if debated hypothesis in cognitive neuroscience—and therefore if linguistic relations were carried forward during the course of the sentence in working memory, there is a reasonable possibility that we could uncover indices of these relations in neural measures.

2. NEURAL CORRELATES OF LINGUISTIC RELATIONS: THREE CANDIDATES

One classic approach to psycholinguistic and neurolinguistics research on language has been to define some metric of structural complexity over sentences, and then to cast a wide net for any behavioral or neural dependent measure that correlates with this metric. In principle, this can be attempted without any explicit hypotheses about the parsing model—in other words, without any hypotheses about what operations occur in what sequence with what memory architecture such that the structural relations can be incrementally, accurately, and efficiently deduced from a serial input stream in real time. Although such a data-driven approach would be logically sound, it has an obvious weakness: it may well be that the processing model that links the input sentence to the output measure is complex enough that we will never be able to identify the true correlations between structure and neural activity without building in some basic hypotheses about it. For example, one strategy for parsing a sentence, known as 'bottom-up', is to record the incoming string until an end-of-sentence signal is encountered, and only then to compute the linguistic relations that support sentence meaning. An alternative parsing strategy, known as 'top-down', is to infer the relations incrementally by making well-informed guesses about what the rest of the string will contain. It should be clear from these extremes that the timecourse of neural activity associated with computing and representing linguistic relations could be wildly different depending on which parsing strategy humans actually use.

The rest of the chapter is therefore organized around three candidate types of neural correlates for linguistic relations (Fig. 1) that rely on simple (but not uncontroversial) hypotheses about the processing model. First, we can assume simple, 'word-by-word' incrementality, and then look for the neural correlates of the process of initially instantiating a local linguistic relation between two elements in the time-window immediately after the second element—for example, investigating the neural activity following *eggs* in the context of the phrase *chocolate eggs*. The second and third candidates discussed here are neural correlates of aspects of the structured memory representation—the processes involved in carrying forward these relational information over time—given particular hypotheses about the human memory architecture and its neural implementation. The second section

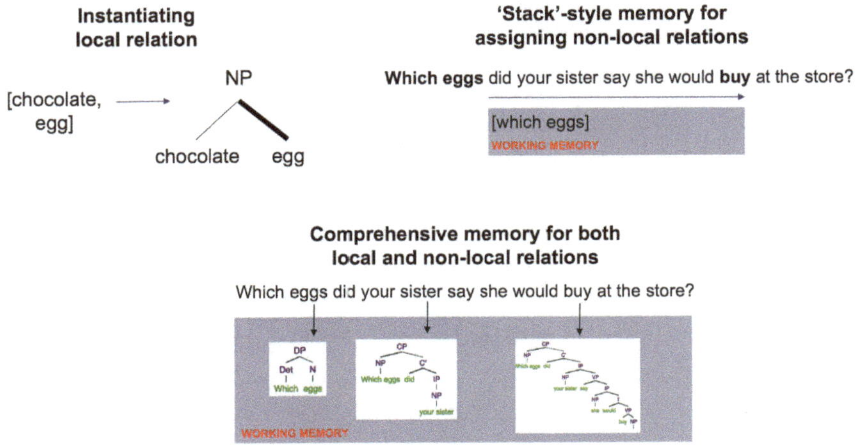

Figure 1 Three candidate types of neural correlates for linguistic relations

focuses on a classic conception of syntactic working memory that emphasizes the information which must be carried forward in cases of non-local relations in order to determine whether the input string could have been generated by the listener's grammar or not. The third section looks at a broader conception of memory for structure, which is at least the relational information which must be carried forward to ensure successful interpretation of the sentence, and at most a full structural analysis of the input, which might be useful for recovery if it turns out that part of the sentence was misanalysed.

2.1 Instantiating Local Linguistic Relations

The first aspect of the neural implementation of linguistic structure that I will consider is exactly the moment-to-moment neural processes that result in a unified syntactic, semantic, and conceptual representation for word sequences presented in a context that encourages linguistic structuring. While in subsequent sections I will examine how these structured representations may be carried forward in time over seconds or minutes, in this first section the question is, what kind of neural activity is associated with establishing these relational representations in the first place?

Perhaps the most straightforward approach to answering this question is simply to look for extra brain activity engaged at the moment when linguistic combination happens, relative to when it does not. In a highly influential paper Bemis and Pylkkänen (2011), recorded MEG responses while participants read simple two word phrases (*red boat*), looking for evidence of activity at *boat* reflecting the instantiation of syntactic or semantic relations between the two words. They compared this response to several controls designed not to elicit the instantiation of syntactic or semantic relations: *xkq boat*, in which the first stimulus was a consonant string that was not associated with a word or concept and so could not be linguistically related to the second, and *cup boat*, where the task instructions were designed to encourage processing of the sequence as an unrelated list of two nouns rather than a phrasal compound. In MEG source localization analyses, Bemis and Pylkkänen showed an increased response to the noun in the phrasal context by around 250 ms in left anterior temporal cortex, an effect that they soon replicated in other experiments (Bemis & Pylkkänen, 2012, 2013a, 2013b). In some of these studies Bemis and Pylkkänen also observed increased activity for the phrasal condition in ventromedial prefrontal cortex (Bemis & Pylkkänen, 2011) and angular gyrus (Bemis & Pylkkänen, 2012), although effects in these other regions have been slightly later and less reliable across studies. We have replicated these basic observations in EEG, showing that the same paradigm elicits differential responses within 180—400 ms after a noun is presented in a phrasal vs. non-phrasal context, in the form of more negative potentials for the noun in a phrasal context (Neufeld et al., 2016). We also observed an even earlier combinatorial effect, which we tentatively attributed to predictive structure-building operations.

This approach towards identifying neural correlates of combinatoric relations has thus seemed to bear fruit. However, further work suggests that at least one of these responses is better attributed to non-linguistic, conceptual combination processes rather than linguistic ones. Initially, Bemis and Pylkkänen argued that the increased anterior temporal activity indexed the computation of linguistic (specifically, syntactic) relations between the adjective and the noun, because the response was relatively early and was not observed for non-linguistic combination such as stimuli using images (Bemis & Pylkkanen, 2013a, 2013b). However, later studies cast doubt on this interpretation because they showed that not all cases of linguistic combination elicited an increased anterior temporal response; rather, there was a strong dependence on the conceptual properties of the nouns being composed (Westerlund & Pylkkänen, 2014; Zhang & Pylkkänen, 2015). For example, while *red boat* elicited an increased response relative to *xkq boat*, *blue canoe* did not elicit an increased response relative to *xkq canoe*, and while *tomato dish* elicited an increased response relative to *qptg dish*, *vegetable dish* did not elicit an increased response relative to *qptg dish*. Based on this pattern, Pylkkänen et al. suggested that the increased anterior temporal response was specifically tied to increases in semantic 'specificity'. This can be understood as restriction in the size of the set of objects that are denoted by the phrase (e.g., the set of tomato dishes is much smaller than the set of dishes, while the set of vegetable dishes is still reasonably large in comparison) or in terms of diagnosticity (in the phrase, *red boat*, the color adjective is critical for distinguishing the denoted item from other items in the broader boat category, whereas canoes already have a number of features that distinguish them from other boats). This hypothesis can be related to a broader literature on semantic memory and conceptual combination that is beyond the scope of the current chapter (e.g. Patterson, Nestor, & Rogers, 2007; Baron, Thompson-Schill, Weber, & Osherson, 2010; Boylan, Trueswell, & Thompson-Schill, 2017). However, several puzzles remain for the semantic specificity account. First, it is not yet clear how exactly to operationalize specificity or diagnosticity such that *red boat* > *boat* in this property but *vegetable dish* = *dish* (Zhang & Pylkkänen note that it is easier to quantify relative specificity for noun modifiers than adjectives). Second, the linking hypothesis between specific interpretation processes and the increase in anterior temporal activity at ~200 ms remains somewhat unclear—in other words, why exactly is there greater or more synchronized neural activity in anterior temporal cortex at 200 ms for phrases that denote more specific concepts?

In an EEG replication of the original Bemis and Pylkkänen (2011) work, we noted that the timing and distribution of the 'red boat' effect bear some resemblance to standard N400 effects in EEG (Neufeld et al., 2016); this also fits with prior localization work showing that some anterior temporal regions contribute to N400 effects (Van Petten & Luka, 2006; Lau et al., 2008, 2014; Lau, Weber, Gramfort, Hämäläinen, & Kuperberg, 2014). If this resemblance indeed reflects a common source, we could draw on current theories about N400 effects to inform the linking hypothesis for the 'red boat' effect. One family of theories suggests that increased N400 responses reflect the increased activation of conceptual and lexical memory networks (e.g. Kutas & Federmeier, 2000; Lau et al., 2008; Federmeier & Laszlo, 2009), and that this network activation is reduced, resulting in smaller N400 responses, when the prior context predicts or primes the subsequent input. In the context of this framework, the 'red boat' effect might reflect increased attention to or retrieval of conceptual features associated with the individual words, in the service of deriving inferences about the intended message or of retrieving exemplars of the entity or event described from episodic memory. For example, out of context, seeing the word *boat* may trigger retrieval of some minimal conceptual information to the effect of 'vehicle for transport in water', but in the context of *I started to deflate my boat*, *boat* may additionally trigger retrieval of episodic memories about inflatable boats I have seen, which may afford inferences about the likely appearance, approximate size, and uses of the boat. It is important to note that, like the specificity hypothesis above, this explanation would attribute the proximate cause of the 'red boat' effect to a non-combinatorial computation (activation of stored conceptual features) that is indirectly modulated by combinatorial contexts. Similarly, while fMRI studies consistently observe increased activity in anterior temporal cortex for structured sentences over lists (e.g. Mazoyer et al., 1993; Friederici, Meyer, & von Cramon, 2000; Vandenberghe, Nobre, & Price, 2002; Rogalsky & Hickok, 2008), this may not indicate that anterior temporal cortex instantiates the combinatorial relations itself, but rather that sentence structure increases the activation and retrieval of individual conceptual features (see Wilson et al., 2014; for more discussion).

Recent work in fMRI has made use of similar paradigms, where two-word and three-word phrases are compared to unstructured lists (Graves, Binder, Desai, Conant, & Seidenberg, 2010; Zaccarella & Friederici 2015a,b; Matchin, Hammerly, & Lau, 2017; Schell, Zaccarella, & Friederici, 2017), and have reported increased activity for phrases across a variety of brain regions associated with language processing such as anterior and posterior inferior frontal cortex, posterior superior temporal sulcus, anterior temporal cortex, and angular gyrus. While an existing fMRI literature had often reported comparisons between sentences and word lists, these newer studies on short phrases are much better-positioned to identify activity associated with forming linguistic relations than most of the prior fMRI literature on syntactic processing. This is because the poor temporal resolution of fMRI means that activity from forming the relations on the fly cannot be distinguished from activity associated with other components of sentence processing such as working memory and discourse updating, or post-stimulus task performance. However, even though the studies of shorter phrases represent an improvement in this respect, the impact of post-stimulus task performance on neural responses is still an ongoing concern and may explain why results have varied across these studies.

Although current evidence suggests that the 'red boat' effect reflects conceptual processing that may not be strictly combinatorial, other neural response differences for two-word combination may well have a linguistic source. Solving a few remaining methodological challenges may help to bring broader acceptance to this approach, which remains a simple and powerful way of identifying neural indices of instantiating both conceptual and linguistic relations with time-sensitive EEG/MEG measures. First, there are often concerns about nuisance differences in the 'control', non-combinatorial condition. The control has often replaced the first word of the phrase with a consonant string (*xkq boat*), but does this impact the response to *boat* in unexpected ways, particularly given the participant's task of matching a subsequent image probe to the prior linguistic input? We should work to develop alternative control conditions that may be subject to fewer concerns. In one recent study (Lau, Neufeld, & Idsardi, in prep.), we used identical two-noun sequences in both conditions, but manipulated the relations applied to the input through the use of task instructions, where participants were either encouraged to read the sequence as a list (*pizza, flower*) or as a novel compound (*pizza flower*). Another possibility would be to use a wholly non-linguistic placeholder for the first part of the control trial, which in the experimental context, may seem more natural to participants than consonant strings (e.g. ### ### *boat* vs. ### *red boat*). Second, many of the prior studies used two-word phrases that, because of their morphology, could not begin a well-formed sentence of English (**Red boat is approaching in the distance*). It is not clear what effects, if any, this should have on the extent to which linguistic relations will be automatically applied to the input, but it might be desirable for future studies to emphasize two-word sequences that could have begun a well-formed sentence (e.g. *red boats*).

A very different approach to this problem is illustrated by the work of Jonathan Brennan and others, which I will term the 'naturalistic neuro-computational' approach (Brennan, 2016). Rather than conducting a categorical comparison between cases that require combination and cases that do not, Brennan creates a model of the number and kind of combinatorial operations that might be hypothesized to occur at each word of a naturalistic connected text or discourse. Neural activity is then recorded while participants read or listen to the input, and the neural correlates of combinatorial operations are discovered by looking for correlations between the neural activity time-locked to each word and the combinatorial operations hypothesized to occur at each word. For example, in Brennan and Pylkkänen (2017), the MEG source localization estimate for each word of the input is correlated to the number of parsing steps that would be invoked by a left-corner parsing strategy (a popular intermediate between fully top-down and fully bottom-up) to instantiate the syntactic relations between the input word and the prior material. They find that the number of parse steps correlates with left anterior temporal activity between 350 and 500 ms after a word is presented. In other work, Brennan and colleagues use a similar approach to compare different models of parsing strategies and grammar, by asking which model results in the best correlation with neural data (Brennan, Stabler, Van Wagenen, Luh, & Hale, 2016; see also Nelson et al. 2017 for a similar approach using ECoG data). More recently Brodbeck, Presacco, and Simon (2018), have introduced a novel linear kernel estimation approach for MEG data that makes it possible to directly compare estimated spatiotemporal response functions for acoustic features, lexical properties, and combinatorial operations such as semantic composition.

The benefits of the naturalistic neuro-computational approach over the two-word phrase paradigm discussed above is that neural responses are being measured as participants are engaged in what is presumably much more natural language comprehension processes—listening to a continuous narrative—than in more controlled experiments (Brennan, 2016). The challenges are of course exactly in this lack of control; there are many strong internal correlations between linguistic factors—such as function words being shorter and more frequent as well as playing particular kinds of roles in syntactic relations—and so analysis of this kind of data must fully model all of these nuisance variables or risk misattributing variability in the neural response to the wrong factor.

Finally, a new methodology introduced by Nai Ding, David Poeppel and colleagues may provide yet another way of extracting neural responses associated with combinatorial operations (Ding et al., 2016). In what I will term the **constituent-rate** paradigm, different phrases associated with exactly the same structural relations are presented at a constant and unbroken rate. For example, in their first experiments in Mandarin and English they used an unbroken sequence of four syllable sentences with an [adjective-noun]-[verb-noun] structure, where all syllables were exactly the same length and no extra pauses were inserted between sentences (e.g. *dry fur rubs skin sick ducks hate ponds* ...). The critical innovation is that a time-frequency analysis is applied to the neural response across the whole sequence of sentences. From their discussion, Ding and colleagues appear to be particularly interested in investigating whether oscillatory activity is used to support short-term memory encoding of structure (as discussed further in the last section of this chapter). However, this method would also pick out activity associated with combinatorial operations that occur at particular moments in the sentence. For example, if a combinatorial operation were invoked whenever a noun is combined with a modifier or a verb, then this operation would happen after the presentation of each of the four nouns in the string *dry fur rubs skin sick ducks hate ponds*, such that if each word had a duration of 250 ms, the corresponding neural activity would show a peak at 2 Hz. Ding and colleagues indeed show reliable peaks in the neural time-frequency data at the sentence rate and at the NP-VP rate in MEG (Ding et al., 2016) and EEG (Ding et al., 2017). Although it may seem a bit convoluted to use a time-frequency measure if the underlying parameter of interest is the *evoked* response associated with instantiating relations, the advantage of the methodology proposed by Ding et al. is that it requires no overt response from participants and has a very fast run-time (15—20 min), while the stereotyped nature of the stimuli means that there is less variability of the kind associated with the naturalistic neuro-computational approach. This gives the Ding et al. methodology the potential to be used to study combinatorial operations in populations that may not be able to perform a task or sit through a long recording session, such as infants and patients, or during sleep (Makov et al., 2017).

Here I summarize the takeaways from this first section. Probably the most well-established generalization thus far about rapid neural correlates of combinatorial operations is the large amount of evidence accrued by Liina Pylkkänen's group that the second word of a two word phrase often elicits increased activity in left anterior temporal cortex at around 250 ms, relative to the same word in non-combinatorial contexts. However, their evidence indicates that this activity reflects computations associated with *conceptual* combination, rather than the instantiation of linguistic relations associated with syntactic phrases or logical forms, and this approach still awaits broader exploration and validation by the field. Other approaches have intriguing early results that might be taken to suggest correlates of linguistic combination, but similarly need a stronger base of evidence before clear conclusions can be drawn.

There are several promising possibilities for making more progress in this area. First, as noted above, advances in the design of two-word phrase experiments could increase their chances of finding responses associated with linguistic structure building. Second, more experiments could take a similar approach to exploring structure-building operations at a slightly higher level, by contrasting the response to lists of phrases with full sentences with time-sensitive neurophysiological measures (Matchin, Brodbeck, Hammerly, & Lau, submitted). There are a number of experimental contexts in which participants find unstructured lists to be a fairly natural mode of input, and researchers could be more fully exploiting these contexts to create non-combinatorial controls for which we have good linking hypotheses.

2.2 Classic Conceptions of Syntactic Working Memory

Gallistel and King (2011) emphasize the importance of distinguishing between something like procedural memory, or 'learning', and something like declarative memory, or 'memory'. Their point is that the neural wiring changes that facilitate computational *procedures* when they are frequently applied—for example when we get better at two-digit multiplication with practice—may be very different from the changes that allow us to stably preserve symbolic *representations* in memory—for example when we hold the number '8' in memory while adding together two other numbers. In the first section of the chapter, I explored research aimed at the neural correlates of the procedures for identifying and forming structured relations between two linguistic elements. The rest of this chapter examines the search for neural indices associated with carrying forward these relations in memory after they are formed.

This second section reviews the neuroimaging literature that was classically grouped under the heading of investigations of 'syntactic working memory'. However, to understand why a few phenomena received most of the focus in this literature, it is important first to recognize that there is a lot of relational information that the combinatorial systems under investigation may not actually *need* to carry forward in memory—just in the way that, if I'm computing a running sum that starts with '2 + 4 + 1 + ... ' I can carry forward only the current sum '7', without needing to also retain the inputs and functions that gave me that sum. For example, imagine an utterance that begins *Your red boat is* .... In order to compute the intended meaning of this phrase, the listener must indeed instantiate particular syntactic and semantic relations between *red* and *boat*, and between *your* and the phrase *red boat*. But for the purpose of assigning the correct structural relations to the rest of the sentence, all that needs to be carried forward in memory is just the fact that the sentence began with a noun phrase. For this purpose it doesn't matter whether there was adjectival modification relations inside the noun phrase or not; e.g., there is no rule in English grammar that says something like 'intransitive verbs can follow simple noun phrases, but only transitive verbs can follow noun phrases that contain adjectives'. Similarly, for interpretive purposes, what was initially a novel combinatorial object with relations between the meanings of the three words, can be converted in the discourse model into something like a simple pointer to a long-term memory representation of the particular beloved red boat that the listener has owned for years, and this can be what is carried forward in memory rather than the initial relational object. Therefore, it is easy to imagine a system where many of the on-the-fly syntactic and semantic relations do not have to be carried forward in memory in the same way that the addition relation between 2 and 4 does not have to be carried forward in memory after the result is computed. And note that in this system, there might be no point in time where the brain would have a representation of the data structure depicted by a full syntactic tree diagram—it might be that by the time the verb phrase was being computed, the relations within the noun phrase would have been discarded.

Because of the uncertainty about whether all relational information is maintained in memory throughout the sentence, then, the classic neuroimaging literature on syntactic working memory has had a particular focus on the cases of non-local relations, where immediate conversion of relational information to non-relational representations appears less straightforward. Wh-dependencies such as *I wonder which bird Thomaz looked carefully at* are one such case. Syntactically and interpretively, there appears to be a relation between the leading wh-phrase *which bird* and the verb phrase *looked carefully at*. But since these phrases are not adjacent to each other, and since they can in fact be indefinitely far apart (*I wonder which bird you claimed Thomaz looked carefully at*), forming this relation intuitively seems to require carrying relational information forward in memory across the intervening material. Note that it is easy to detect ungrammaticality at the end of the sentence if the object of the verb phrase is missing (*Thomaz looked carefully at*); the fact that the sentence becomes good in the context of a wh-phrase indicates that its presence must be available in memory, and crucially not just its presence but its structural relations with other parts of the input (as illustrated by the ungrammaticality of **I heard a bird from which country Thomaz looked carefully at*). The parser also needs to retain information about the syntactic properties of the wh-phrase, and not just the fact that it is a wh-phrase in general (*To whom/who does that belong? /* **Whom/***Who does that belong?*). Finally, much behavioral psycholinguistic work has established that comprehenders do not process such dependencies with a 'reactive' strategy where the wh-phrase is forgotten until it is noted that an argument is missing, but indeed use a 'proactive' strategy for computing the dependency that indicates that the presence of the wh-phrase is remembered throughout the subsequent material (Stowe, 1986; Traxler & Pickering, 1996; see Phillips & Wagers, 2007 for review). Therefore, wh-dependencies have been seen as a paradigmatic case of syntactic working memory demands.

How might this kind of syntactic working memory be neurally implemented? One classic idea is that it involves sustained neural activity across the duration of the linguistic dependency. Highly influential single-cell recording studies in primates (Fuster & Alexander, 1971; Funahashi, Bruce, & Goldman-Rakic, 1989) showed that during a working memory task where information needed to be maintained over a delay, certain prefrontal neurons persistently fired across the delay period while the task-relevant representation had to be maintained, and that this sustained firing appeared to be specific to which representation was being held in memory. More recently, researchers have argued that sustained neural activity is not critical for all short-term memory representations, but more selectively for those in the focus of attention (Lewis-Peacock, Drysdale, Oberauer, & Postle, 2012; LaRocque, Lewis-Peacock, Drysdale, Oberauer, & Postle, 2013; Rose et al., 2016). Either way, these domain-general models provide motivation for perhaps the most easily testable hypothesis about how syntactic dependencies are encoded across time, where a key prediction is increased neural activity across the duration of the dependency, compared to a control case without an extended dependency.

Consistent with this hypothesis, for decades there have been numerous ERP reports of a sustained anterior negativity (SAN) across the timecourse of wh-dependencies relative to control conditions (Kluender & Kutas, 1993; King & Kutas, 1995; Fiebach, Schlesewsky, & Friederici, 2002; Ueno & Kluender, 2003; Phillips, Kazanina, & Abada, 2005). Some of these studies have contrasted object relative clauses (*The reporter who the senator harshly attacked*) with subject relative clauses (*The reporter who harshly attacked the senator*), where the dependency between *the reporter* and the verb can be almost immediately satisfied in the subject relative clause, but is lengthened by the intervening noun phrase in the object relative clause (Fiebach et al., 2002; King & Kutas, 1995). Other studies have contrasted long-distance wh-dependencies (*The lieutenant knew which accomplice the detective hoped that the shrewd witness would recognize …*) with conditions that contain only local dependencies (*The lieutenant knew that the detective hoped that the shrewd witness would recognize …*) (Phillips et al., 2005). Because these differences sustain even through positions of the sentence where there is no possibility that the dependency could be completed, these effects are hard to explain as a series of targeted memory retrievals, but are exactly the kind of data that a sustained neural activity model of working memory would predict (Sprouse et al., in prep). A number of fMRI studies have also demonstrated increased hemodynamic response in left inferior frontal cortex for the processing of long-distance dependencies relative to controls (e.g. Fiebach, Schlesewsky, Lohmann, Von Cramon, & Friederici, 2005; Santi & Grodzinsky, 2007), although the lack of timecourse information makes it difficult to determine whether this increase reflects sustained differences through the dependency. These data have sometimes been taken to suggest that sustained activity in left inferior frontal cortex reflects necessary machinery for computing non-local structural dependencies, and forms part of the explanation for the uniqueness of human language relative to other species (Friederici, 2012; Fitch, 2014).

However, this longstanding account is also not without problems. On the theoretical side, the last several decades have seen a number of prominent models in which the memory architecture for sentence processing does not require active, sustained maintenance of prior material in a dedicated short-term memory buffer (McElree, Foraker, & Dyer, 2003; Lewis & Vasishth, 2005), although they do include an attentional component that could perhaps be invoked to account for the sustained negativities. On the empirical side there are also various puzzles. Fiebach et al. (2002) observe a SAN for object vs. subject relative clauses even though in German, the linear distance spanned by the wh-dependency is the same in both; their fMRI results associating left inferior frontal cortex with this activity (Fiebach et al. 2005) have the same property. This could mean that what is carried forward in memory is *predicted* syntactic positions; in the subject condition, only a verb need be predicted to host the wh-phrase as a subject, while in the object condition, an object position needs to be predicted in addition to the (transitive) verb itself. However Fiebach et al. (2002), also fail to observe a SAN for sentences with shorter dependencies (… *who on Tuesday the doctor called*) even though the SAN had a relatively rapid onset in sentences with longer dependencies (in … *who on Tuesday afternoon after the accident the doctor called* … the SAN onsets at the word *afternoon*). Phillips et al. (2005) also observe differences between short and long distance dependencies, where the negativity is more posteriorly distributed in the short distance case. In a more recent study Yano and Koizumi (2018), find that the sustained anterior negativity observed for object-scrambled dependencies in Japanese disappears when a more supportive discourse context is provided. In fMRI Matchin, Sprouse, and Hickok (2014), failed to replicate effects of wh-dependency distance in left inferior frontal cortex. These findings are unexpected if sustained neural activity is necessary for accurate computation of non-local syntactic relations.

One natural means of reconciling the apparently conflicting data in the literature is to postulate that wide-scale sustained neural activity in inferior frontal cortex is not *necessary* for computing non-local syntactic relations, but that the sustained effects that have been observed rather index some optional process that makes these computations faster or more accurate. For example, the sustained anterior negativity observed in ERP studies might reflect subvocal rehearsal in an articulatory loop, which could be optionally invoked by the comprehender in order to facilitate memory retrieval in the context of a non-local dependency. This optional process might be engaged more by certain individuals, and this would explain why previous studies have reported individual differences in the amplitude of the SAN based on performance in memory span tasks (Fiebach et al. 2002) and on comprehension questions (King & Kutas, 1995) (although note that the reported effects are somewhat inconsistent with each other, as King and Kutas observed larger SANs for 'good' comprehenders, and Fiebach et al. reported larger SANs for 'low-span' participants who tended to make more comprehension errors).

It could also be that these effects reflect some non-memory property that is often correlated with long distance dependencies. One possibility that to our knowledge has not yet been seriously pursued is implicit prosody. Prior work by Steinhauer and colleagues (Steinhauer & Friederici, 2001; Steinhauer, 2003) argues that prosodic contours elicit characteristic ERP responses and that these responses can be observed even for text, as readers impose these prosodic contours implicitly. The variability in the SAN could be partially accounted for if dependencies of different lengths result in different (implicit) prosodic boundaries, and if individuals vary in the extent to which they apply implicit prosody in reading.

Overall, I believe that the classic literature on sustained neural effects in non-local structural dependencies deserves renewed attention because of its relevance to current theories of the sentence processing memory architecture and corresponding theories of the neural encoding of non-local structural relations. I think the best starting place is more direct evaluations of whether sustained activity is necessary for immediate resolution of non-local structural relations. This can be approached with manipulations designed to modulate the need for optional facilitatory mechanisms, but probably the most critical component of this work is to include sensitive behavioral or neural measures of the extent to which participants in fact resolved the dependency immediately. The extensive behavioral psycholinguistic literature on the instantiation of non-local structural relations in real time suggests several possible means of assessing this, such as evaluating the degree to which responses indicate immediate surprise when a wh-dependency cannot be resolved at the first verb (Stowe, 1986; Traxler & Pickering, 1996). Another important direction is to investigate sustained effects in long-distance dependencies other than wh-dependencies, where different hypotheses about the underlying mechanism may diverge in their predictions (e.g. Matchin et al., 2014).

2.3 Broader Conceptions of Syntactic Memory

The classic literature on syntactic working memory focused on cases where retaining relational information in memory would be strictly necessary for determining whether or not the sentence is grammatical. However, more recent neurophysiological work has assumed that more than this minimal relational information is carried forward. This could be the case for several reasons.

First, the goal of comprehension is to derive an interpretation from an utterance, not to verify grammaticality, and this may require additional relational information to be carried forward. For example, in a 'right-branching' sentence like '*Thomaz thinks that you invaded the house of your new boss*', a top-down parser can evaluate the grammaticality of the sentence by maintaining nothing in memory at each position of the sentence except a single prediction for the next category. After *Thomaz* is related to a noun phrase position, memory needs to contain the prediction for the verb phrase that will complete the sentence grammatically. After *thinks* is related as the head of that verb phrase, memory needs to contain the prediction for the complement clause that will complete the sentence grammatically. After *that* is related to the complementizer position, memory needs to contain the prediction for the embedded clause that will complete the sentence grammatically, etc. In this way, without retaining any of the prior syntactic relations in memory, a 'grammatical' judgment can be accurately returned if all predictions are fulfilled, and an 'ungrammatical' judgment if they are not. However, the described procedure alone obviously will not be sufficient for computing the interpretation, and it is somewhat unintuitive (though logically possible) to hypothesize that an incomplete meaning like *Thomaz thinks that you invaded* ... can be converted to a wholly non-relational format in the way I suggested above for a noun phrase like *your red boat.*

Second, retaining relational information from previous parts of the sentence may aid in recovery when it turns out that the current structural analysis was incorrect and needs to be revised. Although certain misanalyses are famously difficult to recover from (strong 'garden paths' like *The horse raced past the barn fell*), recovering from many other kinds of misanalyses is often successful and associated with only mild processing cost (weak garden paths like *I know your friend [... likes ice cream]*). This entails either that comprehenders represent multiple alternative structures in parallel, or that they retain information about the prior input form in memory. If the latter, recovery would likely be easier if this information were structured rather than unstructured (so that the parts of the structural analysis that were still correct could be re-used rather than re-computed altogether).

Pallier, Devauchelle, and Dehaene (2011) reported a seminal fMRI study that was relatively unique in testing a specific model of structure maintenance. In their model, inspired by Smolensky and Legendre (2006), structural relations are built incrementally (as in a top-down parsing strategy), but those relations must be actively maintained in memory (inducing metabolic cost) until the phrase or constituent that they contribute to is complete (reminiscent of a bottom-up parsing strategy). This can be illustrated with the right-branching sentence above; since the verb phrase headed by *thinks* isn't completed until *boss*, their model assumes that the relational information for *thinks* and each of the intervening words has to be actively maintained in memory. They further hypothesize that there is a simple function from the amount of relational information to be maintained and metabolic cost as indexed by fMRI. Therefore, this model predicts that neural activity associated with representing syntactic relations should increase monotonically across the duration of a right-branching sentence, but should return to baseline when the sentence or phrase is completed.

In order to test the predictions of this model, Pallier et al. used a **constituency size paradigm**, creating contrasts between 12-word sequences which varied according to their phrasal structure; some sequences were 12-word right-branching sentences such that no phrase was completed until word 12, while other sequences were two 6-word sentences, so that activity should reset after word 6, three 4-word phrases, so that activity should reset after word 4, etc. The results of Pallier et al.'s study were in line with the predictions of their model: the fMRI response increased monotonically with connected phrase length in left inferior frontal, temporal, and parietal regions. In order to try to distinguish whether these effects reflected the representation of syntactic vs. interpretive relations, they also included a 'jabberwocky' manipulation in which content words were replaced by meaningless pseudowords (*I tosseive that you should begapt the tropufal of your tew viroate*). For jabberwocky materials, manipulating the length of connected structure modulated activity in the left inferior frontal cortex and the left posterior superior temporal sulcus only, leading Pallier et al. to hypothesize that these two regions maintain an ongoing representation of the syntactic relations between elements of an incomplete phrase. In later work they report electrocorticography (ECoG) data from patients in further support of this model (Nelson et al., 2017).

Although these results are intriguing and important, the Pallier et al. active maintenance model conflicts with the recent theories of memory discussed above in which even short-term memory representations are encoded passively without the need for ongoing neural activity. Given this ongoing debate, it is important to consider and rule out alternative explanations for the Pallier et al. results. In Matchin et al. (2017), we discuss several aspects of the design that could motivate alternative accounts of the data: Pallier et al. did not control lexical content across conditions, so that e.g. differences in the proportion of function words across conditions could have contributed to differences in the neural response, and because participants did not know whether an upcoming trial would be a full sentence or a list of phrases, some of the effects observed might have reflected disengagement or surprise upon the realization that a connected structure could not be formed from the input. We addressed these concerns in a partial replication of the constituency-size paradigm, contrasting scrambled lists, two-word phrases, and six-word sentences. Although we failed to find differences between the lists and the two-word phrases as the Pallier et al. model predicted, this could be because of the brief duration across which relations must be maintained in forming two-word phrases. However, we also point out that constituency-size effects could index the computation of syntactic predictions, rather than the representation of preceding syntactic relations.

A more straightforward means of testing the Pallier et al. model or others like it is through the use of temporally sensitive neurophysiological measures such as EEG and MEG. The Pallier et al. model makes very clear predictions about how the neural response should evolve over time: neural activity should systematically increase with each new morpheme across the duration of an incomplete phrase. However, the poor temporal resolution of fMRI measures means that this prediction could be only weakly evaluated in their original study. In a recent EEG study (Lau & Liao, 2018) we used a noun phrase coordination paradigm that allowed a tight lexical and semantic match between conditions with different constituent sizes. Here we compared trials in which coordinated noun phrases were presented (*new books and shiny floors*) with trials in which the same two noun phrases were presented as an unstructured list (*new books ### shiny floors*). Because in the coordination condition the constituent spans all five words of the trial, the Pallier et al. model predicts that in this condition should be observed increased neural activity across the second noun phrase associated with the maintenance of the structural relations from the first noun phrase. In two EEG experiments we saw exactly this pattern, with a sustained anterior negativity observed throughout the second noun phrase for the coordination condition relative to the list condition. However, we did not observe analogous effects for jabberwocky phrases (*blargal cloffs and slithy toves*), raising the possibility that these sustained differences reflect interpretive processes rather than the maintenance of structural relations. Fedorenko et al. (2016) report somewhat consistent data from intracranial ECoG recordings in patients. They compared the amplitude of high-gamma activity for participants reading sentences and word lists, and showed that across many electrode sites, high-gamma amplitude increased steadily across sentences but not word lists. They also show no corresponding increase for structured but meaningless jabberwocky sentences, leading them to argue that the response is a correlate of the representation of sentence-level meaning and not syntactic structure. Also potentially related are results of an MEG time-frequency analysis by Bastiaansen, Magyari, and Hagoort (2010), which showed an increase in amplitude in the beta frequency band (13−18 Hz) across the course of well-formed sentences, which was not observed for scrambled sentences, and which appeared to be truncated in stimuli that began as well-structured sentences prior to a word category violation.

Ding et al. (2016) implicitly appeal to a different mechanism for encoding prior structure. Their **constituent rate** paradigm involves repeating simple structures like sentences with an [[Adj N] [V N]] structure at a constant rate and evaluating the extent to which increases in MEG or EEG power are observed at the phrasal or sentence frequencies. As they do observe such increases, they gesture to the idea that oscillatory synchronization may support the encoding of structural relations. Although this model is not yet fully worked out (see Martin & Doumas, 2017 for a preliminary proposal), it is worth noting that proposals in which synchronous oscillations encode syntactic relations (e.g. that the syntactic relation between *red* and *boat* is encoded by synchronization of their firing) over the course of the sentence are thus implicitly assuming that this relational information needs to be maintained over time, vs. being rapidly transformed to a non-relational format. However, as discussed earlier, it is as yet unclear whether the time-frequency effects observed in the constituent rate paradigm are in fact due to differences in oscillatory properties across time, rather than reflecting event-related responses associated with the initial encoding of a relation (Zhou, Melloni, Poeppel, & Ding, 2016; Frank & Yang, 2018).

To sum up the results reviewed in this section, both Pallier et al. (2011) and Ding et al. (2016) suggest intriguing proposals about how structural relations are carried forward in memory through sustained neural activity, but initial evidence for these proposals from EEG, MEG, and fMRI are mixed. Even if these proposals turn out to be incorrect, however, a critical contribution of this work is to draw attention to questions about memory for structure that have largely been missing from the classic literature on syntactic working memory, that is: beyond what is minimally needed to confirm that a sentence is grammatically well-formed, what kind of structural relations need to be maintained in memory for interpretative and reanalysis purposes? More neurophysiological work is needed using constituent size and constituent rate paradigms of these types, looking for sustained differences in activity associated with structural properties of the prior input, rather than or in addition to the brief event-related manipulations that currently represent the vast majority of electrophysiological research on syntactic processing.

3. CONCLUSION

In this chapter, I have reviewed a body of past and present work aimed at discovering how the brain initially encodes structured relations in linguistic input and how the brain carries forward this relational information across the course of a sentence when needed. We can now see promising avenues for gaining traction on this problem, with the introduction of creative new paradigms for investigating neural responses to structural relations. I would argue that the most prominent obstacles for real progress are (a) the relatively small set of researchers currently working on this problem and (b) the insufficient synergy between the sub-disciplines of syntax, semantics, psycholinguistics, cognitive neuroscience, and computational linguistics that I believe is needed to devise good paradigms and draw appropriate conclusions. With respect to (a), we need to create a broader awareness that studying the neural bases of linguistic structure speaks to big-picture questions about neural computation, and can go far beyond studies that just record neural responses to ungrammatical sentences. With respect to (b), we need to create better interdisciplinary training approaches so that researchers in this area don't waste time reinventing the wheel or pursuing dead-end paths, but rather accelerate progress by building upon a solid foundation of existing well-established generalizations on linguistic relations and human sentence processing.

REFERENCES

Baron, S. G., Thompson-Schill, S. L., Weber, M., & Osherson, D. (2010). An early stage of conceptual combination: Superimposition of constituent concepts in left anterolateral temporal lobe. *Cognitive Neuroscience, 1*(1), 44–51.

Bastiaansen, M., Magyari, L., & Hagoort, P. (2010). Syntactic unification operations are reflected in oscillatory dynamics during on-line sentence comprehension. *Journal of Cognitive Neuroscience, 22*(7), 1333–1347.

Bemis, D. K., & Pylkkänen, L. (2011). Simple composition: A magnetoencephalography investigation into the comprehension of minimal linguistic phrases. *Journal of Neuroscience, 31*(8), 2801–2814.

Bemis, D. K., & Pylkkänen, L. (2012). Basic linguistic composition recruits the left anterior temporal lobe and left angular gyrus during both listening and reading. *Cerebral Cortex, 23*(8), 1859–1873.

Bemis, D. K., & Pylkkänen, L. (2013a). Flexible composition: MEG evidence for the deployment of basic combinatorial linguistic mechanisms in response to task demands. *PLoS One, 8*(9), e73949.

Bemis, D. K., & Pylkkanen, L. (2013b). Combination across domains: An MEG investigation into the relationship between mathematical, pictorial, and linguistic processing. *Frontiers in Psychology, 3*, 583.

Boylan, C., Trueswell, J. C., & Thompson-Schill, S. L. (2017). Relational vs. attributive interpretation of nominal compounds differentially engages angular gyrus and anterior temporal lobe. *Brain and Language, 169*, 8–21.

Brennan, J. (2016). Naturalistic sentence comprehension in the brain. *Language and Linguistics Compass, 10*(7), 299–313.

Brennan, J. R., & Pylkkänen, L. (2017). MEG evidence for incremental sentence composition in the anterior temporal lobe. *Cognitive Science, 41*, 1515–1531.

Brennan, J. R., Stabler, E. P., Van Wagenen, S. E., Luh, W. M., & Hale, J. T. (2016). Abstract linguistic structure correlates with temporal activity during naturalistic comprehension. *Brain and Language, 157*, 81–94.

Brodbeck, C., Presacco, A., & Simon, J. Z. (2018). Neural source dynamics of brain responses to continuous stimuli: Speech processing from acoustics to comprehension. *NeuroImage, 172,* 162–174.

Ding, N., Melloni, L., Zhang, H., Tian, X., & Poeppel, D. (2016). Cortical tracking of hierarchical linguistic structures in connected speech. *Nature Neuroscience, 19*(1), 158.

Ding, N., Melloni, L., Yang, A., Wang, Y., Zhang, W., & Poeppel, D. (2017). Characterizing neural entrainment to hierarchical linguistic units using electroencephalography (EEG). *Frontiers in human neuroscience, 11,* 481.

Federmeier, K. D., & Laszlo, S. (2009). Time for meaning: Electrophysiology provides insights into the dynamics of representation and processing in semantic memory. *Psychology of Learning and Motivation, 51,* 1–44.

Fedorenko, E., Scott, T. L., Brunner, P., Coon, W. G., Pritchett, B., Schalk, G., & Kanwisher, N. (2016). Neural correlate of the construction of sentence meaning. *Proceedings of the National Academy of Sciences of the United States of America, 113*(41), E6256–E6262.

Fiebach, C., Schlesewsky, M., & Friederici, A. (2002). Separating syntactic memory costs and syntactic integration costs during parsing: The processing of German WH-questions. *Journal of Memory and Language, 47*(2), 250–272.

Fiebach, C. J., Schlesewsky, M., Lohmann, G., Von Cramon, D. Y., & Friederici, A. D. (2005). Revisiting the role of Broca's area in sentence processing: Syntactic integration versus syntactic working memory. *Human Brain Mapping, 24*(2), 79–91.

Fitch, W. T. (2014). Toward a computational framework for cognitive biology: Unifying approaches from cognitive neuroscience and comparative cognition. *Physics of Life Reviews, 11*(3), 329–364.

Frank, S. L., & Yang, J. (2018). Lexical representation explains cortical entrainment during speech comprehension. *PLoS One, 13*(5), e0197304.

Friederici, A. D. (2012). The cortical language circuit: From auditory perception to sentence comprehension. *Trends in Cognitive Sciences, 16*(5), 262–268.

Friederici, A. D., Meyer, M., & von Cramon, D. Y. (2000). Auditory language comprehension: An event-related fMRI study on the processing of syntactic and lexical information. *Brain and Language, 74*(2), 289–300.

Funahashi, S., Bruce, C. J., & Goldman-Rakic, P. S. (1989). Mnemonic coding of visual space in the monkey's dorsolateral prefrontal cortex. *Journal of Neurophysiology, 61*(2), 331–349.

Fuster, J. M., & Alexander, G. E. (1971). Neuron activity related to short-term memory. *Science, 173*(3997), 652–654.

Gallistel, C. R., & King, A. P. (2011). *Memory and the computational brain: Why cognitive science will transform neuroscience.* John Wiley & Sons.

Graves, W. W., Binder, J. R., Desai, R. H., Conant, L. L., & Seidenberg, M. S. (2010). Neural correlates of implicit and explicit combinatorial semantic processing. *NeuroImage, 53*(2), 638–646.

Jarvella, R. J. (1971). Syntactic processing of connected speech. *Journal of Memory and Language, 10*(4), 409.

King, J. W., & Kutas, M. (1995). Who did what and when? Using word-and clause-level ERPs to monitor working memory usage in reading. *Journal of Cognitive Neuroscience,* 7(3), 376–395.

Kluender, R., & Kutas, M. (1993). Subjacency as a processing phenomenon. *Language & Cognitive Processes, 8*(4), 573–633.

Kutas, M., & Federmeier, K. D. (2000). Electrophysiology reveals semantic memory use in language comprehension. *Trends in Cognitive Sciences, 4*(12), 463–470.

LaRocque, J. J., Lewis-Peacock, J. A., Drysdale, A. T., Oberauer, K., & Postle, B. R. (2013). Decoding attended information in short-term memory: An EEG study. *Journal of Cognitive Neuroscience, 25*(1), 127–142.

Lau, E., & Liao, C. H. (2018). Linguistic structure across time: ERP responses to coordinated and uncoordinated noun phrases. *Language, Cognition and Neuroscience, 33*(5), 633–647.

Lau, E. F., Phillips, C., & Poeppel, D. (2008). A cortical network for semantics:(de) constructing the N400. *Nature Reviews Neuroscience, 9*(12), 920.

Lau, E. F., Weber, K., Gramfort, A., Hämäläinen, M. S., & Kuperberg, G. R. (2014). Spatiotemporal signatures of lexical–semantic prediction. *Cerebral Cortex, 26*(4), 1377–1387.

Lewis-Peacock, J. A., Drysdale, A. T., Oberauer, K., & Postle, B. R. (2012). Neural evidence for a distinction between short-term memory and the focus of attention. *Journal of Cognitive Neuroscience, 24*(1), 61–79.

Lewis, R. L., & Vasishth, S. (2005). An activation-based model of sentence processing as skilled memory retrieval. *Cognitive Science, 29*(3), 375–419.

Makov, S., Sharon, O., Ding, N., Ben-Shachar, M., Nir, Y., & Golumbic, E. Z. (2017). Sleep disrupts high-level speech parsing despite significant basic auditory processing. *Journal of Neuroscience*, 0168-17.

Martin, A. E., & Doumas, L. A. (2017). A mechanism for the cortical computation of hierarchical linguistic structure. *Plos Biology, 15*(3), e2000663.

Matchin, W., Hammerly, C., & Lau, E. (2017). The role of the IFG and pSTS in syntactic prediction: Evidence from a parametric study of hierarchical structure in fMRI. *Cortex, 88*, 106–123.

Matchin, W., Sprouse, J., & Hickok, G. (2014). A structural distance effect for backward anaphora in Broca's area: An fMRI study. *Brain and Language, 138*, 1–11.

Mazoyer, B. M., Tzourio, N., Frak, V., Syrota, A., Murayama, N., Levrier, O., ... Mehler, J. (1993). The cortical representation of speech. *Journal of Cognitive Neuroscience, 5*(4), 467–479.

McElree, B., Foraker, S., & Dyer, L. (2003). Memory structures that subserve sentence comprehension. *Journal of Memory and Language, 48*(1), 67–91.

Murphy, G. L., & Shapiro, A. M. (1994). Forgetting of verbatim information in discourse. *Memory & Cognition, 22*(1), 85–94.

Nelson, M. J., El Karoui, I., Giber, K., Yang, X., Cohen, L., Koopman, H., ... Dehaene, S. (2017). Neurophysiological dynamics of phrase-structure building during sentence processing. *Proceedings of the National Academy of Sciences, 114*(18), E3669–E3678.

Neufeld, C., Kramer, S. E., Lapinskaya, N., Heffner, C. C., Malko, A., & Lau, E. F. (2016). The electrophysiology of basic phrase building. *PLoS One, 11*(10), e0158446.

Pallier, C., Devauchelle, A.-D., & Dehaene, S. (2011). Cortical representation of the constituent structure of sentences. *Proceedings of the National Academy of Sciences, 108*(6), 2522–2527.

Patterson, K., Nestor, P. J., & Rogers, T. T. (2007). Where do you know what you know? The representation of semantic knowledge in the human brain. *Nature Reviews Neuroscience, 8*(12), 976.

Phillips, C., Kazanina, N., & Abada, S. (2005). ERP effects of the processing of syntactic long-distance dependencies. *Cognitive Brain Research, 22*(3), 407–428.

Phillips, C., & Wagers, M. (2007). Relating structure and time in linguistics and psycholinguistics. In *Oxford handbook of psycholinguistics* (pp. 739–756).

Rogalsky, C., & Hickok, G. (2008). Selective attention to semantic and syntactic features modulates sentence processing networks in anterior temporal cortex. *Cerebral Cortex, 19*(4), 786–796.

Rose, N. S., LaRocque, J. J., Riggall, A. C., Gosseries, O., Starrett, M. J., Meyering, E. E., & Postle, B. R. (2016). Reactivation of latent working memories with transcranial magnetic stimulation. *Science, 354*(6316), 1136–1139.

Sachs, J. S. (1967). Recognition memory for syntactic and semantic aspects of connected discourse. *Attention, Perception, & Psychophysics, 2*(9), 437–442.

Santi, A., & Grodzinsky, Y. (2007). Working memory and syntax interact in Broca's area. *NeuroImage, 37*(1), 8–17.

Schell, M., Zaccarella, E., & Friederici, A. D. (2017). Differential cortical contribution of syntax and semantics: An fMRI study on two-word phrasal processing. *Cortex, 96*, 105–120.

Smolensky, P., & Legendre, G. (2006). *The harmonic mind: From neural computation to optimality-theoretic grammar.* MIT Press.

Steinhauer, K. (2003). Electrophysiological correlates of prosody and punctuation. *Brain and Language, 86*(1), 142–164.

Steinhauer, K., & Friederici, A. D. (2001). Prosodic boundaries, comma rules, and brain responses: The closure positive shift in ERPs as a universal marker for prosodic phrasing in listeners and readers. *Journal of Psycholinguistic Research, 30*(3), 267–295.

Stowe, L. A. (1986). Parsing WH-constructions: Evidence for on-line gap location. *Language & Cognitive Processes, 1*(3), 227–245.

Traxler, M. J., & Pickering, M. J. (1996). Plausibility and the processing of unbounded dependencies: An eye-tracking study. *Journal of Memory and Language, 35*(3), 454–475.

Ueno, M., & Kluender, R. (2003). Event-related brain indices of Japanese scrambling. *Brain and Language, 86*(2), 243–271.

Van Petten, C., & Luka, B. J. (2006). Neural localization of semantic context effects in electromagnetic and hemodynamic studies. *Brain and Language, 97*(3), 279–293.

Vandenberghe, R., Nobre, A. C., & Price, C. J. (2002). The response of left temporal cortex to sentences. *Journal of Cognitive Neuroscience, 14*(4), 550–560.

Westerlund, M., & Pylkkänen, L. (2014). The role of the left anterior temporal lobe in semantic composition vs. semantic memory. *Neuropsychologia, 57*, 59–70.

Wilson, S. M., DeMarco, A. T., Henry, M. L., Gesierich, B., Babiak, M., Mandelli, M. L., Miller, B. L., & Gorno-Tempini, M. L. (2014). What role does the anterior temporal lobe play in sentence-level processing? Neural correlates of syntactic processing in semantic variant primary progressive aphasia. *Journal of Cognitive Neuroscience, 26*(5), 970–985.

Yano, M., & Koizumi, M. (2018). Processing of non-canonical word orders in (in) felicitous contexts: Evidence from event-related brain potentials. *Language, Cognition and Neuroscience*, 1–15.

Zaccarella, E., & Friederici, A. D. (2015a). Reflections of word processing in the insular cortex: a sub-regional parcellation based functional assessment. *Brain and language, 142*, 1–7.

Zaccarella, E., & Friederici, A. D. (2015b). Merge in the human brain: A sub-region based functional investigation in the left pars opercularis. *Frontiers in Psychology, 6*, 1818.

Zhang, L., & Pylkkänen, L. (2015). The interplay of composition and concept specificity in the left anterior temporal lobe: An MEG study. *NeuroImage, 111*, 228–240.

Zhou, H., Melloni, L., Poeppel, D., & Ding, N. (2016). Interpretations of frequency domain analyses of neural entrainment: Periodicity, fundamental frequency, and harmonics. *Frontiers in Human Neuroscience, 10*, 274.

CHAPTER SIX

Familial Sinistrality and Language Processing

Chia-lin Lee[*,§,¶,||]
[*]Graduate Institute of Linguistics, National Taiwan University, Taiwan
[§]Department of Psychology, National Taiwan University, Taiwan
[¶]Graduate Institute of Brain and Mind Sciences, National Taiwan University, Taiwan
[||]Neurobiology and Cognitive Neuroscience Center, National Taiwan University, Taiwan
E-mail: chialinlee@ntu.edu.tw

Contents

1. Introduction 144
2. Language-Related Processing as a Function of Familial Sinistrality 145
   2.1 Familial Sinistrality and Language Lateralization 145
   2.2 Familial Sinistrality and Language Processing 147
3. Theories of Familial Sinistrality and Language 148
   3.1 The Relational/Holistic Model 148
   3.2 The Declarative/Procedural Model 150
   3.3 Commonalities Between the Views 152
4. New Considerations for Familial Sinistrality and Language Processing 153
   4.1 Right Hemisphere Capability for Language 153
   4.2 Bi-hemispheric Syntactic Processing 157
   4.3 Bi-lateralization of Language 158
      4.3.1 Inter-hemispheric Transmission Time Delay *158*
      4.3.2 Inter-hemispheric Inhibition *159*
      4.3.3 Right Hemisphere Compensation *161*
      4.3.4 Future Directions *163*
   4.4 Sensitivity to Syntactic Information 164
5. Conclusions 168
References 169

Abstract

An increasing number of studies have revealed striking within- and between-individual differences in language processing even in healthy native speakers, suggesting higher heterogeneity of the neurological and cognitive mechanisms involved in language processing than typically assumed. One of the prominent explanatory factors for this variability that has been highlighted in the literature is familial sinistrality (i.e., whether or not one has left handed biological relatives). Presence and absence of familial sinistrality has been associated with different consequences after unilateral brain lesions

Psychology of Learning and Motivation, Volume 68
ISSN 0079-7421
https://doi.org/10.1016/bs.plm.2018.08.005

and different response patterns in language comprehension and memory tasks, as assessed with behavioral, neuroimaging, and electrophysiological measures. While one of the consensuses regarding familial sinistrality is its association with reduced language asymmetry, there are still open questions regarding the nature and the neurological basis of this tendency toward bi-lateralization. As is reviewed in this chapter, familial sinistrality provides a window for examining the possible variations of how neural mechanisms are coordinated across the hemispheres, and this literature suggests that the orchestration of neurological and cognitive mechanisms involved in successful language processing may be more dynamic and adaptive within and across individuals than previously thought.

1. INTRODUCTION

Language, with its special status, has been given tremendous attention in cognitive science, with impressive numbers of theoretical accounts put forward to try to capture the mechanisms of language processing. Most of the models draw on empirical work wherein a high degree of cross-individual homogeneity is assumed, which has inadvertently led to the impression that one unified model of language processing proceeds uniformly among healthy proficient native speakers. However, language processing enlists a wide array of cognitive-neural resources so as to rapidly associate sensory stimuli with long-term memory and to integrate the incoming information and the message gleaned from the preceding context in a timely fashion. As such, small quantitative differences in any aspect of processing could potentially amount to qualitative differences in the whole process. Indeed, emerging data has revealed striking within- and between-individual differences in language processing even in healthy native speakers (e.g., Grey, Tanner, & van Hell, 2017; Hancock & Bever, 2013; Kos, van den Brink, & Hagoort, 2012; Nieuwland, 2014; Tanner & Van Hell, 2014). These findings suggest more diverse and dynamic configurations of the neurological and cognitive mechanisms involved in successful language processing than has often been supposed.

The present review will focus on familial sinistrality, a trait that has been repeatedly noted in association with individual variation in language processing, and will discuss the implications of this literature for views of language lateralization. Familial sinistrality – whether an individual has at least one left-handed blood relative – has been recognized by increasing numbers of studies as an important dimension along which processing and hemispheric dominance of language can differ (e.g., Chiarello, Vazquez, Felton, & Leonard, 2013; Grey et al., 2017; Kos et al., 2012; Lee & Federmeier, 2015; Tanner & Van Hell, 2014; Tzourio-Mazoyer, Simon, et al., 2010). Based on large-scale questionnaire surveys, individuals with familial sinistrality account for roughly 30% to 40% of the total population and half of the right-handers—the main population tested in the psycholinguistics and neurolinguistics literature (Bever, 2013; Spiegler & Yeni-Komshian, 1983). A better understanding of how familial sinistrality may modulate the language processing system thus has the potential to bring about a more thorough understanding of the mechanisms involved in successful language processing.

In this chapter, I will first review the literature showing links between familial sinistrality and language processing. Next, I will review prominent theoretical views about how familial sinistrality interacts with the language system. In the second part of this chapter I will then discuss some recent data that has called for reconsideration of the existing accounts.

2. LANGUAGE-RELATED PROCESSING AS A FUNCTION OF FAMILIAL SINISTRALITY

2.1 Familial Sinistrality and Language Lateralization

One of the earliest links between familial sinistrality and language performance comes from lesion studies. It has been noted that left hemisphere damage is less likely to cause language abnormalities in right-handers with familial sinistrality (FS+) than in right-handers without familial sinistrality (FS−) (Hécaen, De Agostini, & Monzon-Montes, 1981). When language deficits do occur following left hemisphere strokes, FS+ individuals also tend to recover more quickly regardless of their own handedness (Hécaen & Sauguet, 1971; Luria, 1970). By contrast, FS+ individuals are more likely to suffer from language deficits subsequent to right hemisphere strokes (Brown & Hécaen, 1976; Joanette, Lecours, Lepage, & Lamoureux, 1983; Subirana, 1958). These data have led to views associating the trait of FS with reduced degrees of leftward "cerebral dominance" for language processing, with lower left hemisphere dominance and greater right-hemispheric involvement (Annett, 1994; Hardyck & Petrinovich, 1977; Hécaen et al., 1981; McKeever & VanDeventer, 1977).

Data from neurologically intact participants corroborate the view that language is more bilaterally represented for FS+ individuals. For example, Kee, Bathurst, and Hellige (1983) reported that solving word anagrams interferes with right-handed finger tapping more for right-handed FS− than for right-handed FS+ individuals, suggesting that processes used in solving anagram problems are more bilaterally distributed for FS+ individuals. Furthermore, using dichotic listening or tachistoscopic tasks to measure hemispheric asymmetry, research has found that right-handed FS+ individuals have significantly smaller asymmetries in performance across presentation ears or visual fields than do right-handed FS− individuals (Hines & Satz, 1971; Piazza, 1980; Springer & Searleman, 1978; also see Zurif & Bryden, 1969; and Lishnian & McMeekan, 1977 for related findings from FS+ left-handed participants). In addition, the incidence of FS+ was also noted to be much higher in individuals with atypical language lateralization than individuals with typical language lateralization (Satz, Fennell, & Jones, 1969).

Functional magnetic resonance imaging data have led to similar conclusions. For example, Szaflarski et al. (2002), studying left-handed and ambidextrous people, found that FS+ individuals showed significantly less left-lateralized brain activation for a semantic categorization task than did FS− individuals. Also, among FS+ left-handers there were a significantly higher number of people showing right-hemispheric language lateralization than within FS− left-handers (Hund-Georgiadis, Lex, Friederici, & Cramon, 2002). Furthermore, FS+ right-handers with weak manual preference strength showed no left hemisphere dominance while listening to a story in their native language, despite a leftward functional asymmetry found in FS+ right-handers with strong manual preference and FS− right-handers regardless of the strength of their hand preference (Tzourio-Mazoyer, Petit, et al., 2010). Note, however, that associations between atypical langauge lateralization and FS+ are not always reliably found (Behavioral: Briggs & Nebes, 1976; Functional transcranial Doppler-ultrasonography: Knecht, Dräger, et al., 2000; fMRI: Springer et al., 1999; Orsini, Satz, Soper, & Light, 1985).

Findings of reduced functional asymmetry among FS+ individuals are corroborated by structural brain imaging data also showing reduced asymmetries in FS+ individuals, specifically in language-related brain areas (Steinmetz, Volkmann, Jäncke, & Freund, 1991; Szaflarski, Holland, Schmithorst, & Byars, 2006; Tzourio-Mazoyer, Petit, et al., 2010; Tzourio-Mazoyer, Simon, et al., 2010). For example, Tzourio-Mazoyer and colleagues (2010), using brain magnetic resonance imaging, found that FS+ individuals had reduced leftward asymmetry in gray matter, which was hypothesized to reflect reduced synaptic pruning in the right hemisphere. Moreover, they found that the surface area of a cortical region linked to phonological processing, the "planum temporale" (PT), decreased proportionally with the number of left-handed relatives, but only in the left hemisphere, thus reducing the typical left-greater-than-right asymmetry typically seen for this area. The strong association between FS and reduced or atypical language lateralization patterns in the above-reviewed studies has thus led to FS being treated as one of the most prominent predictors for cerebral organization, among other potentially interacting factors such as strength of manual preference, sex, and head size (e.g., Searleman, Tweedy, & Springer, 1979; Tzourio-Mazoyer, Petit, et al., 2010).

2.2 Familial Sinistrality and Language Processing

In addition to influencing language lateralization, FS has also been shown to affect patterns of language processing. One behavioral trend that has been noted in the literature is that, relative to individuals without familial sinistrality background, FS+ individuals seem to access lexical information more readily and are better at understanding sentences presented under difficult reading/listening conditions. For example, FS+ individuals understood a short essay better when the story was presented with each word alternating between the ears than when the whole story was presented consistently to one ear, while the converse was true for FS− individuals (Bever, Carrithers, Cowart, & Townsend, 1989). In addition, relative to FS− individuals, FS+ individuals were faster at recognizing auditory probe words from a just-heard sentence fragment (Bever, Carrithers, & Townsend, 1987; Townsend, Carrithers, & Bever, 2001), and showed superior performance in a visual word-by-word sentence-reading task (Bever et al., 1987).

FS+ individuals are also found to be less affected by manipulations of syntactic structure during language processing. For example, Bever et al. (1987) found that when participants listened to a two-clause sentence along with a tone and were asked to locate the position of the tone, FS− participants were more accurate when the tone occurred between clauses than within either clause. However, FS+ participants' performance was not affected by the position of the tone relative to clause boundaries. In addition, in a task wherein participants needed to report whether a probe word appeared in a previously-heard, complex sentence, FS− adult right-handers took more time to respond to words occurring later in the sentence, whereas FS+ adult right-handers were overall faster and were not affected by the serial position of the probe word (Townsend et al., 2001). Finally, FS+ individuals were faster than FS− individuals when asked to self-pace through sentences word by word, a manner of reading that does not highlight the syntactic structure of the sentence, while the reversed pattern was observed when the sentence was presented clause by clause (Bever et al., 1987).

FS has also been found to affect language acquisition. It has been found that, at least for sign language acquisition, FS+ was associated with an earlier-ending critical period for mastery of sign language proficiency than was observed for FS− deaf children. Whereas FS+ deaf children showed reliably lower sign language proficiency when acquiring sign language after age of 8, a similar drastic drop of learning outcome was not seen in FS− deaf children (Ross & Bever, 2004). Note that despite the differential sensitivity observed for specific aspects of language as a function of FS background, there are usually no reliable differences in global verbal performance measures (e.g., Tzourio-Mazoyer, Petit, et al., 2010), corroborating the idea that successful comprehension can be achieved via different pathways.

3. THEORIES OF FAMILIAL SINISTRALITY AND LANGUAGE

Below, I summarize two prominent theoretical views about language processing that are of high relevance to how FS may affect the coordination of neurological and cognitive resources during language processing (Bever, 2013; Hancock & Bever, 2013; Ullman, 2004).

3.1 The Relational/Holistic Model

Building on the assumption that cerebral functional lateralization is primarily a consequence of an information bottleneck across the corpus callosum (Ringo, Doty, Demeter, & Simard, 1994), Hancock and Bever (2013) suggested that cortical asymmetry of the signal-to-noise ratio (SNR)—the ratio of stimulus-induced brain responses to the non-induced fluctuations around the signal—could give rise to cerebral lateralization of language as well as motor functions. Hancock and Bever (2013) propose that the source of the cortical asymmetry of SNR can be traced back to hemispheric differences in the modulatory action of dopamine activity (especially D2 activity) in the cortico-striatal-thalamic loop. This view is consistent with findings showing greater D2 activity in the left than in the right striatum in normal right handers (Larisch et al., 1998; Vernaleken et al., 2007) and the association between dopaminergic activity and motor lateralization (e.g., hand preference in humans: Mohr, Landis, Bracha, & Brugger, 2003;

Mohr & Lievesley, 2007). That the major candidate gene for handedness (LRRTM1) is also notably expressed in the human striatum, implicating its potential interaction with the dopaminergic synapses there, also lends additional support for the claim by Hancock and Bever (2013).

Hancock and Bever (2013) propose that language may be preferentially lateralized to the hemisphere that has a computationally more optimal cortical SNR. With inter-hemispheric inhibition over the course of development, a slightly greater advantage in computational power in the left hemisphere from the outset could gradually lead to the superiority of the left hemisphere for language (Hancock & Bever, 2013). As a consequence, the dominant left hemisphere takes on the 'relational' processing that requires handling multiple mental representations at the same time and manipulating the relation among these representations, and the non-dominant right hemisphere is left associated with the 'holistic' processing that involves the direct association of a mental representation with a stimulus and response (Bever, 1975, 1980; also see; Hughlings Jacksons, 1878).

Familial sinistrality is an important element in this framework, as it suggests the possibility of a genetic basis for how language is used and represented. Consistent with earlier views from lesion work (e.g., Luria, 1970), in this framework, FS+ individuals are also thought to represent and process language more bilaterally. It is suggested that this reduced language lateralization in FS+ individuals could be due to more equilateral dopamine D2 activity in FS+ people, which in turn could lead to changes in the asymmetry of cortical SNR (Hancock & Bever, 2013). For example, an increase in noise in the left hemisphere could lead to reduced reliance on and specialization of the left hemisphere; alternatively, bilateral increases in noise levels could result in a reduced degree of emergent functional lateralization (Hancock, 2013). Critically, however, Bever et al. suggest that the bilateral representation of language in FS+ individuals may be specific to lexical knowledge (Bever, 2013; Hancock & Bever, 2013). The rationale is that syntactic processing is computationally demanding, and thus is relegated exclusively to the computationally more powerful left hemisphere. By comparison, lexical acquisition and processing is computationally less demanding, and thus lexical information is more likely to be represented and processed in the right hemisphere in addition to the left hemisphere. A less lateralized language system in FS+ individuals thus allows for more right hemisphere representation and processing of the simpler aspects of language—lexical processing.

In brief, the view of Bever and colleagues holds that while lexical processing is more bilateral in FS+ right-handers than in FS− right-handers, syntactic processing should be lateralized to the left hemisphere regardless of FS background. Having a more widespread representation of individual lexical items allows FS+ individuals to access each word more readily and to process language with less dependence on syntactic structural processing but more contact with other conceptual knowledge than would be the case for FS− individuals (Bever et al., 1987). Moreover, Bever and colleagues also propose that FS− and FS+ individuals differ in whether they 'attend' to the activities of one hemisphere or the other. While FS+ individuals comprehend by initially extracting individual words and their associations, FS− individuals give greater attention to syntactic organization and comprehend by determining the syntactic relations between words within sentences (Townsend et al., 2001). As such, FS− and FS+ individuals may also differ in the extent to which they encode linguistic information in the two hemispheres. While FS− individuals may encode linguistic information in the left hemisphere in the form of sequence representations, FS+ individuals may encode it in the right hemisphere, emphasizing words and meaning.

3.2 The Declarative/Procedural Model

With a co-optation view of language, Ullman proposed that language depends importantly on previously existing neurobiological systems regardless of whether these systems have subsequently become further specialized phylogenetically or ontogenetically for language (Ullman, 2004, 2016). In particular, important aspects of the human language system, such as learning and retaining more idiosyncratic or episodic aspects of language (e.g., lexical knowledge, irregular morphological forms, idioms, proverbs, and the context in which a word has been encountered or learned) could be co-opted from the declarative memory system whereas sequencing and more rule-like aspects of language (e.g., word segmentation, grammatical rules, morphosyntax, non-lexical semantics, and phonological constraints) critically involve the procedural memory system (Babcock, Stowe, Maloof, Brovetto, & Ullman, 2012; Kidd, 2012).

These two memory systems have been found to engage distinct neural correlates. The learning phase of declarative memory has been thought to involve regions such as hippocampus, temporal lobes (in particular medial temporal lobe, or MTL, structures, such as the entorhinal, perirhinal, and parahippocampal cortices, Stark & Stark, 2016), as well as structures connected to the MTL (including the fornix and various diencephalic structures and tracts, Squire & Wixted, 2011). Accessing those acquired declarative memories has been shown to involve different cortical regions depending on the type of knowledge (e.g., objects and tools; Martin, 2007; Squire & Wixted, 2011). The procedural memory system, on the other hand, is thought to involve a network of brain structures rooted in frontal/basal-ganglia circuits (including frontal premotor areas and Broca's area, particularly BA 44) and the cerebellum, and is also thought to be dopamine-dependent (Ashby, Turner, & Horvitz, 2010).

Based on findings showing that these two memory systems can, to some extent, acquire similar or analogous knowledge or skills, albeit in different forms of representation (Poldrack & Packard, 2003; Ullman, 2004), it is proposed that many aspects of language can have representations coexisting in both systems redundantly even though they are usually preferentially linked to one system over the other (Ullman, 2004, 2016). For example, more productive, rule-like aspects of grammar that are normally thought to involve procedural memory can be learned or processed via the declarative memory with explicit rules, analogical generalization across chunk representations of regular morphological forms, or shallow parsing of sentences based on lexical-semantic knowledge (e.g., Hamrick, 2014; Wong, Ettlinger, & Zheng, 2013). Similarly, sequences, rules, and categories could be learned implicitly in procedural memory or more explicitly in declarative memory (e.g., Alegre & Gordon, 1999; Clahsen & Felser, 2006; Hamrick, 2014; Hartshorne & Ullman, 2006; Lieberman, Chang, Chiao, Bookheimer, & Knowlton, 2004). Under this redundancy hypothesis, declarative memory can at least partly support most if not all aspects of language that are normally subserved by procedural memory (Ullman, 2016).

In addition to memory- or learning-related factors (e.g., explicit/implicit instruction, task types, learning materials, etc.) that could lead to more reliance on one system over the other, these two memory systems are also hypothesized to interact with each other in compensatory or competitive ways. As such, one memory system could be preferentially relied on or even enhanced (the 'seesaw effect') due to relative dysfunction of the other system. For example, individuals with known abnormalities of procedural memory (e.g., Specific Language Impairment, dyslexia, autism, or obsessive-compulsive disorder) have been shown to rely more on the intact declarative memory system and use means such as chunking or explicit rules to compensate (Ullman & Pullman, 2015). Alternatively, one memory system could be downplayed as a result of the competition or inhibition from the other stronger system. Evidence for these between-system interactions has been seen in humans as well as animals (Poldrack & Packard, 2003; Ullman, 2004). For example, while estrogen is found to enhance declarative memory, it is also found to inhibit procedural memory (Ullman, 2004).

In this framework, left-handedness (self or familial), has been noted to be associated with a declarative memory advantage (Ullman, 2016). In addition, as the procedural system is considered to be more dominant in the left hemisphere (Ullman, 2001), left-handedness (self or familial) – and thus, as already described, a reduced tendency toward left-hemisphere dominance for language and motor functions – is hypothesized to be associated with less efficient procedural memory. Therefore, while individuals with left hemisphere dominant language are expected to significantly engage procedural memory-related activity during morpho-syntactic processing, FS+ individuals are thought to rely more on declarative memory-related lexical/semantic processing for morpho-syntactic information, in order to make up for their less efficient procedural memory ability (Ullman et al., under revision as cited in Grey et al., 2017).

3.3 Commonalities Between the Views

Despite fundamental differences in perspective between the two frameworks reviewed above, there are some critical commonalities, especially with regard to how FS influences language processing. For example, both accounts hypothesize a reduced leftward asymmetry for cortical representations of language in FS+ individuals. In addition, both frameworks assume a similar 'resource allocation' arrangement across the two hemispheres regardless of FS background. For example, both views assume that syntactic or procedural-memory-based processing is specialized to or dominant in the left hemisphere, whereas lexical aspects of processing can be done by both hemispheres. With these common assumptions, both frameworks account for the language processing differences associated with FS by suggesting that FS+ individuals are more bilateral in language processing, specifically in the sense that they rely more on the right hemisphere for lexical-associative processing. On these views, it is not always clear whether FS+ individuals' greater reliance on the right hemisphere is compensatory in nature, as this greater reliance could be due to more limited structural processing in the left hemisphere or, instead, a bias (e.g., in attention) toward lexical associative processing happening in the right hemisphere. Nevertheless, both views would predict that FS+

individuals engage more lexical-conceptual processing in response to syntactic information and show less structural processing that is based in the left hemisphere (Bever, 2013; Hancock & Bever, 2013; Ullman, 2004).

Some recent studies measuring electrophysiological data have provided evidence in line with some aspects of these views. At the same time, these studies also demonstrated result patterns that differ in important ways from what would have been predicted based on these views. These findings thus present the need to re-consider some fundamental assumptions with regard to how FS influences language processing. These differences are discussed in turn in what follows.

4. NEW CONSIDERATIONS FOR FAMILIAL SINISTRALITY AND LANGUAGE PROCESSING

4.1 Right Hemisphere Capability for Language

While the consequences of left hemisphere brain injury for language functions are often devastating, homologous right hemisphere lesions do not tend to compromise language to the same extent (e.g., Broca, 1861; Luria, 1970; also see Martin, 2003; and Vigneau et al., 2011 for reviews). These findings have led to the general impression that there is an exclusive link between the left hemisphere and verbal ability. However, the neurobiological basis of language has been noted to be less homogenous than typically thought, and it has further been shown that inter-individual variation in language lateralization is primarily driven by differences in right hemisphere activation rather than differences in left hemisphere activation (Seghier, Kherif, Josse, & Price, 2011). For example, with blood oxygenation level-dependent (BOLD)-fMRI responses recorded during auditory story comprehension, Tzourio-Mazoyer, Petit, et al. (2010) and Tzourio-Mazoyer, Simon, et al. (2010) found that the distribution of hemispheric functional lateralization for language is, as expected, skewed toward leftward hemispheric asymmetry. However, unlike hemispheric functional lateralization for spatial attention, which has a Gaussian distribution, the distribution of hemispheric asymmetries for language is more variable and has a long tail for rightward hemispheric asymmetry. Similar variability in language lateralization, showing subpopulations of right-handers without a leftward asymmetry, has also been noted in a study using functional transcranial Doppler ultrasonography (fCTD) (Knecht, Deppe, et al., 2000). These findings underscore the linguistic potential of the right hemisphere and suggest that, at least in some cases, the right hemisphere can take the lead in language processing. These findings are also consistent with the literature showing that patients with left hemispherectomy, and hence a lone right hemisphere, can grow up to be linguistically normal (Curtiss et al., 2001; Devlin et al, 2003; de Bode, Chanturidze, Mathern, & Dubinsky, 2015) and show normal developmental patterns (Curtiss & Shaeffer, 1997).

Even in cases wherein the left hemisphere is the dominant one, the right hemisphere has been found to be more linguistically capable than previously thought, as a growing literature using a wide range of methods has suggested a significant role for the right hemisphere in prosodic or paralinguistic aspects of speech processing and for meaning comprehension, especially in cases when non-literal or secondary meanings need to be considered, and context information used flexibly (see Federmeier, 2007; Federmeier, Wlotko, & Meyer, 2008; and Lindell, 2006 for comprehensive reviews). Nevertheless, because evidence for right hemisphere sensitivity to syntactic processing is scarcer, syntax, the core aspect thought to underlie the unique productivity of human language, is still generally thought to be specialized to the left hemisphere and unavailable to the right hemisphere, at least in right-handed individuals.

However, some studies have found evidence for right hemisphere involvement in syntactic processing. For example, the right hemisphere is sensitive to at least some syntactic cues, such as those establishing expectations for syntactic word class or number agreement (Arambel & Chiarello, 2006; Liu, Chiarello, & Quan, 1999). Right hemisphere activation has also been noted for healthy young adults in several brain imaging studies of syntactic processing (Just, Carpenter, Keller, Eddy, & Thulborn, 1996; Moro et al., 2001; Service, Helenius, Maury, & Salmelin, 2007). Increased right hemisphere activity has also been noted for healthy older adults in syntactic processing as well as other cognitive domains (Cabeza, 2002), and has sometimes been considered compensatory (Tyler et al., 2010). Finally, the right hemisphere has been suggested to serve a facilitative role for beginning learners with regard to sequential learning or processing. For example, Loui, Li, and Schlaug (2011), found positive correlations between performance during learning of a pitch-based artificial grammar and white matter connectivity in right hemisphere analogs of language areas (e.g., right inferior frontal gyrus, right middle temporal gyrus, and temporal-parietal junction). Similarly, Qi, Han, Garel, San Chen, and Gabrieli (2015), found that outcomes for Mandarin learning after an intensive course were positively correlated with white matter integrity (measured with fractional anisotropy or FA) in the right superior longitude fasciculus and the right inferior longitude fasciculus assessed before the course.

It is unclear, however, in the above-mentioned cases, whether the right hemisphere handles syntactic information in a way similar to the structural analysis in the left hemisphere, or instead, as suggested by some studies, if the right hemisphere is processing syntactic information via more lexical-semantic or association based mechanisms. Indeed, some studies have noted that the right hemisphere showed better performance on syntactic judgments when the grammatical distinctions were lexically marked as opposed to morphologically marked (e.g., "the fish is/are eating" vs. "the fish eats/eat"; Zaidel, 1983; also see Gazzaniga & Hillyard, 1971 for relevant discussions). Furthermore, it has been found that when required to recognize a melody, experienced musicians, who are able to analyze the internal relations between a sequence of musical notes, rely more on the left hemisphere, whereas naïve listeners, who tend to perceive the melody as a whole, rely more on the right hemisphere (Bever & Chiarello, 1974). The latter case in particular is consistent with the idea that the declarative memory system can pick up grammatical information more quickly, perhaps through the means of chunking, before the procedural memory masters sequential relations (Ullman, 2016). It is against this backdrop that findings showing more bilaterally represented language in FS+ individuals are taken to suggest that FS+ individuals rely more on the right hemisphere for processes that are more lexical-associative in nature (Bever et al., 1989; Hancock & Bever, 2013; Townsend et al., 2001).

The hypothesis that the right hemisphere may process syntactic information via more lexically based mechanism has gained partial support. Using separable, functionally well-characterized electrophysiological indices of lexico-semantic and syntactic processes, Lee and Federmeier (2015) assessed right-handers' Event-Related Potential (ERP) responses while they viewed and judged the grammaticality of English two-word phrases. These phrases consisted of a centrally presented one-word syntactic cue ('to' for verbs; 'the' for nouns) followed by a laterally presented verb or noun, matching or mismatching the syntactic expectancy. Their results showed that, for right-handers overall, syntactic processing, as indexed by P600 responses previously reported to be elicited by various other kinds of syntactic violations (Hagoort, Brown, & Groothusen, 1993; Osterhout & Holcomb, 1992), was only found with initial presentation to the left hemisphere (via the right visual field). Their results showed that the right hemisphere was sensitive to syntactic violations, in the sense that grammatical and ungrammatical targets projected to the left visual field and preferentially processed by the right hemisphere elicited reliably different ERP responses. This right hemisphere grammaticality effect, however, was manifested on the N400 component that has been considered a reflection of the activation states of a distributed, multimodal semantic network (see Kutas & Federmeier, 2011 for a comprehensive review). These right hemisphere N400 grammaticality effects thus suggest that, for right-handers as a group, the right hemisphere may respond to syntax mainly via patterns of lexical association. These lateralization patterns for ERP responses to word class violations have since then been replicated with Mandarin materials in young Mandarin-speaking right-handers with a similar design (Weng and Lee, submitted). Data from Lee and Federmeier (2015) and Weng and Lee (submitted) are re-plotted in Fig. 1.

Critically, however, further analysis suggests that the nature of right hemisphere sensitivity to syntactic information is qualitatively moderated by familial sinistrality. Despite the group-level trend that P600 responses were only reliable with left hemisphere biased (right visual field)

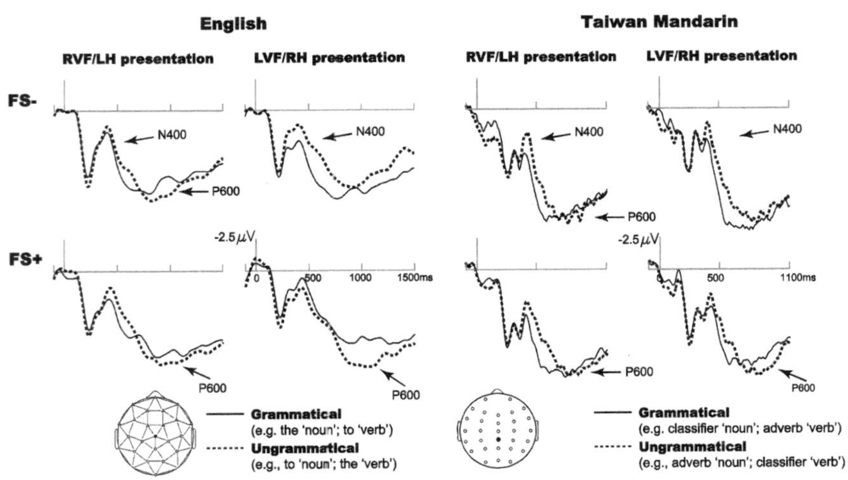

Figure 1 Grand-average event-related potentials (ERPs) of the N400 and P600 differences between grammatical and ungrammatical phrases for participants without familial sinistrality (FS−; top panels) and with familial sinistrality (FS+; bottom panels). Data from native speakers of English reading English phrases (Lee & Federmeier, 2015) and native speakers of Taiwan Mandarin reading Taiwan Mandarin phrases (Weng and Lee, submitted) are shown on the left and right panels respectively. The graphs present the ERPs recorded at a representative electrode site (*black dot*). Negative voltage is plotted upward.

presentation, some individuals – in particular FS+ individuals – elicited P600 responses with right hemisphere biased (left visual field) presentation as well (Lee & Federmeier, 2015; also see Weng and Lee, submitted, for a replication in Taiwan Mandarin) (Fig. 1). As such, analyses that grouped participants according to their FS background showed that syntactic processing (as indexed by the P600 grammaticality effect) is unilaterally manifested in the left hemisphere for FS− right-handers, but bilaterally for FS+ right-handers. Thus, while data from FS− participants supports traditional views of the functional lateralization of syntactic processing, results from FS+ participants provide compelling evidence that the right hemisphere is capable of engaging syntactic processes that are qualitatively similar to those found in the left hemisphere.

4.2 Bi-hemispheric Syntactic Processing

It is important to note in the data reviewed above that, although FS+ right handers show reduced functional asymmetry, mirroring many findings in the literature as already discussed, this reduced asymmetry arises because of left hemisphere-like syntactic processing in the right hemisphere, rather than decreased left hemisphere processing capabilities. This is different from predictions that associate familial sinistrality with less effective information access in the left hemisphere (Bever, 2013; Bever et al., 1989; Hancock & Bever, 2013) or a less efficient procedural system due to a less dominant left hemisphere (Ullman, 2004). Contrary to these predictions, the findings in Lee and Federmeier (2015) and Weng and Lee (submitted) showed that, like FS− right-handers, FS+ right-handers were able to elicit reliable P600 responses to grammatical errors presented to the right visual field, indicating comparable left hemisphere syntactic capability for right-handers across the familial sinistrality background, at least for syntactic processing pertaining to word class information.

Furthermore, different from the hypothesis that FS+ individuals may rely on the right hemisphere to process syntactic information via lexical-associative- based mechanisms, P600 responses are bilaterally available in FS+ right-handers (Lee & Federmeier, 2015; Weng and Lee, submitted). This suggests that, while FS+ right-handers indeed engage more bilateral resources for syntactic processing, what critically differentiates brain response patterns between FS+ and FS− right-handers is the additional structural analysis indexed by P600s in the right hemisphere. While verifications using a variety of syntactic constructions will be needed to back up this observation, these findings make an important case that bi-hemispheric syntactic processing in FS+ individuals is possible and highlight the importance of taking heterogeneity in brain functional mapping into account when considering the influence of FS on language processing.

4.3 Bi-lateralization of Language

Evidence for bi-hemispheric syntactic processing, and especially for cases in which right hemisphere processing is qualitative similar to that in the left hemisphere, has subsequently been observed in different healthy populations using similar visual half field presentation methods. For example, consistent with findings in the aging literature showing that cognitive processes that are unilateral in young adults often tend to engage more bilateral processes in healthy older adults (Cabeza, 2002), it has also been found that P600 responses to grammatical errors that are unilaterally manifested in FS− young right-handers are bilaterally reliable in FS− older right-handers (Chen, Chen, & Lee, 2017; Leckey & Federmeier, 2017). A right-hemisphere-based P600 grammaticality effect has also been observed in second language learners, but only in those of intermediate to low proficiency as opposed to learners of high proficiency (Chen et al., 2018).

Together, these data provide an interesting opportunity to examine the factor(s) that may potentially tune right hemisphere activity and contribute to language lateralization. Considering the differences in neurobiological resources and language experiences among these populations, it is possible that multiple independent or interactive factors jointly determine right hemisphere involvement in processing. Below, I discuss some factors that have been proposed to account for language lateralization in relation to recent data about FS influences on language processing.

4.3.1 Inter-hemispheric Transmission Time Delay

Inter-hemispheric communication delays through the corpus callosum as a result of increased brain size have been deemed a critical factor leading to functional lateralization (Aboitiz, López, & Montiel, 2003; Ringo et al., 1994). In this view, inter-hemispheric transmission time delays, which could be on the order of tens of milliseconds, impose a critical constraint on inter-hemispheric communication and can be functionally deleterious, especially for tasks, like language, that require rapid information transmission. These processing limitations due to transmission time delay are thus thought to promote the development of brain lateralization that helps keep required processes within one hemisphere (Ringo et al., 1994). The 'callosal distance hypothesis' has received support from neuroanatomical evidence. For example, it has been found that behavioral asymmetries can be predicted by overall skull size, with larger brains showing more lateralization for language processing (Tzourio-Mazoyer, Petit, et al., 2010).

Along the line of the 'callosal distance hypothesis', it has been suggested that the relative size of the corpus callosum could influence the amount of information transfer between the two hemispheres, with a larger corpus callosum encouraging more interhemispheric transfer of information (Christman, 1993; Christman, 2001; Honda, 1982; McKeever & Hoff, 1983; Moscovitch & Smith, 1979; Potter & Graves, 1988; but see; Clarke, Lufkin, & Zaidel, 1993). Individuals with a larger corpus callosum could thus be more efficient at inter-hemispheric communication and may then have the ability to processing language more bilaterally. Indeed, it has been found that callosal size, in some callosal segments in particular, is larger in individuals that are reportedly less lateralized in anatomy or function (Aboitiz, 1992; Dorion et al., 2000; Witelson, 1989). A larger corpus callosum has been associated with left handedness (Burke & Yeo, 1994; Cowell, Allen, Kertesz, Zalatimo, & Denenberg, 1994; Habib et al., 1991; Kertesz, Polk, Howell, & Black, 1987), which, in turn, was associated with better interhemispheric transfer performance (Potter & Graves, 1988). Differences in corpus callosum size per se have yet to be documented for FS+ individuals (Tzourio-Mazoyer, Petit, et al., 2010; Tzourio-Mazoyer, Simon, et al., 2010). However, it has been suggested that, in comparison with FS− individuals, FS+ individuals may have a greater degree of interhemispheric interaction, as they show superior episodic memory, which is thought by some to based in interhemispheric processing (Christman & Propper, 2001; but see Piccirilli, Finali, & Sciarma, 1989). Similar arguments for superior interhemispheric communication may be made for language processing in FS+ individuals based on findings showing FS+ individuals' superior performance in comprehending an essay when the words were perceived monaurally with alternating ear (Hancock & Bever, 2013).

4.3.2 Inter-hemispheric Inhibition

In addition to interhemispheric integration, interhemispheric inhibition has also been suggested to be a mechanism contributing to underlying brain asymmetry. For example, it has been suggested that left hemisphere-equivalent language capabilities exist in the right hemisphere but are oftentimes masked by transcallosal interhemispheric inhibition from the dominant left hemisphere (Chiarello & Maxfield, 1996; Karbe et al., 1998), as parallel and independent processes within each hemisphere could help address the demands of coordinating and integrating task-specific activities generated from the processing streams in each hemisphere (Banich & Belger, 1990; Weissman & Banich, 2000; Chiarello & Maxfield, 1996; Cook & Beech, 1990; Hoptman & Davidson, 1994). This view is supported by findings showing increased activation in the right hemisphere homologs of left hemisphere language areas after left hemisphere strokes, which presumably release the inhibition from the left hemisphere (Hamilton, Chrysikou, & Coslett, 2011). In addition, utilizing noninvasive brain stimulation such as Transcranial Magnetic Stimulatoin (TMS) and Transcranial Direct Current Stimulation (tCDS) as a therapy for aphasia, Hamilton and colleagues suggest that, in some cases with left-hemisphere lesions, suppressing the less efficient or dysfunctional neural activity in the right hemisphere may reduce deleterious noise to the overall function of the language network and facilitates the recovery process of the left hemisphere (Hamilton et al., 2011).

Related to this hypothesis, recent data have shown suggestive evidence that the magnitude of right hemisphere P600 responses to grammatical errors is moderated by individuals' interhemispheric inhibition ability. Weng and Lee (submitted) assessed young FS− and FS+ right-handers' interhemispheric inhibition ability using a behavioral bilateral flanker task (Eriksen & Eriksen, 1974; Spencer & Banich, 2005), in which participants attended to and judged the direction of a horizontal arrow presented in a pre-specified visual field. Meanwhile, another arrow or a square appeared in the non-attended visual field, yielding congruent, incongruent, and neutral conditions. Inhibition on the right hemisphere (presumably from the left hemisphere) was measured by subtracting the reaction times on neutral trials from those on incongruent trials in the attend-right visual field condition. The results showed no reliable differences in the size of flanker effect between young FS+ and FS− right-handers at the group level. However, correlational analysis showed an association between more effective ability to inhibit distractor information projected to the right hemisphere and a smaller right hemisphere P600 grammaticality effect for young FS+ right-handers. Young FS− right-handers, a population that shows much fewer right-hemisphere P600 responses than FS+ right-handers in general, did not show this correlation in this study, but showed a similar correlation in a follow-up study wherein the flanker task was sped up (Yeh, Yen, Hsu, & Lee, 2018).

These findings thus corroborated the role of inter-hemispheric inhibition in moderating left-hemispheric equivalent syntactic processing in the right hemisphere among right-handers. That interhemispheric inhibition ability did not differ reliably as a function of FS needs to be further confirmed with a larger sample size. However, these extant data suggest that if FS+ and FS− right-handers are statistically equivalent in terms of interhemispheric inhibition, inhibition alone may not be sufficient in explaining the tendency of bilateralization in FS+ individuals. Instead, the association between inhibition ability and right hemisphere P600 found in these studies may reflect a common variance among right-handers that is easier to capture in young FS+ individuals due to higher variability of right hemispheric responses in this group, which could be captured in FS− individuals so long as more variance in the inhibition index could be pulled out by task demand.

4.3.3 Right Hemisphere Compensation

Reduced language lateralization due to increased right hemisphere participation has sometimes been linked to the idea of compensation when the processing is considered challenging for the individuals. Age-related increases in the tendency to show bilateral (vs. asymmetric) activation patterns have been documented for syntactic processing (Tyler et al., 2010) as well as in other cognitive domains (Cabeza, 2002; Park & Reuter-Lorenz, 2009). With a word monitoring task, Tyler et al. (2010) found that healthy older adults showed an increase of activity in right hemisphere language regions proportional to the degree of age-related atrophy in homologous left hemisphere regions. This bi-lateralized pattern has been interpreted as compensatory in nature, helping to preserve syntactic processing in older adults (Tyler et al., 2010). The nature of the increased right-hemisphere activity, however, was not clear in this study.

Based on the ERP results reported in Lee and Federmeier (2015), at least for processing of syntactic word class, there could be subtypes of bilateral activation, indicating greater lexical-semantic activity (reflected by the N400 grammaticality effect) or greater structural-based analysis (reflected by the P600 grammaticality effect) in the right hemisphere, and behavioral consequences of bi-lateralization could differ as a function of the type of right hemisphere involvement (see Meyer & Federmeier, 2010 for an example of how different bilateral activation patterns are differentially linked to meaning selection abilities among older adults). Leckey and Federmeier (2017) investigated this issue using the visual half field and word class violation paradigm from Lee and Federmeier (2015). Their results showed reliable P600 grammaticality effects in healthy older FS− right-handers with presentation to both visual fields. These results thus suggest that the highly lateralized pattern of syntactic processing seen in young FS− right-handers can shift to a much more bilateral pattern over the course of normal aging, with left-hemisphere-like syntactic processing becoming more available in the right hemisphere (Leckey & Federmeier, 2017). This finding of bilateral P600 grammaticality effects in older adults was replicated testing older FS− right-handers comprehending their native language (Chen et al., 2017), Taiwan-Mandarin. Note, however, that in this study, the older adults were still reliably less accurate than their young counterparts (Chen et al., 2017), suggesting that compensatory efforts may not always be completely successful.

Shifts away from left hemisphere dominance, with the right hemisphere contributing in a left hemisphere-like manner, have also been observed in the context of language learning. For example, a right-hemisphere biased P600 grammaticality effect was found in intermediate to low-proficiency, but not high-proficiency, second language learners (Chen et al., 2018). A similar association between P600 grammaticality effects in the right hemisphere and lower language learning performance was also noted in an artificial grammar learning study (Yeh et al., 2018). Yeh et al. (2018) used an artificial grammar learning design adapted from Gómez (2002), in which participants learned a new artificial grammar by listening monaurally to a set of 3-element strings generated according to pre-determined rules (which were not made explicit to participants). After training, they then judged the grammaticality of another set of monaurally presented strings. The data showed an overall bilateral P600 grammaticality effect, with the degree of P600 left-lateralization associated with judgment accuracy. A median-split based on behavioral accuracy showed that for learners with higher judgment accuracy, P600 grammaticality effects were reliable only in the right ear presentation condition and thus were lateralized to the left hemisphere. However, for learners with lower judgment accuracy, P600 grammaticality effects were reliable with both right-ear and left-ear presentations, replicating the link between right hemisphere P600s and lower language learning outcome (Chen et al., 2018). Critically, when participants were tested on syntactic word class based grammaticality in their native language, both sub-groups showed left-lateralized P600 effects, reinforcing the relation between the right-hemisphere P600 grammaticality effect and learning and processing of a *new* syntactic regularity.

The findings reviewed above raise the question of whether left-hemisphere-like effects in the right hemisphere are compensatory in nature or, instead, provide competition that is potentially deleterious to language processing (e.g., interferes with the ability to acquire language patterns). To adjudicate between these two hypotheses, Yen, Tang, Hsu, and Lee (2018), used a design similar to that in Yeh et al. (2018), but focused on ERP grammaticality effects from successful learners only. In addition, instead of the salient non-adjacent regularity trained and tested in Yeh et al. (2018), Yen and colleagues trained and tested salient (i.e., easier to learn) versus less salient (i.e., harder to learn) syntactic regularities, with the saliency of regularity manipulated as a between-subjects factor. With matched high accuracy between the two groups, their results showed that successful learners exhibited left-lateralized P600s in the salient condition but bilateral P600 effects in the less salient condition. These results thus disfavor the interference account and provide support for the compensatory account by suggesting that the right hemisphere P600 responses might reflect extra neural resources being devoted to aiding challenging syntactic learning or processing. These findings are also in line with past findings suggesting that while interhemispheric interaction can be inefficient for tasks that are simple (Belger & Banich, 1998; Fling et al., 2010), it can be beneficial for tasks that are complex and demanding (Scalf, Banich, Kramer, Narechania, & Simon, 2007; Weissman & Banich, 2000). As such, lateralized or more bilateral language processing may be optimal configurations for different scenarios even within the same individual.

What about the bi-hemispheric P600 effect seen in FS+ right-handers — is this also a compensatory response? Although there are no statistical differences between the left hemisphere P600 responses in FS− and FS+ right-handers, it is possible that, as proposed in Bever et al.'s and Ullman's accounts (Bever, 2013; Ullman, 2016), syntactic processing is more challenging for FS+ right-handers. If so, then the P600 responses from the right hemisphere may be compensatory, allowing FS+ individuals to achieve levels of behavioral grammaticality judgment performance comparable to those in FS− individuals (Lee & Federmeier, 2015; Weng and Lee, submitted). This hypothesis, however, could be further bolstered with independent evidence for lower syntactic sensitivity in the same group of FS+ right-handers.

4.3.4 Future Directions

Although the underlying mechanisms and the functional significance of bi-lateralized language processing associated with familial sinistrality are still not clear, the above reviewed studies point out promising directions for future investigations. As multiple factors can contribute to bilateral syntactic processing, it is clearly important to control for factors known to be relevant for language lateralization or to investigate multiple factors concurrently so as to obtain a clearer picture of how these factors may interact. For example, some have found that the reduced leftward language lateralization (as assessed by a story-listening task) associated with FS is only seen among individuals with weak manual preference, but not among those with strong manual preference (Tzourio-Mazoyer, Petit, et al., 2010). It would also be informative to include an index for behavioral performance or task difficulty, as right hemisphere involvement could be beneficial or deleterious in different situations. A parallel trend can be seen in rehabilitation work with aphasic patients. Although right hemisphere recruitment has been found to be beneficial in the recovery from acute or subacute left hemisphere lesions, right hemisphere recruitment during the recovery process of chronic aphasia was found to have a negative correlation with the restoration of damaged region in the left hemisphere (Hamilton et al., 2011). These data in turn suggest that the right hemisphere may often possess left-like competencies with language processing, but only manifests those when the dominant hemisphere cannot carry the burden alone (Karbe et al., 1998).

4.4 Sensitivity to Syntactic Information

The studies reviewed in Section 2.2 highlight that FS background affects behavioral response patterns for experimental manipulations pertaining to syntactic structures. For example, when asked to judge whether a phrasal probe was related to a preceding sentence in meaning, both FS− and FS+ groups were faster to respond to main-clause probes than to subordinate-clause probes, but the difference across experimental conditions was larger for the FS− group than the FS+ group (Bever et al., 1987; see Hancock & Bever, 2013 for a review of other result patterns). Behavioral patterns associated with FS+ like this were once thought to be analogous to response patterns from agrammatic aphasics (Bever et al. 1987) and were sometimes taken further to implicate that FS+ participants were 'less sensitive' to syntactic structures and, instead, relied more on lexical-semantic processing. However, in some cases, FS+ individuals' smaller effect sizes were associated with faster overall responses (e.g., Townsend et al., 2001); hence it is not always clear whether smaller response differences to syntactic manipulations reflect superior or less effective processing.

Along these lines, some recent electrophysiological studies have found that, relative to FS− right-handers, FS+ individuals are less likely to elicit P600 responses to morphosyntactic violations (Grey et al., 2017; Kos et al., 2012; Tanner & Van Hell, 2014), and, in some cases, tend to elicit N400 grammaticality effects instead (Grey et al., 2017; Tanner & Van Hell, 2014). For example, Tanner and Van Hell (2014), using stimuli with basic morphosyntactic violations in subject-verb agreement (e.g., 'The clerk at the clothing boutique *was/were* severely underpaid and unhappy') and tense marking (e.g., 'The crime rate was *increasing/increase* despite the growing police force'), found that most of their right-handed participants elicited P600 grammaticality effects to these errors. However, a substantial number of participants showed an additional N400 grammaticality effect, resulting in a biphasic N400-P600 effect pattern, or showed just the N400 grammaticality effect alone. This inter-individual variability was not accounted for by variation in cognitive abilities such as working memory, cognitive control, lexical processing speed or language proficiency, but was reliably associated with participants' FS background (which explained approximately 19% of the variance). Specifically, while FS− right-handers showed monophasic P600 responses to both types of morphosyntactic violations, FS+ right-handers were more likely to show biphasic N400-P600 effects. Although no direct comparisons were made regarding the effect sizes of P600 responses as a function of FS, when effect sizes for both the N400 and P600 effects were taken into consideration to compute response dominance, FS+ right-handers showed reliably lower response dominance for the P600 effects than did FS− right-handers. Building on results in Tanner and Van Hell (2014), Grey et al. (2017), additionally tested left-handers, and discovered that left-handers also showed similar response patterns to FS+ right-handers, with biphasic N400-P600 effect patterns and a response dominance index for the P600 effect that was statistically equivalent to that of FS+ right-handers but was marginally lower than that of FS− right-handers.

These results seem to present a contradictory picture from studies suggesting bi-hemispheric syntactic processing in FS+ individuals. Specifically, contrary to findings showing P600 grammaticality effects that were robustly available in both hemispheres in FS+ right-handers (Lee & Federmeier, 2015; Weng and Lee, submitted), the above studies suggest lower syntactic sensitivity and less available P600-based structural analysis in FS+ right-handers (Grey et al., 2017; Tanner & Van Hell, 2014). However, there are some important differences in methodology among these studies that may shed light on the seemingly inconsistent findings. First, Lee and Federmeier (2015) used the visual half field paradigm to preferentially present stimuli to one hemisphere only, whereas the studies that found reduced P600 grammaticality effects in FS+ right-handers used central presentation (Grey et al., 2017; Tanner & Van Hell, 2014). What each hemisphere can do when it preferentially processes an input versus what happens when input is made available to both hemispheres in parallel may well be different. For example, it is possible that, for FS+ right-handers the hemisphere that initially receives the input will proceed with the structural analysis and arrive at the final output (P600 response) with relatively little interference, as processing in the other syntax-capable hemisphere is delayed or even blocked, because it gets no direct stimulation. However, with central presentation, qualitatively similar structural analyses may be initiated in both hemispheres concurrently, yielding non-identical and potentially conflicting output that impedes the formation of the (joint) P600 response. Under this hypothesis, the central and lateral data do not contradict each other, but rather jointly corroborate the inhibition view described in Section 5.3.2.

Another potential explanation for the mixed results is additional individual variability within or across the samples in these studies. Left-handedness (own or familial) has been shown to interact with other factors in modulating the amplitude of the P300 (Hancock, 2013; Polich, 2007), a component that has been argued to share neurological resources with P600 (Kutas, Federmeier, & Urbach, 2014). Thus it may be informative to examine this issue by measuring individual variation in more domain-general characteristics. Relative to right-handers, left-handers have been found to elicit larger and earlier P300s at anterior and central electrode sites, a pattern that has been attributed to the larger corpus callosum and consequently better interhemispheric communication generally observed for this population (J. Polich & Hoffman, 1998; Polich, 2007). The trait of FS in particular, has also been found to affect P300 responses (P3b in particular). Hancock (2013), using a classical auditory oddball task, reported an interactive effect of FS and sex on the amplitude of P300, with FS+ males eliciting smaller P300s in the posterior region and a trend toward larger P300s in the anterior region compared to FS− males. This effect of FS, however, was not seen in females. Hancock's finding was based on a relatively small group (24 FS+ and 12 FS− participants balanced for gender). As P3b has also been shown to be sensitive to various other demographic and cognitive abilities (e.g., background knowledge: Carrión & Bly, 2008; gender: Conroy & Polich, 2007;

fitness: Pontifex, Hillman, & Polich, 2009; brain amyloid load: Bennys et al., 2017), the issue of whether the FS effect on P3b in males reported by Hancock (2013) could be explained by differences in other demographic or cognitive characteristics needs to be addressed. These studies, however, suggest the possibility that interaction between FS and other traits (e.g., manual preference and gender) may also contribute to the mixed results (Lee & Federmeier, 2015; Weng and Lee, submitted; Grey et al., 2017; Tanner & Van Hell, 2014). They also point out the interesting possibility of a wider influence from the FS trait on human cognition, and a means of further testing the influence of familial sinistrality on hemispheric specialization and interactions between the hemispheres.

Yet another possible explanation for the mixed findings is that different aspects of the syntactic system were probed across these studies. Whereas the visual half field studies that found bi-hemispheric P600 grammaticality effects in FS+ right-handers tested syntactic word class information (Lee & Federmeier, 2015; Weng and Lee, submitted), studies that used central presentation and found reduced P600 grammaticality effects in FS+ right-handers tested more morphological aspect of syntactic knowledge such as subject-verb agreement and tense marking (Grey et al., 2017; Tanner & Van Hell, 2014). It is possible that FS differentially affects the recruitment of neural resources needed for processing syntactic word class expectation versus those used to process subject-verb agreement violations. Notably, P600 grammaticality effects elicited by subject-verb agreement have been shown to co-vary with the Late Positive Complex (LPC) effect elicited by semantic anomalies, which has also been suggested to be affected by the trait of FS (Kos et al., 2012). The LPC is a broad posterior positivity subsequent to N400 responses, thus sharing similar latency and topographic distribution with the P600. However, whereas the P600 has been associated with syntactic processing, the LPC has been linked to semantics and, especially, to participants explicitly noting the presence of a semantic anomaly, and perhaps trying to repair it (Van Petten & Luka, 2012). The elicitation of an LPC to semantic anomalies is quite variable and does not seem to straightforwardly hinge on stimulus factors such as sentence constraint or working memory capacity. However, Kos et al. (2012), observed a suggestive role for FS in accounting for the substantial inter-individual variation. Kos et al. (2012) noted that, among the 14 FS+ individuals tested in their study, 11 of them did not elicit LPC responses to outright semantic violations (e.g., 'Whipped cream tastes *anxious* and creamy'). Interestingly, mean amplitudes of this LPC semantic anomaly effect correlated with those of the P600 grammaticality effect to subject-verb agreement errors ($r = .37$, e.g., 'The spoiled child *throws/throw* the toy on the floor.'). Kos et al. did not discuss the proportion of FS+ individuals who did not show the P600 grammaticality effects. However, this study suggests the possibility that the tendency for smaller P600 response dominance in FS+ right-handers might be more likely to be observed for aspects of syntactic processing that load more strongly on lexico-semantic analysis (Grey et al., 2017; Tanner & Van Hell, 2014).

The review above provides some hypotheses that are difficult to assess with extant data in the literature, but could be tested with studies that systematically compare data from the same set of closely matched FS− and FS+ right-handers across different presentation methods and a variety of morphosyntactic manipulations.

5. CONCLUSIONS

A phenomenon that has been receiving increasing attention is the striking within- and between-individual differences that can be found for language processing even in healthy native speakers, suggesting the need to re-consider the typical assumption that there is high homogeneity for language processing mechanisms. In view of this, the present article focused on familial sinistrality (FS), a trait that has been widely noted in association with individual variability in language processing. As reviewed in this chapter, studies using neuropsychological, behavioral, neuroimaging, and electrophysiological techniques have indicated a special role for FS in moderating the allocation and lateralization of language processing resources. Studies using visual half field presentation in particular confirmed the hypothesis of reduced leftward language asymmetry in FS+ individuals. Furthermore, these data extended this hypothesis by showing that, at least in the case of processing syntactic word class information, the reduced language lateralization associated with FS+ individuals is driven by more (qualitatively similar) activation in the right hemisphere. These findings also point out that the right hemisphere is capable of eliciting left-hemispheric equivalent syntactic processing, which echoes the view from Bradshaw and Nettleton (1981) suggesting that differences in function between the hemispheres tend to be quantitative rather than qualitative. These findings thus broaden the possible interpretations of bi-lateralization and hemispheric interaction for language processing. This chapter also reviewed several driving forces for language lateralization that have been proposed in the literature, in relation to findings from FS+ individuals. This review highlights that language lateralization is not a uniform phenomenon but is, instead, dynamic. In addition, direction and degree of language lateralization can be jointly determined by multiple factors such as genetic features (e.g., familial handedness history), neurobiological features (e.g., distribution of the dopaminergic system), demographic features (e.g., age and gender), experience (e.g., language learning), test materials (e.g., saliency of the syntactic rules), and so forth. This calls for caution when interpreting and generalizing functional significance over similar lateralization patterns, as the final configurations for language processing (e.g., unilateral vs. bi-lateral) could result from different contributions of these modulating factors and could therefore be advantageous or deleterious depending on the resources and tasks at hand. Jointly, the findings reviewed here endorse the view that successful language processing can be achieved via multiple different configurations of neurological and cognitive mechanisms that are adaptive to available cognitive resources and task demand over time.

REFERENCES

Aboitiz, F. (1992). Brain connections: Interhemispheric fiber systems and anatomical brain asymmetries in humans. *Biological Research, 25*(2), 51–61.

Aboitiz, F., López, J., & Montiel, J. (2003). Long distance communication in the human brain: Timing constraints for inter-hemispheric synchrony and the origin of brain lateralization. *Biological Research, 36*(1), 89–99. https://doi.org/10.4067/S0716-97602003000100007.

Alegre, M., & Gordon, P. (1999). Frequency effects and the representational status of regular inflections. *Journal of Memory and Language, 40*(1), 41–61. https://doi.org/10.1006/jmla.1998.2607.

Annett, M. (1994). Handedness as a continuous variable with dextral shift: Sex, generation, and family handedness in subgroups of left- and right-handers. *Behavior Genetics, 24*(1), 51–63. https://doi.org/10.1007/BF01067928.

Arambel, S. R., & Chiarello, C. (2006). Priming nouns and verbs: Differential influences of semantic and grammatical cues in the two cerebral hemispheres. *Brain and Language, 97*(1), 12–24. https://doi.org/10.1016/j.bandl.2005.07.003.

Ashby, F. G., Turner, B. O., & Horvitz, J. C. (2010). Cortical and basal ganglia contributions to habit learning and automaticity. *Trends in Cognitive Sciences, 14*(5), 208–215. https://doi.org/10.1016/j.tics.2010.02.001.

Babcock, L., Stowe, J. C., Maloof, C. J., Brovetto, C., & Ullman, M. T. (2012). The storage and composition of inflected forms in adult-learned second language: A study of the influence of length of residence, age of arrival, sex, and other factors★. *Bilingualism: Language and Cognition, 15*(4), 820–840. https://doi.org/10.1017/S1366728912000053.

Banich, M., & Belger, A. (1990). Interhemispheric interaction: how do the hemispheres divide and conquer a task? *Cortex, 26*, 77–94.

Belger, A., & Banich, M. T. (1998). Costs and benefits of integrating information between the cerebral hemispheres: A computational perspective. *Neuropsychology, 12*(3), 380–398. https://doi.org/10.1037/0894-4105.12.3.380.

Bennys, K., Gabelle, A., Berr, C., De Verbizier, D., Andrieu, S., Vellas, B., ... MAPT-DSA Study group. (2017). Cognitive event-related potential, an early diagnosis biomarker in Frail Elderly subjects: The ERP-MAPT-PLUS Ancillary study. *Journal of Alzheimer's Disease: JAD, 58*(1), 87–97. https://doi.org/10.3233/JAD-161012.

Bever, T. G. (1975). Psychologically real grammar emerges because of its role in language acquisition. *Developmental Psycholinguistics: Theory and Application*, 63–75.

Bever, T. G. (1980). Broca and Lashley were right: Cerebral dominance is an accident of growth. In D. Kaplan, & N. Chomsky (Eds.), *Biology and language* (pp. 186–232). Cambridge, MA: MIT Press.

Bever, T. G. (2013). The biolinguistics of language universals—the next years. In M. Sanz, I. Laka, & M. K. Tanenhaus (Eds.), *Language down the garden path: The cognitive and biological basis for linguistic structures* (pp. 385–406). Oxfor University Press.

Bever, T. G., Carrithers, C., Cowart, W., & Townsend, D. J. (1989). Language processing and familial handedness. In A. Galaburda (Ed.), *From neurons to reading*. Cambridge, MA: MIT Press.

Bever, T. G., Carrithers, C., & Townsend, D. J. (1987). A tale of two brains; or, the sinistral quasimodularity of language. *Cognitive Science Society*, 764–773.

Bever, T. G., & Chiarello, R. J. (1974). Cerebral dominance in musicians and nonmusicians. *Science, 185*(4150), 537–539.

Bradshaw, J. L., & Nettleton, N. C. (1981). The nature of hemispheric specialization in man. *Behavioral and Brain Sciences, 4*(1), 51–63. https://doi.org/10.1017/S0140525X00007548.

Briggs, G. G., & Nebes, R. D. (1976). The effects of handedness, family history and sex on the performance of a dichotic listening task. *Neuropsychologia, 14*(1), 129–133. https://doi.org/10.1016/0028-3932(76)90014-2.

Broca, P. (1861). Remarques sur le siège de la faculté du langage articulé, suivies d'une observation d'aphémie (perte de la parole). *Bulletin et Memoires de La Societe Anatomique de Paris, 6*, 330–357.

Brown, J. W., & Hécaen, H. (1976). Lateralization and language representation. *Neurology*, (26), 183.

Burke, H. L., & Yeo, R. A. (1994). Systematic variations in callosal morphology: The effects of age, gender, hand preference, and anatomic asymmetry. *Neuropsychology, 8*(4), 563–571. https://doi.org/10.1037/0894-4105.8.4.563.

Cabeza, R. (2002). Hemispheric asymmetry reduction in older adults: The HAROLD model. *Psychology and Aging, 17*(1), 85–100. https://doi.org/10.1037/0882-7974.17.1.85.

Carrión, R. E., & Bly, B. M. (2008). The effects of learning on event-related potential correlates of musical expectancy. *Psychophysiology, 45*(5), 759–775. https://doi.org/10.1111/j.1469-8986.2008.00687.x.

Chen, P.-H., Chen, M.-H., & Lee, C. L. (2017). Inter-hemispheric communication ability modulates the laterality of P600 responses to syntactic category violations in healthy young and older adults. *Psychophysiology, 54*(Suppl. 1), S73.

Chen, P.-H., Yeh, C., Lu, C., Hsieh, S.-K., Chou, T.-L., Su, L. I.-W., & Lee, C. L. (2018). Multiple scaffolding mechanisms for L2 syntactic processing–An Event-Related Potential study. *Journal of Chinese Language Teaching, 15*(2), 63–94.

Chiarello, C., & Maxfield, L. (1996). Varieties of interhemispheric inhibition, or how to keep a good hemisphere down. *Brain and Cognition, 30*(1), 81–108. https://doi.org/10.1006/brcg.1996.0006.

Chiarello, C., Vazquez, D., Felton, A., & Leonard, C. M. (2013). Structural asymmetry of anterior insula: Behavioral correlates and individual differences. *Brain and Language, 126*(2), 109–122. https://doi.org/10.1016/j.bandl.2013.03.005.

Christman, S. (1993). Handedness in musicians: Bimanual constraints on performance. *Brain and Cognition, 22*(2), 266–272. https://doi.org/10.1006/brcg.1993.1038.

Christman, S. D. (2001). Individual differences in stroop and local-global processing: A possible role of interhemispheric interaction. *Brain and Cognition, 45*(1), 97–118. https://doi.org/10.1006/brcg.2000.1259.
Christman, S. D., & Propper, R. E. (2001). Superior episodic memory is associated with interhemispheric processing. *Neuropsychology, 15*(4), 607–616. https://doi.org/10.1037/0894-4105.15.4.607.
Clahsen, H., & Felser, C. (2006). Grammatical processing in language learners. *Applied PsychoLinguistics, 27*(1), 3–42. https://doi.org/10.1017/S0142716406060024.
Clarke, J. M., Lufkin, R. B., & Zaidel, E. (1993). Corpus callosum morphometry and dichotic listening performance: Individual differences in functional interhemispheric inhibition? *Neuropsychologia, 31*(6), 547–557. https://doi.org/10.1016/0028-3932(93)90051-Z.
Conroy, M. A., & Polich, J. (2007). Normative variation of P3a and P3b from a large sample. *Journal of Psychophysiology, 21*(1), 22–32. https://doi.org/10.1027/0269-8803.21.1.22.
Cook, N. D., & Beech, A. R. (1990). The Cerebral Hemispheres and Bilateral Neural Nets. *International Journal of Neuroscience, 52*(3-4), 201–210. https://doi.org/10.3109/00207459009000522.
Cowell, P. E., Allen, L. S., Kertesz, A., Zalatimo, N. S., & Denenberg, V. H. (1994). Human corpus callosum: A stable mathematical model of regional neuroanatomy. *Brain and Cognition, 25*(1), 52–66. https://doi.org/10.1006/brcg.1994.1022.
Curtiss, S., de Bode, S., & Mathern, G. W. (2001). Spoken Language Outcomes after Hemispherectomy: Factoring in Etiology. *Brain and Language, 79*(3), 379–396. https://doi.org/10.1006/brln.2001.2487.
Curtiss, Susan & Jeannette Shaeffer. (1997). In E. Hughes (Ed.), *Syntactic development in children with hemispherectomy: The INFL-system*. Somerville, MA: BUCLD 21 Proceedings.
de Bode, S., Chanturidze, M., Mathern, G. W., & Dubinsky, S. (2015). Literacy after cerebral hemispherectomy: Can the isolated right hemisphere read? *Epilepsy & Behavior, 45*, 248–253. https://doi.org/10.1016/j.yebeh.2015.01.007.
Devlin, A. M., Cross, J. H., Harkness, W., Chong, W. K., Harding, B., Vargha-Khadem, F., & Neville, B. G. R. (2003). Clinical outcomes of hemispherectomy for epilepsy in childhood and adolescence. *Brain, 126*(3), 556–566. https://doi.org/10.1093/brain/awg052.
Dorion, A. A., Chantôme, M., Hasboun, D., Zouaoui, A., Marsault, C., Capron, C., & Duyme, M. (2000). Hemispheric asymmetry and corpus callosum morphometry: A magnetic resonance imaging study. *Neuroscience Research, 36*(1), 9–13.
Eriksen, B. A., & Eriksen, C. W. (1974). Effects of noise letters upon the identification of a target letter in a nonsearch task. *Perception & Psychophysics, 16*(1), 143–149. https://doi.org/10.3758/BF03203267.
Federmeier, K. D. (2007). Thinking ahead: The role and roots of prediction in language comprehension. *Psychophysiology, 44*(4), 491–505. https://doi.org/10.1111/j.1469-8986.2007.00531.x.
Federmeier, K. D., Wlotko, E. W., & Meyer, A. M. (2008). What's "right" in language comprehension: Event-related potentials reveal right hemisphere language capabilities. *Language and Linguistics Compass, 2*(1), 1–17. https://doi.org/10.1111/j.1749-818X.2007.00042.x.
Fling, B. W., Walsh, C. M., Bangert, A. S., Reuter-Lorenz, P. A., Welsh, R. C., & Seidler, R. D. (2010). Differential Callosal Contributions to Bimanual Control in Young and Older Adults. *Journal of Cognitive Neuroscience, 23*(9), 2171–2185. https://doi.org/10.1162/jocn.2010.21600.
Gazzaniga, M. S., & Hillyard, S. A. (1971). Language and speech capacity of the right hemisphere. *Neuropsychologia, 9*(3), 273–280. https://doi.org/10.1016/0028-3932(71)90022-4.
Gómez, R. L. (2002). Variability and detection of invariant structure. *Psychological Science, 13*(5), 431–436. https://doi.org/10.1111/1467-9280.00476.

Grey, S., Tanner, D., & van Hell, J. G. (2017). How right is left? Handedness modulates neural responses during morphosyntactic processing. *Brain Research, 1669*, 27–43. https://doi.org/10.1016/j.brainres.2017.05.024.

Habib, M., Gayraud, D., Oliva, A., Regis, J., Salamon, G., & Khalil, R. (1991). Effects of handedness and sex on the morphology of the corpus callosum: A study with brain magnetic resonance imaging. *Brain and Cognition, 16*(1), 41–61. https://doi.org/10.1016/0278-2626(91)90084-L.

Hagoort, P., Brown, C., & Groothusen, J. (1993). The syntactic positive shift (sps) as an erp measure of syntactic processing. *Language & Cognitive Processes, 8*(4), 439–483. https://doi.org/10.1080/01690969308407585.

Hamilton, R. H., Chrysikou, E. G., & Coslett, B. (2011). Mechanisms of aphasia recovery after stroke and the role of noninvasive brain stimulation. *Brain and Language, 118*(1–2), 40–50. https://doi.org/10.1016/j.bandl.2011.02.005.

Hamrick, P. (2014). Recognition memory for novel syntactic structures. *Canadian Journal of Experimental Psychology/Revue Canadienne de Psychologie Expérimentale, 68*(1), 2–7. https://doi.org/10.1037/cep0000002.

Hancock, R. (2013). *Dynamic properties of dopamine asymmetry: A basis for functional lateralization* (Doctoral dissertation). The University of Arizona. Retrieved from https://repository.arizona.edu/handle/10150/299105.

Hancock, R., & Bever, T. G. (2013). Genetic factors and normal variation in the organization of language. *Biolinguistics,* 7(0), 75–95.

Hardyck, C., & Petrinovich, L. F. (1977). Left-handedness. *Psychological Bulletin, 84*(3), 385–404. https://doi.org/10.1037/0033-2909.84.3.385.

Hartshorne, J. K., & Ullman, M. T. (2006). Why girls say "holded" more than boys. *Developmental Science, 9*(1), 21–32. https://doi.org/10.1111/j.1467-7687.2005.00459.x.

Hécaen, H., De Agostini, M., & Monzon-Montes, A. (1981). Cerebral organization in left-handers. *Brain and Language, 12*(2), 261–284. https://doi.org/10.1016/0093-934X(81)90018-3.

Hécaen, H., & Sauguet, J. (1971). Cerebral dominance in left-handed subjects. *Cortex,* 7(1), 19–48. https://doi.org/10.1016/S0010-9452(71)80020-5.

Hines, D., & Satz, P. (1971). Superiority of right visual half-fields in right-handers for recall of digits presented at varying rates. *Neuropsychologia, 9*(1), 21–25. https://doi.org/10.1016/0028-3932(71)90057-1.

Honda, H. (1982). Effects of handedness on the inhibition of reaction time in double stimulation situations. *Japanese Psychological Research, 24*(1), 43–47. https://doi.org/10.4992/psycholres1954.24.43.

Hoptman, M. J., & Davidson, R. J. (1994). How and why do the two cerebral hemispheres interact? *Psychological Bulletin, 116*(2), 195–21.

Hughlings Jacksons, J. (1878). On affections of speech from disease of the brain. *Brain, 1,* 304–330.

Hund-Georgiadis, M., Lex, U., Friederici, A. D., & Cramon, D. Y. (2002). Non-invasive regime for language lateralization in right and left-handers by means of functional MRI and dichotic listening. *Experimental Brain Research, 145*(2), 166–176. https://doi.org/10.1007/s00221-002-1090-0.

Joanette, Y., Lecours, A. R., Lepage, Y., & Lamoureux, M. (1983). Language in right-handers with right-hemisphere lesions: A preliminary study including anatomical, genetic, and social factors. *Brain and Language, 20*(2), 217–248. https://doi.org/10.1016/0093-934X(83)90043-3.

Just, M. A., Carpenter, P. A., Keller, T. A., Eddy, W. F., & Thulborn, K. R. (1996). Brain activation modulated by sentence comprehension. *Science (New York, NY), 274*(5284), 114–116.

Karbe, H., Thiel, A., Weber-Luxenburger, G., Herholz, K., Josef, K., & Heiss, W.-D. (1998). Brain plasticity in poststroke aphasia: What is the contribution of the right hemisphere? *Brain and Language*, (64), 215–230.

Kee, D. W., Bathurst, K., & Hellige, J. B. (1983). Lateralized interference of repetitive finger tapping: Influence of familial handedness, cognitive load and verbal production. *Neuropsychologia, 21*(6), 617–624. https://doi.org/10.1016/0028-3932(83)90059-3.

Kertesz, A., Polk, M., Howell, J., & Black, S. E. (1987). Cerebral dominance, sex, and callosal size in MRI. *Neurology, 37*(8), 1385–1388.

Kidd, E. (2012). Implicit statistical learning is directly associated with the acquisition of syntax. *Developmental Psychology, 48*(1), 171–184. https://doi.org/10.1037/a0025405.

Knecht, S., Deppe, M., Dräger, B., Bobe, L., Lohmann, H., Ringelstein, E.-B., & Henningsen, H. (2000). Language lateralization in healthy right-handers. *Brain, 123*(1), 74–81. https://doi.org/10.1093/brain/123.1.74.

Knecht, S., Dräger, B., Deppe, M., Bobe, L., Lohmann, H., Flöel, A., … Henningsen, H. (2000). Handedness and hemispheric language dominance in healthy humans. *Brain, 123*(12), 2512–2518. https://doi.org/10.1093/brain/123.12.2512.

Kos, M., van den Brink, D., & Hagoort, P. (2012). Individual variation in the late positive complex to semantic anomalies. *Frontiers in Psychology, 3*, 318. https://doi.org/10.3389/fpsyg.2012.00318.

Kutas, M., & Federmeier, K. D. (2011). Thirty years and counting: Finding meaning in the N400 component of the event-related brain potential (ERP). *Annual Review of Psychology, 62*, 621–647. https://doi.org/10.1146/annurev.psych.093008.131123.

Kutas, M., Federmeier, K. D., & Urbach, T. P. (2014). The "negatives" and "positives" of prediction in language. In M. S. Gazzaniga (Ed.), *The cognitive neurosciences* (5th ed., pp. 649–656). Cambridge: MIT Press.

Larisch, R., Meyer, W., Klimke, A., Kehren, F., Vosberg, H., & Müller-Gärtner, H. W. (1998). Left-right asymmetry of striatal dopamine D2 receptors. *Nuclear Medicine Communications, 19*(8), 781–787.

Leckey, M., & Federmeier, K. D. (2017). Age-related shifts in hemispheric dominance for syntactic processing. *Psychophysiology, 54*(12), 1929–1939. https://doi.org/10.1111/psyp.12970.

Lee, C. L., & Federmeier, K. D. (2015). It's all in the family brain asymmetry and syntactic processing of word class. *Psychological Science, 26*(7), 997–1005. https://doi.org/10.1177/0956797615575743.

Lieberman, M. D., Chang, G. Y., Chiao, J., Bookheimer, S. Y., & Knowlton, B. J. (2004). An event-related fMRI study of artificial grammar learning in a balanced chunk strength design. *Journal of Cognitive Neuroscience, 16*(3), 427–438. https://doi.org/10.1162/089892904322926764.

Lindell, A. K. (2006). In your right mind: Right hemisphere contributions to language processing and production. *Neuropsychology Review, 16*(3), 131–148. https://doi.org/10.1007/s11065-006-9011-9.

Lishnian, W. A., & McMeekan, E. R. L. (1977). Handedness in relation to direction and degree of cerebral dominance for language. *Cortex, 13*(1), 30–43. https://doi.org/10.1016/S0010-9452(77)80051-8.

Liu, S. R. A., Chiarello, C., & Quan, N. (1999). Hemispheric sensitivity to grammatical cues: Evidence for bilateral processing of number agreement in noun phrases. *Brain and Language, 70*(3), 483–503. https://doi.org/10.1006/brln.1999.2185.

Loui, P., Li, H. C., & Schlaug, G. (2011). White matter integrity in right hemisphere predicts pitch-related grammar learning. *Neuroimage, 55*(2), 500–507. https://doi.org/10.1016/j.neuroimage.2010.12.022.

Luria, A. R. (1970). *Traumatic aphasia: Its syndromes, psychology and treatment*. Walter de Gruyter.

Martin, R. C. (2003). Language processing: Functional organization and neuroanatomical basis. *Annual Review of Psychology, 54*(1), 55–89. https://doi.org/10.1146/annurev.psych.54.101601.145201.

Martin, A. (2007). The representation of object concepts in the brain. *Annual Review of Psychology, 58*, 25–45. https://doi.org/10.1146/annurev.psych.57.102904.190143.

McKeever, W. F., & Hoff, A. L. (1983). Further evidence of the absence of measurable interhemispheric transfer time in left-handers who employ an inverted handwriting posture. *Bulletin of the Psychonomic Society, 21*(4), 255–258. https://doi.org/10.3758/BF03334702.

McKeever, W. F., & VanDeventer, A. D. (1977). Visual and auditory language processing asymmetries: Influences of handedness, familial sinistrality, and sex. *Cortex, 13*(3), 225–241. https://doi.org/10.1016/S0010-9452(77)80033-6.

Meyer, A. M., & Federmeier, K. D. (2010). Event-related potentials reveal the effects of aging on meaning selection and revision. *Psychophysiology, 9999*(9999). https://doi.org/10.1111/j.1469-8986.2010.00983.x.

Mohr, C., Landis, T., Bracha, H. S., & Brugger, P. (2003). Opposite turning behavior in right-handers and non-right-handers suggests a link between handedness and cerebral dopamine asymmetries. *Behavioral Neuroscience, 117*(6), 1448–1452.

Mohr, C., & Lievesley, A. (2007). Test-retest stability of an experimental measure of human turning behaviour in right-handers, mixed-handers, and left-handers. *Laterality, 12*(2), 172–190. https://doi.org/10.1080/13576500601051580.

Moro, A., Tettamanti, M., Perani, D., Donati, C., Cappa, S. F., & Fazio, F. (2001). Syntax and the brain: Disentangling grammar by selective anomalies. *Neuroimage, 13*(1), 110–118. https://doi.org/10.1006/nimg.2000.0668.

Moscovitch, M., & Smith, L. C. (1979). Differences in neural organization between individuals with inverted and noninverted handwriting postures. *Science (New York, N.Y.), 205*(4407), 710–713.

Nieuwland, M. S. (2014). "Who's he?" Event-related brain potentials and unbound pronouns. *Journal of Memory and Language, 76*, 1–28. https://doi.org/10.1016/j.jml.2014.06.002.

Orsini, D. L., Satz, P., Soper, H. V., & Light, R. K. (1985). The role of familial sinistrality in cerebral organization. *Neuropsychologia, 23*(2), 223–232. https://doi.org/10.1016/0028-3932(85)90106-X.

Osterhout, L., & Holcomb, P. J. (1992). Event-related brain potentials elicited by syntactic anomaly. *Journal of Memory and Language, 31*(6), 785–806. https://doi.org/10.1016/0749-596X(92)90039-Z.

Park, D. C., & Reuter-Lorenz, P. (2009). The adaptive brain: Aging and neurocognitive scaffolding. *Annual Review of Psychology, 60*, 173–196. https://doi.org/10.1146/annurev.psych.59.103006.093656.

Piazza, D. M. (1980). The influence of sex and handedness in the hemispheric specialization of verbal and nonverbal tasks. *Neuropsychologia, 18*(2), 163–176. https://doi.org/10.1016/0028-3932(80)90062-7.

Piccirilli, M., Finali, G., & Sciarma, T. (1989). Negative evidence of difference between right- and left- handers in interhemispheric transfer of information. *Neuropsychologia, 27*(7), 1023–1026. https://doi.org/10.1016/0028-3932(89)90079-1.

Poldrack, R. A., & Packard, M. G. (2003). Competition among multiple memory systems: Converging evidence from animal and human brain studies. *Neuropsychologia, 41*(3), 245–251.

Polich, J. (2007). Updating P300: An integrative theory of P3a and P3b. *Clinical Neurophysiology: Official Journal of the International Federation of Clinical Neurophysiology, 118*(10), 2128–2148. https://doi.org/10.1016/j.clinph.2007.04.019.

Polich, J., & Hoffman, L. D. (1998). P300 and handedness: On the possible contribution of corpus callosal size to ERPs. *Psychophysiology, 35*(5), 497–507.

Pontifex, M. B., Hillman, C. H., & Polich, J. (2009). Age, physical fitness, and attention: P3a and P3b. *Psychophysiology, 46*(2), 379–387. https://doi.org/10.1111/j.1469-8986.2008.00782.x.

Potter, S. M., & Graves, R. E. (1988). Is interhemispheric transfer related to handedness and gender? *Neuropsychologia, 26*(2), 319–325.

Qi, Z., Han, M., Garel, K., San Chen, E., & Gabrieli, J. D. E. (2015). White-matter structure in the right hemisphere predicts Mandarin Chinese learning success. *Journal of Neurolinguistics, 33*, 14–28. https://doi.org/10.1016/j.jneuroling.2014.08.004.

Ringo, J. L., Doty, R. W., Demeter, S., & Simard, P. Y. (1994). Time is of the essence: A conjecture that hemispheric specialization arises from interhemispheric conduction delay. *Cerebral Cortex (New York, N.Y.: 1991), 4*(4), 331–343.

Ross, D. S., & Bever, T. G. (2004). The time course for language acquisition in biologically distinct populations: Evidence from deaf individuals. *Brain and Language, 89*(1), 115–121. https://doi.org/10.1016/S0093-934X(03)00308-0.

Satz, P., Fennell, E., & Jones, M. B. (1969). Comments on: A model of the inheritance of handedness and cerebral dominance. *Neuropsychologia,* 7(1), 101–103. https://doi.org/10.1016/0028-3932(69)90049-9.

Scalf, P. E., Banich, M. T., Kramer, A. F., Narechania, K., & Simon, C. D. (2007). Double take: Parallel processing by the cerebral hemispheres reduces attentional blink. *Journal of Experimental Psychology: Human Perception and Performance, 33*(2), 298–329. https://doi.org/10.1037/0096-1523.33.2.298.

Searleman, A., Tweedy, J., & Springer, S. P. (1979). Interrelationships among subject variables believed to predict cerebral organization. *Brain and Language,* 7(3), 267–276. https://doi.org/10.1016/0093-934X(79)90023-3.

Seghier, M. L., Kherif, F., Josse, G., & Price, C. J. (2011). Regional and hemispheric determinants of language laterality: Implications for preoperative fMRI. *Human Brain Mapping, 32*(10), 1602–1614. https://doi.org/10.1002/hbm.21130.

Service, E., Helenius, P., Maury, S., & Salmelin, R. (2007). Localization of syntactic and semantic brain responses using magnetoencephalography. *Journal of Cognitive Neuroscience, 19*(7), 1193–1205. https://doi.org/10.1162/jocn.2007.19.7.1193.

Spencer, K. M., & Banich, M. T. (2005). Hemispheric biases and the control of visuospatial attention: An ERP study. *BMC Neuroscience, 6*(1), 51. https://doi.org/10.1186/1471-2202-6-51.

Spiegler, B. J., & Yeni-Komshian, G. H. (1983). Incidence of left-handed writing in a college population with reference to family patterns of hand preference. *Neuropsychologia, 21*(6), 651–659.

Springer, J. A., Binder, J. R., Hammeke, T. A., Swanson, S. J., Frost, J. A., Bellgowan, P. S., … Mueller, W. M. (1999). Language dominance in neurologically normal and epilepsy subjects: A functional MRI study. *Brain: a Journal of Neurology, 122*(Pt 11), 2033–2046.

Springer, S. P., & Searleman, A. (1978). The ontogeny of hemispheric specialization: Evidence from dichotic listening in twins. *Neuropsychologia, 16*(3), 269–281. https://doi.org/10.1016/0028-3932(78)90020-9.

Squire, L. R., & Wixted, J. T. (2011). The cognitive neuroscience of human memory since H.M. *Annual Review of Neuroscience, 34*, 259–288. https://doi.org/10.1146/annurev-neuro-061010-113720.

Stark, S. M., & Stark, C. E. L. (2016). Chapter 67 - introduction to memory. In G. Hickok, & S. L. Small (Eds.), *Neurobiology of language* (pp. 841–854). San Diego: Academic Press. Retrieved from https://www.sciencedirect.com/science/article/pii/B9780124077942000675.

Steinmetz, H., Volkmann, J., Jäncke, L., & Freund, H.-J. (1991). Anatomical left-right asymmetry of language-related temporal cortex is different in left- and right-handers. *Annals of Neurology, 29*(3), 315–319. https://doi.org/10.1002/ana.410290314.

Subirana, A. (1958). The prognosis in aphasia in relation to cerebral dominance and handedness. *Brain, 81*, 415.

Szaflarski, J. P., Binder, J. R., Possing, E. T., McKiernan, K. A., Ward, B. D., & Hammeke, T. A. (2002). Language lateralization in left-handed and ambidextrous people: fMRI data. *Neurology, 59*(2), 238–244.

Szaflarski, J. P., Holland, S. K., Schmithorst, V. J., & Byars, A. W. (2006). fMRI study of language lateralization in children and adults. *Human Brain Mapping, 27*(3), 202–212. https://doi.org/10.1002/hbm.20177.

Tanner, D., & Van Hell, J. G. (2014). ERPs reveal individual differences in morphosyntactic processing. *Neuropsychologia, 56*, 289–301. https://doi.org/10.1016/j.neuropsychologia.2014.02.002.

Townsend, D. J., Carrithers, C., & Bever, T. G. (2001). Familial handedness and access to words, meaning, and syntax during sentence comprehension. *Brain and Language, 78*(3), 308–331. https://doi.org/10.1006/brln.2001.2469.

Tyler, L. K., Shafto, M. A., Randall, B., Wright, P., Marslen-Wilson, W. D., & Stamatakis, E. A. (2010). Preserving syntactic processing across the adult Life span: The modulation of the frontotemporal language system in the context of age-related atrophy. *Cerebral Cortex, 20*(2), 352–364. https://doi.org/10.1093/cercor/bhp105.

Tzourio-Mazoyer, N., Petit, L., Razafimandimby, A., Crivello, F., Zago, L., Jobard, G., ... Mazoyer, B. (2010). Left hemisphere lateralization for language in right-handers is controlled in part by familial sinistrality, manual preference strength, and head size. *Journal of Neuroscience, 30*(40), 13314–13318. https://doi.org/10.1523/JNEUROSCI.2593-10.2010.

Tzourio-Mazoyer, N., Simon, G., Crivello, F., Jobard, G., Zago, L., Perchey, G., ... Mazoyer, B. (2010). Effect of familial sinistrality on planum temporale surface and brain tissue asymmetries. *Cerebral Cortex, 20*(6), 1476–1485. https://doi.org/10.1093/cercor/bhp209.

Ullman, M. T. (2001). The declarative/procedural model of lexicon and grammar. *Journal of Psycholinguistic Research, 30*(1), 37–69. https://doi.org/10.1023/A:1005204207369.

Ullman, M. T. (2004). Contributions of memory circuits to language: The declarative/procedural model. *Cognition, 92*(1), 231–270. https://doi.org/10.1016/j.cognition.2003.10.008.

Ullman, M. T. (2016). Chapter 76 - the declarative/procedural model: A neurobiological model of language learning, knowledge, and use. In G. Hickok, & S. L. Small (Eds.), *Neurobiology of language* (pp. 953–968). San Diego: Academic Press. Retrieved from https://www.sciencedirect.com/science/article/pii/B9780124077942000766.

Ullman, M. T., Grey, S., Lovelett, J. T., Gelfand, M. P., Litcofsky, K. A., Pullman, M. Y., ... Walenski, M. (2018). *The influence of handedness on language: Storage versus composition differences in left- and right-handers* (under revision).

Ullman, M. T., & Pullman, M. Y. (2015). A compensatory role for declarative memory in neurodevelopmental disorders. *Neuroscience and Biobehavioral Reviews, 51*, 205–222. https://doi.org/10.1016/j.neubiorev.2015.01.008.

Van Petten, C., & Luka, B. J. (2012). Prediction during language comprehension: Benefits, costs, and ERP components. *International Journal of Psychophysiology, 83*(2), 176–190. https://doi.org/10.1016/j.ijpsycho.2011.09.015.

Vernaleken, I., Weibrich, C., Siessmeier, T., Buchholz, H.-G., Rösch, F., Heinz, A., ... Gründer, G. (2007). Asymmetry in dopamine D(2/3) receptors of caudate nucleus is lost with age. *Neuroimage, 34*(3), 870–878. https://doi.org/10.1016/j.neuroimage.2006.10.013.

Vigneau, M., Beaucousin, V., Hervé, P.-Y., Jobard, G., Petit, L., Crivello, F., ... Tzourio-Mazoyer, N. (2011). What is right-hemisphere contribution to phonological, lexico-semantic, and sentence processing? Insights from a meta-analysis. *Neuroimage, 54*(1), 577–593. https://doi.org/10.1016/j.neuroimage.2010.07.036.

Weissman, D. H., & Banich, M. T. (2000). The cerebral hemispheres cooperate to perform complex but not simple tasks. *Neuropsychology, 14*(1), 41–59. https://doi.org/10.1037/0894-4105.14.1.41.

Weng, Y.- L., & Lee, C.- L. Reduced lateralization for syntactic category processing is associated with weaker interhemispheric inhibition: evidence from healthy young right-handers with familial sinistrality background. Submitted.

Witelson, S. F. (1989). Hand and sex differences in the isthmus and genu of the human corpus callosum. A postmortem morphological study. *Brain: A Journal of Neurology, 112*(Pt 3), 799–835.

Wong, P. C. M., Ettlinger, M., & Zheng, J. (2013). Linguistic grammar learning and DRD2-TAQ-IA polymorphism. *PLoS One, 8*(5). https://doi.org/10.1371/journal.pone.0064983.

Yeh, C., Yen, H.-H., Hsu, H.-J., & Lee, C.-L. (2018). Right hemisphere syntactic processing suppressed for native language is seen for a newly learned grammar: Evidence from event-related potentials. *Presented at the International Max Planck Research School for the language sciences conference.*

Yen, H.-H., Tang, L., Hsu, H. J., & Lee, C.-L. (2018). Finding syntactic regularities in the right brain: An event-related potential investigation of artificial grammar learning. *Presented at the 58th annual meeting of the Society for Psychophysiological Research.*

Zaidel, E. (1983). On multiple representations of the lexicon in the brain—the case of two hemispheres. In M. Studdert-Kennedy (Ed.), *Psychobiology of language*. Cambridge: MIT Press.

Zurif, E. B., & Bryden, M. P. (1969). Familial handedness and left-right differences in auditory and visual perception. *Neuropsychologia,* 7(2), 179–187. https://doi.org/10.1016/0028-3932(69)90015-3.

CHAPTER SEVEN

How Special Is Language Production? Perspectives From Monitoring and Control

Nazbanou Nozari*,§,[1]
*Department of Neurology, Johns Hopkins University, Baltimore, MD, United States
§Department of Cognitive Science, Johns Hopkins University, Baltimore, MD, United States
[1]Corresponding author. E-mail: nozari@jhu.edu

Contents

1. The Special Place of Language 180
   1.1 Functional Taxonomy of Cognitive Processes: Where Is Language? 180
   1.2 What Is a Module? 181
      1.2.1 Fast and Mandatory Processing, With Limited Central Access to Mental Representations *181*
      1.2.2 Informational Encapsulation *182*
      1.2.3 Domain-Specificity *184*
   1.3 Language Production in Fodor's Taxonomy 185
   1.4 Summary and a Way Forward 189
2. Language Production and the Need for Control 191
   2.1 Control at the Level of Word Production 192
   2.2 Control Beyond Single Word Production 194
3. Assessing the Need for Control: The Primary Job of a Monitoring System 195
   3.1 Production-External Monitoring 195
   3.2 Semi-Production-External Monitoring 196
   3.3 Production-Internal Monitoring 198
   3.4 Monitoring as Decision Making 199
4. Domain-Generality of Monitoring and Control in Language Production 202
   4.1 Domain-Generality of Computations 202
   4.2 Domain Generality of Neural Underpinning 203
   4.3 Domain Generality of Functional Adjustment in Control 204
5. Conclusion: How Special Is Language Production? 207
Acknowledgments 209
References 209

Abstract

In his seminal essay, "The Modularity of Mind" Fodor (1983), presents arguments in favor of language comprehension as a special module along with other input processing

Psychology of Learning and Motivation, Volume 68
ISSN 0079-7421
https://doi.org/10.1016/bs.plm.2018.08.006

systems. His view on language production is less clear. In this chapter, I first demonstrate that language production and comprehension are quite similar when evaluated in light of Fodor's criteria for modules: both meet a subset of those criteria in that their behavior resembles automatic processing; neither, however, is informationally encapsulated. This partial conformity with the criteria for specialized modules, leaves the question "How special is language production?" unanswered. I will then propose that this question can be answered by re-examining the origin of what resembles the behavior of an automatic system. I will argue that language production is, in fact, an efficiently monitored and controlled system, and that such monitoring and control mechanisms are shared between language production and other systems. These domain-general mechanisms, however, operate on domain-specific representations, creating specialized monitoring-control loops that can be selectively trained and selectively damaged.

1. THE SPECIAL PLACE OF LANGUAGE

As a uniquely human ability, language processing has always had a special place in cognition; so special in fact that it has been considered by many to be altogether separate from cognition. Most of the philosophical arguments, as well as empirical research, on the special nature of language processing has concerned language *comprehension*, though. My goal in this chapter is to evaluate how special language *production* is with regard to the rest of cognition. I will start with an overview of Fodor's seminal essay, "The Modularity of Mind" (Fodor, 1983), to introduce a theoretical framework within which it would be possible assess how "special" a cognitive system is, and also to apply the same framework that has been employed to evaluate the specialness of language comprehension to language production. In doing so, I will re-evaluate some of Fodor's assumptions and claims, and discuss a possible extension of his views to language production. I will then evaluate those views in light of the evidence from empirical studies of language production.

1.1 Functional Taxonomy of Cognitive Processes: Where Is Language?

Fodor recognizes two general families of cognitive systems: *modules* and *central processes*. This taxonomy is function-based: the function of modules is input analysis, whereas the function of central processes is the fixation of belief. In Fodor's view, what we believe depends on the input we receive plus information about how reliable that input is, but how we process an input is largely insensitive to what we believe. Where does language fall in this taxonomy? There are two options: (1) to group language with central processes of thought, and contrast that with perceptual processes, e.g., vision, audition, etc. (2) To group language with perceptual systems, consider them all input systems, and contrast them with central processes such as thought and problem solving.

Since the representational nature of language and its similarity to thought in that sense can hardly be denied, there would have to be a strong motivation for choosing the second classification. Fodor argues that such motivation exists. Language comprehension and other perceptual processes form a natural kind, he argues, because they have something fundamental in common: they are all modules.

1.2 What Is a Module?

According to Fodor, a system can be considered modular if it has all (or most) of the key characteristics that follow. Modular systems (1) are fast and mandatory and allow only limited central access to mental representations, i.e., they meet the criteria for automaticity; (2) are informationally encapsulated; and (3) are domain-specific.

1.2.1 Fast and Mandatory Processing, With Limited Central Access to Mental Representations

These three properties were discussed separately by Fodor, but I have grouped them together as all of them are indicators of the "automaticity" of a system. Fodor argues that language comprehension is mandatory because we cannot help but understand the meaning of what is spoken in our native language, even when we do not intend to listen to a conversation or when we are instructed to focus on the acoustic properties of the input instead of its meaning (Marslen-Wilson & Tyler, 1981). On this ground, Fodor has compared language comprehension to a reflex, i.e., a system with no reliance on cognitive control operations. Our first interpretation of a sentence may not always be correct, but Fodor's critical point is that at least some meaning is always extracted online, quickly, and independent of listeners' intentions. This position is compatible with evidence showing that listeners quickly fixate not only the picture whose name they just heard (e.g., lock), but also semantically-related pictures (e.g., key), showing quick and involuntary access to the (extended) semantics of heard words (Yee & Sedivy, 2006).

By "limited central access to mental representations", Fodor means that conscious awareness of intermediate cognitive operations is limited. For example, upon hearing an utterance, a host of phonological, lexical, syntactic, and semantic processes are at work, but we are only aware of the final result: comprehension of the meaning of the sentence. We do not have access to the acoustic features of the words we heard, and we cannot accurately report how many lexical items or syntactic trees have been activated during the parsing of the sentence. This lack of awareness is a desirable property for a cognitive system, because, as Fodor explains, "the world is, in general, considerably more stable than are its projections onto the surfaces of transducers. Constancies correct for this, so that in general percepts correspond to distal layouts better than proximal stimuli do. But, of course, the work of the constancies would be undone unless the central systems which run behavior were required largely to ignore the representations which encode unconnected proximal information." (Fodor, 1983, p. 60).

In summary, language comprehension is fast, mandatory and mostly without conscious awareness of the intermediate processes which map linguistic input to meaning. These three properties make language comprehension a reasonable candidate for an automatic process.

1.2.2 Informational Encapsulation

When is a system informationally encapsulated? In a nutshell, when it does not have access to information from other systems. The tricky question here is what constitutes "other systems". For example, Fodor does not see the McGurk effect (perceiving the same auditory stimulus as different speech sounds depending on the accompanying visual information from mouth movements) as evidence of the influence of information from one system (visual) on another (auditory), because the phenomenon is contained within the "language domain". Similarly, he dismisses evidence of top-down processing within the same system as evidence of the penetrability of modules. Consider the phoneme restoration effect (Warren, 1970): listeners "perceive" phonemes that have been spliced out of familiar words and replaced with noises like coughs. For example, even though they never actually hear an "s" in/legi(cough)lature/, they report having heard "legislature" with a cough in the background. This finding has been taken as evidence of top-down feedback from lexical items to lower-level representations like phonemes. Fodor argues that even if this is true—a claim he disputes on the grounds that guessing is an equally valid explanation—it would not constitute evidence against modularity, because both lexical items and phonemes are part of the language module. In later sections when drawing parallels between comprehension and production, I will discuss evidence showing that lexical and phonological domains are governed by very different principles and operations. Thus evidence of an interaction between the two speaks against their modularity, although such evidence can only speak to the issue within the language domain. But for now, let us ask what kind of evidence Fodor would accept as evidence against the modularity of language processing.

An example of such evidence would be syntactic parsing guided by semantic context or real-world background (Fodor, 1983, p. 77). A large body of research in the '80s and '90s addressed exactly this question. Against a theory of encapsulated syntactic processing—which claims that the first-pass parsing of incoming utterances is purely syntactic and impermeable to other sources of information (e.g., Ferreira & Clifton, 1986; Rayner, Carlson, & Frazier, 1983) — Tanenhaus, Spivey-Knowlton, Eberhard, and Sedivy (1995) showed that even the earliest moments of sentence processing were sensitive to the visual context. They presented participants with sentences such as "Put the apple on the towel in the box" and recorded their eye movements while they viewed a scene with either one or two apples. When a single apple was present, "on the towel" was often interpreted as the destination, a preference that proponents of syntactic encapsulation attributed to syntactic simplicity and the requirements of the argument structure of the verb "put". When, on the other hand, the scene contained two apples (only one of which was on a towel), "on the towel" was interpreted as modifying the noun "apple". The difference between the processing of the one- vs. two-referent scene, which was captured in early looks to objects, cannot be explained by a syntactic module that is insensitive to information provided by other systems. Later studies demonstrated the early influence of non-syntactic information from a broader range of factors such as referential domains constrained by pragmatic factors (e.g., Brown-Schmidt & Tanenhaus, 2008), further challenging the view of language comprehension as an informationally-encapsulated system.

It is important to point out that Fodor had a good reason to insist on modular input analyzers: top-down knowledge cannot easily override perceptual illusions such as the Müller-Lyer illusion or the phoneme restoration effect. But this would only be a problem if one assumes that top-down feedback is stronger than bottom-up input. If, on the other hand, one views input analysis as primarily bottom-up, with some constraints induced by top-down processing, it is possible to accommodate the seemingly contradictory findings. There is some data to support this: we played sentences such as "She will write/see with the sharp pencil."

while participants viewed black and white drawings of a pencil (target), a pen (verb competitor), a syringe (adjective competitor), and an unrelated item that was neither sharp nor used for writing. Results showed that, upon hearing the word "sharp", there were significantly reduced looks to the syringe when the verb was "write" compared to "see". This finding implies that top down processing (in this case the semantic constraints introduced by the verb) had affected the processing of the bottom-up information carried by the word "sharp". Syringes are not used for writing, thus a syringe is not a plausible referent for a sharp object that one writes with. Importantly, however, hearing the verb "write" did not eliminate fixations to the syringe upon hearing "sharp"; there were still significantly more fixations to the syringe than the unrelated distractor, showing that the influence of top down constraint on processing the bottom-up input was only partial (Nozari, Trueswell, & Thompson-Schill, 2016).

In summary, the empirical evidence suggests that language comprehension is not informationally-encapsulated, although top-down feedback does not completely override the influence of bottom-up information.

1.2.3 Domain-Specificity

While the '80s and '90s were the years of intense research on informational encapsulation (i.e., interactivity), the emergence of sophisticated neuroimaging methods shifted the focus to tests of domain-specificity in the late '90s and 2000s. Given its strong ties to cognitive neuroscience, the main question of this line of research is often how specialized certain parts of the brain are for certain operations. Fodor's view of domain-specificity, however, was not primarily neural. For reasons I will discuss in later sections, I believe that it is useful to preserve this distinction. So let us focus for now on Fodor's line of argument for the domain-specificity of language.

The best place to start is to recall what domain-specificity is *not*. Nobody disputes the fact that you cannot hear with your eye or see with your ear. It thus follows trivially that the mechanisms of auditory and visual perception operate on special classes of input. But it does not tell us anything about the nature of such mechanisms or whether they are similar or different. In other words, differences in the physical nature of the input, per se, are not sufficient to argue for the domain-specificity of the operations that process that input. This, however, does not imply that the nature of input is immaterial to the evaluation of domain specificity. In fact, Fodor's argument regarding domain-specificity hinges directly on the nature of the stimulus, but his emphasis is on the complexity of the information required for analyzing the stimuli (as opposed to its physical nature). If such analyses require highly specialized information within a domain, then that domain is "eccentric". Since there are strong arguments for the existence of language universals (e.g., Chomsky, 1975; Pinker & Jackendoff, 2005; cf. Evans & Levinson, 2009) that prevail across human languages and are absent from non-linguistic domains, Fodor concludes that the language domain is highly eccentric.

The argument for domain-specificity is that eccentric domains require specialized operations, and the degree of this specialization is commensurate with the degree of the eccentricity of the domain. But Fodor admits that this inference is loose. While it makes sense to conclude that the existence of specialized operations implies the existence of domain-eccentricity, it does *not* follow that the existence of domain-eccentric information implies the existence of specialized domain-specific operations. The latter must be proven independently. Fodor's solution to this problem is to independently assess the operations of the language faculty in light of the other criteria for modularity. If it is confirmed that language processing fulfills those criteria, this, combined with the eccentricity of the language domain, would enable one to conclude that language processing is domain-specific. As reviewed above, however, language comprehension only meets some of the criteria for modularity.

1.3 Language Production in Fodor's Taxonomy

As mentioned above, the bulk of Fodor's arguments concern language comprehension. He is conspicuously silent on language production. In so far as language concerns motor acts of speech, Fodor states that "It would please me if the kinds of arguments that I shall give for the modularity of input systems proved to have application to motor systems as well. But I don't propose to investigate that possibility here." (Fodor, 1983, p. 42). However, there is good reason to believe that Fodor did not view language production as reducible to a motor module, or believed that the same kind of simple processes that carry out comprehension are successful in production. For example, while defending the utility of automatic associative processing in input models (e.g., the lock-key example discussed earlier), Fodor resists the extension of associative processing to "belief" and "sentence production", arguing that they are products of "judgment" and "planning", respectively, rather than association (Fodor, 1983, p. 81). This and a few other brief allusions to the nature of language production as communicating thoughts and beliefs imply that Fodor viewed language production as better aligned with "thought" (i.e., central processes) than "perception". Similarly Levelt (1989), who views many of the production operations past the level of message generation as automatic, maintains that the first step of language production, i.e., the conceptualization of a message, is a highly-controlled process, far from what can be expected from an automatic module.

While it is understandable why language production might be an ill fit for the "perceptual" category, grouping it with "thought" implies drawing a sharp dichotomy between language production and comprehension. Fodor embraces this dichotomy: in one of his few mentions of language production, he reasons that since communicating one's views through speech requires access to memory and thought, language production cannot be as domain-specific as language comprehension. Differences between language comprehension and production notwithstanding, I will argue in this section that drawing such a sharp distinction between the two, in terms of domain specificity or modularity, is contradictory even to Fodor's own views. The core of the argument is that, when examined carefully, language production and comprehension meet (and fail to meet) the same criteria for modularity.

First, in terms of domain-eccentricity, the exact same universals that apply to the perception of language apply to the production of language: native speakers in their native environment do not apply a different set of syntactic rules to what they produce and what they hear. Moreover, what is perceived during comprehension directly affects what is produced: speakers are prone to producing the same syntactic structure that they have just perceived (Bock, 1986; Pickering & Branigan, 1998). The same is true for the lower levels, acoustic features for perception and articulatory-motor features for production. Even though comprehension and production use different representations at this level, they are still governed by the same rules. For example, in a language like Japanese which does not distinguish between/r/and/l/, both perception and production of these phonemes are affected. Perception-production loops also work at this stage: production of lexical items is altered and shaped by the speaker's perception of those items on prior occasions (Pierrehumbert, 2002). Finally, neuropsychological evidence points to a common semantic system for language comprehension and production. Individuals with semantic dementia do not just lose the ability to name a horse; they can no longer understand what the word "horse" means (Hodges, Patterson, Oxbury, & Funnell, 1992). Given this overlap between the representations in language comprehension and production, their influence on one another, and the identical rules that govern them, it would be strange to consider one an eccentric domain but not the other.

Next, let us turn to Fodor's other criteria for modularity and compare comprehension and production on those. I begin by reviewing the three criteria for automaticity: fast, mandatory and mostly subconscious processing. If we consider comprehension fast, then we must acknowledge that production is also fast, because the two are intimately tied (the normal speaking rate of a neurotypical adult speaker is 2–3 words/second, despite the large number of processes to be completed during sentence production). To argue for the mandatory nature of language comprehension, Fodor appeals to the common phenomenon of involuntary overhearing of others' speech. One could make the same argument for production by appealing to the many cases where speakers blurt out something they should not have. Documentation of (in)famous embarrassing speech errors is evidence of the ballistic nature of the language production system. But one might argue that overhearing is much more common than producing speech errors (about 1 in every 1000 words; Meyer, 1992). Note, however, that the goal is not to claim that comprehension and production are *equally* automatic; rather, that if mandatory processing without control is taken as evidence for automaticity, then such evidence exists in both comprehension and production. Finally, in terms of limited central access to mental representations, production is strikingly similar to comprehension. Speakers are aware of their thoughts and their utterances and little in between. In fact, this lack of awareness of intermediate representations and processes has been taken as evidence to argue for the automaticity of language production past the stage of conceptualization (Levelt, 1989). Language production, then, meets the same criteria for automaticity as language comprehension.

In terms of informational encapsulation, a large body of work has investigated the presence of feedback between the layers of the production system. In contrast to a strictly feed-forward model of language production (e.g., Levelt, Meyer, & Roelofs, 1999), interactive models of word production (e.g., Dell, 1986; Rapp & Goldrick, 2000) posit feedback from lower levels (e.g., phonemes) to higher levels (e.g., words) in the production system. Evidence for such feedback comes from certain patterns of speech errors: for example, phonological errors (e.g., "cap" instead of "cat") are more likely to produce a word than a nonword (e.g., Baars, Motley, & MacKay, 1975; Nozari & Dell, 2009). Similarly, phonological exchanges between the consonants of two words are more likely when the words share a vowel (e.g., "mad back" → "bad mack") compared to when they do not ("mad bike" → "bad mike") both in speaking (Dell, 1986) and in typing (Pinet & Nozari, 2018). In both cases, the effect is observed because of the feedback from segments (phonemes, graphemes, etc.) to lexical items. The overlapping segments (e.g., /k/, /æ/) help activate the competing lexical item (e.g., "cap", "bad"). These items in turn activate their segments, giving them an advantage over other segments that do not enjoy the support of lexical items which have been activated through feedback.

More recently, we have shown further evidence of feedback between segments and lexical items through incremental learning. Producing words in the context of segmentally-related words leads to interference in both spoken and written production (Breining, Nozari, & Rapp, 2016; Nozari, Freund, Breining, Rapp, & Gordon, 2016). In a nutshell, this is why: when "cat" is the target, feedback from its segments activates the competitor "cap", which in turn activates its segment/p/. Once the production of "cat" is completed, incremental learning (e.g., Oppenheim, Dell, & Schwartz, 2010) strengthens the connections between the target segments and their supporting lexical items, while weakening the connections between the non-target segments and their corresponding lexical items. Thus the connection between/p/and "cap" is weakened. This weakening puts "cap" at a disadvantage in the next trial when it becomes the target. This mechanism depends critically on the feedback from segments to words and is also observed during learning of new vocabulary (Breining, Nozari, & Rapp, 2018).

As noted when discussing informational encapsulation in the context of comprehension, Fodor does not accept interactivity within the language system as evidence against encapsulation. But refuting interaction between layers of the production system as evidence against informational encapsulation implies that those layers are not specific enough to be considered special domains by themselves. There is, however, converging evidence pointing to separate systems for lexical and segmental processing in language production. For example, neuropsychological studies of individuals with aphasia have shown the possibility of damage to lexical retrieval processes without impaired segmental encoding and vice versa (Dell, Schwartz, Martin, Saffran, & Gagnon, 1997; Dell, Schwartz, Nozari, Faseyitan, & Coslett, 2013). More importantly, viewing lexical and segmental processing as belonging to the same domain violates Fodor's own criterion for eccentric domains, as these two levels of processing are governed by very different rules. Evidence from speech errors show that lexical (but not segmental) errors tend to preserve their syntactic categories (Garrett, 1975). For example, lexical exchanges on the target "I put the napkin on the table" are much more likely to sound like "I put the *table* on the *napkin*", than "I *napkin* the *put* on the table". However, segmental exchanges are not constrained by syntax. "I nut the papkin on the table" is not far less likely than "I put the tapkin on the nable". Instead, segmental exchanges are more likely to happen over short (e.g., within-phrase) than long distances, compared to lexical exchanges (Garrett, 1975).

These rules are not specific to English (e.g., Del Viso, Igoa, & García-Albea, 1991), and are precisely the kinds of "universals" that Fodor claims make a domain eccentric. If we accept this premise, then lexical and segmental processing are different domains, and the evidence of feedback between them should be taken as evidence against informational encapsulation, at least as far as various parts of the production system are concerned. To take matters one step further, language production is also influenced by extra-linguistic factors. For example, speakers' syntactic choices can be manipulated by cueing objects in the visual scene (Nozari & Omaki, 2018), even when those cues are presented for only 60–80 ms, and none of the participants are aware of them (Gleitman, January, Nappa, & Trueswell, 2007). Moreover, speakers formulate their utterances at least partly based on their addressees' perspectives (e.g., Yoon, Koh, & Brown-Schmidt, 2012). These findings are equivalent of those reported by Tanenhaus et al. (1995) in comprehension: information from the non-linguistic (visual) domain directly influences how a sentence is formulated (in production) or parsed (in comprehension) even though the linguistic information has not changed. It thus seems reasonable to conclude that language production is not informationally encapsulated either with regard to its own sub-systems or with regard to other cognitive systems.

1.4 Summary and a Way Forward

To summarize, if we evaluate the modularity of language production by the same Fodorian criteria as have been applied to language comprehension, the two have striking similarities. Both are fast and show some degree of involuntary processing. Neither allows central access to intermediate mental representations. Both are highly eccentric domains (perhaps with eccentric subdomains) and the evidence suggests that both are permeable to information from other systems. From this, I draw two main conclusions: (1) given the similarities between language comprehension and production, placing them on opposite sides of Fodor's cognitive taxonomy is likely to hinder our understanding of the relationship between language and other cognitive processes. (2) The evidence I reviewed suggests that language processing meets some, but not all, of Fodor's criteria for modularity. The question thus remains: How special is language processing really?

In the remainder of this chapter, I will try to answer this question for language production. The choice of production—and not comprehension—is mainly motivated by the fact that a considerably smaller body of theoretical and empirical work in psycholinguistics has focused on how we produce—as opposed to comprehend—speech. This is, in large part, due to the much greater methodological challenges involved in studying language production: one can have perfect control over the input in order to achieve good experimental control in studies of comprehension, but it is substantially more difficult to control the thoughts and ideas behind speech planning. Methodological issues aside, the above review showed many commonalities between comprehension and production, so the main approach that I will take here can be (and in some cases has been) extended to comprehension. At the same time, Fodor is absolutely correct that production does differ from comprehension in one major respect: it comes from one's own thoughts. Therefore the empirical findings from language comprehension research cannot be readily extended to production. Production must be studied on its own, with the acknowledgment that production and comprehension are intimately connected processes.

Since Fodor's placement of language production in his cognitive taxonomy proved problematic, the first step is to redefine a framework within which the "specialness" of production can be assessed. We can do so while staying faithful to Fodor's general idea: while we agree that language is an eccentric domain, we can ask: Are its operations also specialized to this domain, or are they shared with other domains? If the cognitive processes that mediate language production are specific to the production system, then language production can be considered a special system separate from other cognitive systems. If, on the other hand, these processes are shared with other cognitive systems, then language production is simply part of domain-general cognition, albeit a highly sophisticated part.

The question now is which processes are reasonable targets for such a study. If we choose to focus on operations that are closely tied to language-specific representations, we will soon come to an inevitable dead-end. For example, the sensitivity of lexical errors to grammatical class shows that knowledge of syntactic categories limits lexical selection, e.g., "I put the ——" only accepts a lexical item from the "noun" category. We can now argue that, since the concept of a "noun" is undefined outside of the linguistic domain, then, by definition, a noun-activating mechanism must be specific to language. This is unhelpful. But we can still target the same process from a slightly different angle: when, for example, "table" and "napkin" are both viable candidates for selection, what kinds of processes determine which one is produced? Do these processes have anything in common with processes that adjudicate between response alternatives in other cognitive systems, regardless of the nature of those responses? Are the principles behind detecting competition, resolving competition, and learning to better regulate the system based on this competition shared between language production and other systems?

These questions fall under the general framework of what I call "monitoring and control" in language production, and which I argue provides the soundest test of the "specialness" of the language production system. An obvious objection might be that such processes are "external" to language production, and as such cannot provide insights into the operations of the language production system per se. For example, one could imagine an external monitor that only peeks at the final output of the production system and issues a pass/fail decision. Presumably such a monitor would not reveal much about the internal dynamics of the production system, and would hardly be useful in evaluating how special its internal operations are—or are not—compared to other systems. However, in what follows, I will argue that such a view of monitoring and control in language production is wrong. Monitoring and control are intimately tied to production processes and are guided, in large part, by information generated within the production system itself. The critical question is whether they also have components shared with other systems or not.

2. LANGUAGE PRODUCTION AND THE NEED FOR CONTROL

Earlier I argued that language production is fast, to some degree ballistic, and that speakers are deprived of access to its intermediate representations. In addition to these properties, language production is extremely well practiced, at least by the time one reaches adulthood. These characteristics make production a good candidate for an automatic process (Levelt, 1983), i.e., a process that does not require cognitive control past the level of conceptualization. If this is in fact true, the whole framework I have proposed is invalid. But, before accepting this, it is important to note that the tasks in which automaticity has been claimed based on the above-mentioned criteria often consist of a repetitive procedure with only small deviations from the task routine. Even a complex task like driving from one's home to work involves following the same route over the same roads and turns that one drives every day in the same car, more or less in the same traffic and with the same speed. On the other hand, we almost never repeat the exact same set of sentences in the same order to the same person on two occasions. The highly generative nature of language production invites caution in evaluating its automaticity according to the same criteria that have been applied to tasks that are not nearly as creative. But if production is not automatic, how is it so fast, fluent, and relatively error-free? I will argue below that the key is an efficient monitoring-control loop that constantly evaluates the need for control and deploys the control necessary to keep performance optimal.

2.1 Control at the Level of Word Production

To illustrate why producing a word may require cognitive control, I will present a brief overview of the word production system. While details differ, the backbone of the architecture that I review here is shared between several major computational models of word production (Dell, 1986; Levelt, Meyer, & Roelofs, 1999; Rapp & Goldrick, 2000). It is now accepted that producing a word (i.e., getting from meaning to sound) requires at least two stages of processing: in the first stage, semantic features (e.g., animal, four-legged, pet, meows, etc.) must be mapped onto lexical items (e.g., "cat"; Fig. 1). Because semantic features are shared between several items, this process often activates not only the target, but also semantically-similar lexical items (e.g., "dog"). In systems with cascading and feedback (e.g., Dell, 1986; Rapp & Goldrick, 2000), the activation from lexical items "trickles down" to segments (phonemes in spoken and graphemes in written production), and those segments, in turn, send activation back to the corresponding lexical items. Whether the system has feedback or not, at some point, one of the activated representations at the lexical level is selected for further processing. This marks the end of the first stage of word production.

In the second stage, the selected lexical item is mapped onto its segments. In systems that allow for feedback, activation of segments leads to the activation of segmentally-related words (e.g., "cap"), which in turn activate their own segments to compete with the target's segments. This step also ends with a selection process, which selects segments for their proper positions (segmental encoding). The outcome of this process is passed down to

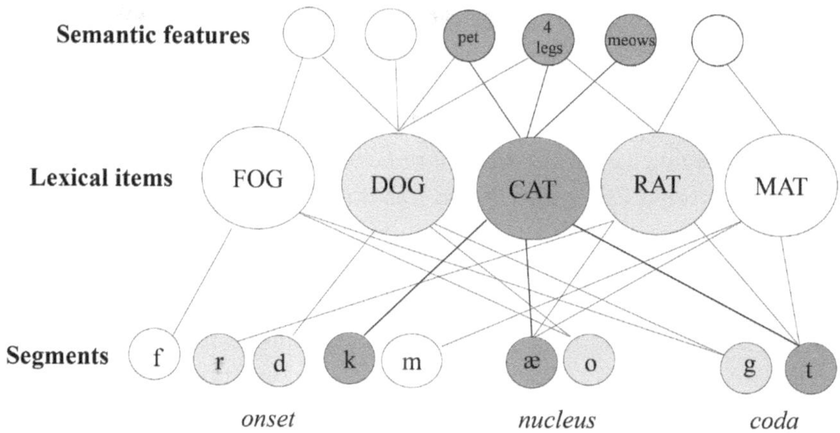

Figure 1 A schematic of the two-step architecture of word production before articulatory-phonetic encoding (e.g., Dell, 1986). The first stage ends by selecting one of the activated lexical items. The second stage ends by selecting one of the segments in each position.

the articulators for articulatory-phonetic encoding, and ultimately overt production.

The need for control arises because, in each stage, the spreading activation processes activate multiple lexical items (cat, dog, etc.) and multiple segments (/t/,/p/, etc.) in each position, but only one representation can be selected at each level (see Fig. 1). Empirical evidence shows that greater activation of competitors makes production more difficult by increasing error rates and/or response latencies: naming pictures is harder in the presence of competing words (e.g., Schriefers, Meyer, & Levelt, 1990) and in the context of other semantically-related (e.g., Belke, Meyer, & Damian, 2005; Schnur et al., 2009) or segmentally-related (Breining et al., 2018, 2016; Nozari et al., 2016) pictures, at least during tasks that elicit repeated naming of a small set of pictures. It is thus reasonable to hypothesize that a control operation which suppresses competing representations in favor of a single response, i.e., inhibitory control, might be at work in word production.

Consistent with this prediction Shao, Roelofs and Meyer (2013), reported a correlation between response latencies in naming pictures in the presence of distractors and participants' inhibitory control abilities as measured by the stop-signal task. Importantly Shao, Roelofs, and Meyer (2012), also reported a correlation between measures of inhibitory control and naming pictures of objects and actions in the absence of overt distractors, demonstrating that the need for control was not specific to cases where competition was enhanced. The correlational nature of this evidence from neurotypical speakers might cast doubt on whether inhibitory control is essential to word production; perhaps it simply helps speed up production through strategic processes that are not really necessary for word production. I recently reported a case study of an individual with aphasia who produced many semantically-related errors, in quick succession, while naming a picture (e.g., "peach, orange, no peach, pineapple!" in response to a picture of a watermelon). Using a battery of linguistic/cognitive tasks I showed that the individual had intact semantic, lexical, and phonological representations and was able to activate lexical items from semantic features. He was, however, selectively impaired on tasks of inhibitory control (e.g., the Simon task), and his performance quickly deteriorated under conditions that increased the activation of competitors (Nozari, 2018). This case provides evidence that selective impairment of inhibitory control creates problems during lexical selection, which in turn implies that selection processes are dependent on inhibitory control.

2.2 Control Beyond Single Word Production

Since conversations are made of sentences and sentences of words, it would not be surprising to find that the need for inhibitory control at the level of word production extends to sentence production and ultimately conversation. Therefore, instead of focusing on what has to be logically true, I will discuss some evidence for the need for control in operations that are at work during sentence production and conversation but are not part of single-word production. Producing sentences consists not only of retrieving words, but also of keeping track of dependencies between different parts. For example, in English—and many other languages—the verb must agree in number with its subject. If the sentence has more than one noun phrase, as in "The lion next to the green snakes ...", sometimes speakers mistakenly produce a verb that agrees with the local noun (snakes) instead of the subject (lion). This phenomenon is called "agreement attraction" (Bock & Miller, 1991), and is most common when a singular subject noun is competing against a plural local noun. We recently showed that increasing the demand for control during the assignment of thematic roles increases the chance of attraction errors (Nozari & Omaki, 2018). Moreover, analysis of individual differences using path modeling showed that individuals with better inhibitory control abilities were less likely to produce attraction errors. At the level of conversations, beyond producing words and sentences, speakers must also tailor their utterances to the referential context and to their listeners' perspectives. Analysis of individual differences has shown that both of these are predicted by speakers' inhibitory control abilities (Trude & Nozari, 2017; Wardlow, 2013).

3. ASSESSING THE NEED FOR CONTROL: THE PRIMARY JOB OF A MONITORING SYSTEM

The findings I reviewed above suggest that inhibitory control plays a role at all levels of language production, from producing single words to conversations. This evidence argues against a strong notion of automaticity in language production, and calls for mechanisms to assess the need for control. Recall that it is the nature of such mechanisms that can shed light on the "specialness" of language production. Production would not be so special if the regulation of its primary processes such as lexical selection depended on the interaction between the production system and other (domain-general) systems which regulate similar selection processes in the non-linguistic domain. This regulation is the job of the monitoring system. Therefore, to understand how special the production system is, we must adjudicate between different accounts of monitoring in language production.

3.1 Production-External Monitoring

The classic view of a speech monitor is a device that captures errors (Levelt, 1983; 1989). Levelt (1983) tasks the comprehension system with this job: a "perceptual loop" monitors the output of the production system (e.g. "cap"), compares it to an intended target ("cat"), and concludes that the speaker has made an error. This proposal is quite elegant, as it proposes that error detection for both self- and other-produced speech is carried out using the comprehension system. If true, such a monitoring mechanism is perfectly compatible with a modular production system, as it operates outside of the production domain and only examines production output once the primary production processes have been completed. However, neuropsychological data suggest a dissociation between monitoring one's own speech and that of others, as well as a double dissociation between comprehension abilities and detecting one's errors (see Nozari, Dell, & Schwartz, 2011 for a review of this evidence). This is not to imply that comprehension plays no role in monitoring. Altered auditory feedback indeed changes speaker's perception of what they have produced (Lind, Hall, Breidegard, Balkenius, & Johansson, 2014). The critical question is whether the comprehension system, by itself, is an efficient enough monitor.

Defining the job of a monitor is a key point here. If one narrows monitoring to the detection of errors, especially after they emerge, the comprehension system is not a bad candidate. But I set up this chapter to show the reader that production monitoring goes far beyond error detection. The design and functioning of the production system, i.e., the interconnected nature of the representations along with the activation of multiple alternatives at each stage of processing, might be expected to make production very error-prone. However, production errors are rare in neurotypical adult speakers because production is efficiently controlled at various stages. Such control can only be deployed efficiently if there is a monitoring mechanism that constantly assesses the need for control. Viewed this way, the primary job of a monitoring system is to *prevent* errors from happening. Catching errors which have already happened is a much smaller part of the job.

In order to achieve this goal, the monitor must work closely in parallel with primary production processes. It thus has to be fast and not require much conscious effort, or it would make production slow, effortful, and disfluent. Importantly, it must have access to various stages of production because, as discussed above, the need for control arises at different levels. A monitor that only has access to the output of the production process is simply too slow to do much good. If we want to accept comprehension as the monitor for the production system, we must define a comprehension system that has access to intermediate representations in production. But as Fodor pointed out, speakers do not have access to the intermediate linguistic processes, so this access must be subconscious. Moreover, the comprehension system should be able to perform a comparison between the representation that is about to be selected and the target, while this comparison remains subconscious. One could try to redefine comprehension to accommodate this definition, but it would be a different process than assumed by the perceptual loop which is "conscious and attentional" and "more or less ... deliberate" (Postma, 2000, p. 115).

3.2 Semi-Production-External Monitoring

A fully production-external monitor that evaluates the intermediate representations of the production system on a constant basis is not very plausible on both theoretical and empirical grounds that I discussed in the previous section. There is, however, one kind of monitoring that can be achieved through an interaction between the production and perception systems (hence I refer to it as semi-production-external) in a fast unconscious manner. Hickok (2012) defines such a monitor as follows: activation of semantic representations during conceptualization of a message activates both production and perception processes. At lower levels, the production and perception systems' representations diverge: we have articulatory-motor representations in production and acoustic representations in perception. Hickok argues that, under normal circumstances, as the motor plan for "cat" gains activation, the perceptual representation of "cat" is gradually suppressed via inhibitory connections. If, however, an erroneous word (e.g., "dog") is sent to the articulator, its motor representation cannot suppress the perceptual representation of "cat", simply because the two are not connected in the system. This leads to persistent activation of the perceptual representation of "cat" which is read as an error alarm by the system.

This account is plausible for several reasons: (a) articulatory-phonetic and acoustic representations are distinct and most likely connected (we can learn to produce /k/ by listening to someone else's production of /k/), (b) perceptual representations in the auditory cortex are activated quickly during production (Tian & Poeppel, 2010), and (c) similar mechanisms are at work in other domains where motor actions have perceptual consequences. But precisely the same reasons that make this kind of monitor a viable model for monitoring motor aspects of speech make it much less of a viable candidate for monitoring at higher levels of language production. There has been much interest in recent years in extending this model (sometimes referred to as "forward model" or "internal model") to monitoring at lexical and syntactic levels of production (e.g., Pickering & Garrod, 2013). Such a model, however, must assume the existence of *two* sets of such representations (i.e., different lexical nodes for production and one for comprehension, equivalent to motor and perceptual representations) and their simultaneous activation during production. I am not aware of any theoretical arguments that motivate double sets of representations at the higher cognitive levels, or of empirical evidence showing that such double sets exist.[a] This is, of course, not proof that they do not exist, but for such a monitoring model to be proposed as a viable alternative at that level, verification of its basic

[a] Note that by "representation" here, I am referring to the nodes (and not the connections) in a network. While the body of knowledge is not stored only in nodes, but also in the connections between nodes that comprise a network, the concept of activation —and inhibition of activation— applies to nodes. Thus a model such as a forward model would require two sets of nodes, and not simply separate connections. It is this duplication of nodes as representations that I question.

assumptions is a necessary first step. Thus, to the degree that control is required at pre-articulatory levels, it is unclear whether this kind of monitoring mechanism provides the right answer.

3.3 Production-Internal Monitoring

Theoretically speaking, the system with the fastest and most complete access to information in the production system is the production system. One should then at least consider the possibility that the production system has a strong hand in monitoring itself. This idea has been floating around since the early '80s (e.g., De Smedt & Kempen, 1987; Laver, 1980; MacKay, 1992; Schlenck, Huber, & Willmes, 1987). The original proposals (e.g., Laver, 1980) still viewed monitoring as a comparative process, in which the produced response was compared to the target response. When situated within a production-internal model, such a comparative process would require the target to have been available in the production system, in which case one might wonder why the correct target was not produced in the first place. A major theoretical development was the proposal of non-comparative monitoring. For example MacKay (1992), proposed that monitoring can still be successful without the monitor having access to the right response. Instead, it would detect a new pattern (e.g., of activation in the network), one not previously experienced by the speaker, as an error signal. This perspective had a groundbreaking implication: it viewed conscious awareness of an error as a *consequence* (and not a precursor) of detecting the error. When the system falls into an unusual state, it triggers an error message which signals to the speaker that something is not right.

The most recent proposal for a production-based monitor is the *conflict-based monitor* (Hanley, Cortis, Budd, & Nozari, 2016; Nozari et al., 2011), which was motivated by a number of properties I have discussed in earlier sections: (1) It is parsimonious in the sense that it uses the information already generated by primary production processes, so no assumptions about a new set of representations are required. Using production-internal information also allows for fast monitoring at all levels of production without privileged access to intermediate representations by an external system. (2) It in non-comparative, i.e., it is blind to the "right response", thus avoiding the problem of producing an error despite having the correct response already at hand. Instead, it uses a certain pattern of information about the state of the production system to deduce the need for additional control. (3) The information it employs is precisely the information needed to deploy control to the production system. When reviewing the need for control at multiple levels of the production system, I pointed out that one of the major reasons—perhaps *the* major reason—for the necessity of control in production is competition between alternative representations at each level (e.g., lexical, segmental, syntactic, etc.). It is thus reasonable to posit that information about such competition is what is being monitored. Specifically, the conflict-based monitor is sensitive to the difference between the activation of two (or more) representations in the same part of the production system (e.g., the lexical layer). If this difference is small, the representations are in high competition with one another. This is exactly the kind of situation in which the deployment of cognitive control is necessary to resolve the competition. Importantly, the same kind of mechanism can determine the need for control in other (non-linguistic) systems as well.

3.4 Monitoring as Decision Making

In the above sections, I argued that the best source of information for a monitor whose goal is to quickly determine the need for control in the production system is information about the level of competition between alternative representations in each part of the system. This information can be quantified as *conflict*, i.e., the inverse of the difference in the activation between two (or more) representations (Nozari et al., 2011). Conflict is high when several items have similar levels of activation, i.e., a little bit of noise can create a high chance for any of them to be selected as the response. If only one response is correct, then the chance of committing an error is proportional to the chance of the competing representations being selected. Conflict is thus a direct index of the *probability* of an error: high conflict equals a high probability of making an error. Note that the system does not need to know what the right response is for this to be informative. When "cat" and "dog" have comparable levels of activation, there is a 50% chance that one of them gets selected. This chance is the same whether "cat" is the actual target or "dog". Moreover, this index allows for *prospective* assessment of error commission; the monitor does not have to wait for selection to be over before it can determine the need for control. It can thus act as a preventive measure to keep the system from making errors, a function I defined earlier as the primary function of the monitoring system.

But here is the hard question: How much conflict is too much conflict and should signal the need for control? This is the kind of question that all non-comparative theories of monitoring—which do not require comparison of a response to the "right" response—must address, as all of them work with some form of information that serves as a proxy for error

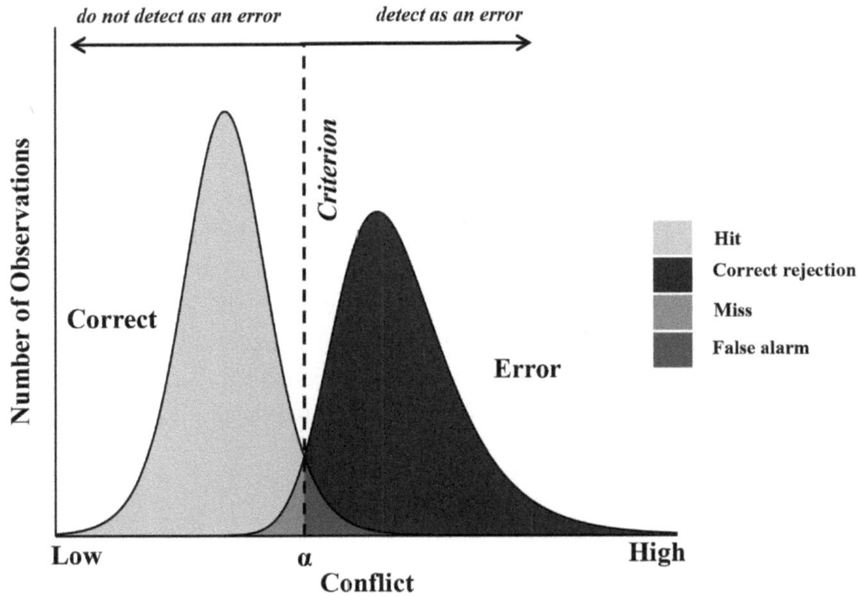

Figure 2 Conflict-based monitoring in the signal detection framework.

probability. I believe the answer is that monitoring is a form of decision making. This is illustrated in Fig. 2, using a framework adopted from the Signal Detection Theory. The two distributions show hypothetical distributions of conflict for correct and error responses. As explained above, error trials are associated with higher levels of conflict (see Nozari et al., 2011 for a computational demonstration of this claim), so the distribution of errors falls to the right of the distribution of correct trials. However, not all high-conflict trials will necessarily lead to errors, because there is always some chance of the correct response being selected from among the activated alternatives and the two distributions thus have some overlap.

The monitor determines which responses to flag as potential errors by setting a criterion somewhere between the two distributions. The location of this criterion is determined by the speaker's goals: if the goal is to produce speech which is free of any errors, the criterion can be placed far to the left so that all errors are caught. Note, however, that this will come at the cost of increased "false alarms". These false alarms can be observed as overt rejection of the correct response in individuals with aphasia (Nozari et al., 2011), but their more common manifestation in neurotypical adult speakers is disfluency (e.g., Postma, Kolk, & Povel, 1990). If, on the other hand, speed and fluency are to be emphasized over accuracy, the criterion would be placed to the far right. When neither speed nor accuracy is to be prioritized over the other, one way to determine the position of the criterion is by allowing it to settle at a point of equilibrium; for example, a point at which the probability that the criterion would have to be moved to the left to avoid a miss would match the probability of it having to be moved to the right to avoid a false alarm (Kac, 1962). Our computational simulations showed that this strategy predicts a detection rate of around 50% with a false alarm rate of <1% (Nozari et al., 2011). This detection rate matches what has been reported in conversational speech (e.g., Nooteboom, 2005) and what we have observed in my lab during typing of single words.

A useful feature of this framework is the possibility of extending it from detecting an error once selection is completed to preventing an error before a response has been selected. We have recently shown that differences in the pattern of responses (longer response latencies and more omission errors vs. shorter response latencies and more commission errors) within the same paradigm can be explained by shifting the position of a criterion that determines whether lexical selection should be allowed to proceed or halted until competition can be resolved, most likely by the intervention of control processes (Nozari & Hepner, in press).

A complete picture is beginning to emerge now: monitoring is a process of decision making, with a flexible criterion, using information generated during production processes to determine the probability of an error. In neurotypical systems, in most cases, this information is used before selection to recruit the control necessary to prevent errors from occurring. Occasionally, however, the relevant information may only be available after selection has been completed. In such cases, the monitor acts as an error detector, but the main function of monitoring remains surveying the system in order to determine the need for control. Such a monitor plays a critical role in regulating and optimizing production processes. For example, when faced with a high-conflict picture naming trial, it can tune the production system to more efficiently resolve conflict in the next high-conflict picture naming trial (e.g., Freund, Gordon, & Nozari, 2016). Importantly, this adaptation from one high-conflict trial to the next seems to persist despite a temporal gap between the two trials and regardless of the nature of the intervening tasks, suggesting the involvement of learning mechanisms in the allocation of control (Freund & Nozari, 2018).

4. DOMAIN-GENERALITY OF MONITORING AND CONTROL IN LANGUAGE PRODUCTION

What I have explained so far can be summarized in a few sentences: language production involves competition at all levels. If unresolved, this competition leads to frequent errors, as is the case with aphasic individuals. However, competition is often successfully resolved because the production system is constantly monitored and control is appropriately applied. Since competition and its resolution are integral parts of the production processes, the operations that affect competition resolution are also part of successfully producing speech. We can now evaluate whether such processes are specialized for language production, or are shared with other systems, and by answering this question determine how "special" language production is relative to the rest of cognition.

In the early parts of this chapter, I reviewed what Fodor considered to be the criteria for computational domain specificity, and alluded to the surge of interest in probing neural domain-specificity in the late '90s and 2000s. Most recently, the investigation of domain-generality has turned to paradigms that assess whether an increased need for control in one task/domain leads to the recruitment of control in another task/domain (see Egner, 2014 for a review). These three approaches have arrived at different answers to the question of the domain-generality of cognitive functions—partly because of methodological variations, but more importantly because the question turns out to have different meanings when viewed from different angles (Nozari & Novick, 2017). I will thus briefly review the evidence for domain-generality from these three angles.

4.1 Domain-Generality of Computations

Competition is everywhere in the cognitive system, and must often be settled in favor of a single winner. It is thus reasonable to assume that the mechanisms that mediate conflict resolution apply across domains. The evidence I reviewed on the role of inhibitory control in various levels of language production lend support to this claim. Analysis of individual differences suggest that individuals' ability to inhibit a response in tasks like stop-signal or go-nogo tasks predicts how quickly and accurately they can name pictures, produce the correct subject-verb agreement, and tailor their utterance to the referential context. While some part of this process may be carried out locally within each system through lateral inhibition between representations, neural evidence suggests that at least some part of the process is mediated by central processes (see below). By the same logic, it is plausible that the need for control is assessed in similar ways across systems that require competition resolution. The conflict-based monitor in speech production was, in fact, inspired by a similar model applied to forced-choice manual responses in a variety of tasks which induce conflict between response alternatives (Botvinick, Braver, Barch, Carter, & Cohen, 2001). Similarly, semi-production-based monitors such as Hickok's (2012) model use the same principle of suppression of a perceptual response by its motor equivalent that has been reported in vision.

In short, similar computational problems exist in language production as in other systems. As such, similar computational solutions are applicable to both, as long as the nature of the representations are compatible with those solutions. A general mechanism of motor control through perceptual suppression is applicable where motor and perceptual representations are concerned in the language system. A general conflict detecting mechanism is applicable where several representations of the same kind are simultaneously competing for selection.

4.2 Domain Generality of Neural Underpinning

A thorough review of the neural underpinnings of language production is beyond the scope of this chapter, but I will review a few key pieces of evidence related to the issue of the domain-generality of monitoring and control processes in speech production. Generally speaking, three techniques have been used to probe this issue: Event-Related Potentials (ERPs), neuroimaging (mostly fMRI) and brain stimulation. The first of these is more concerned with uncovering common neural signatures (if they exist), as opposed to pinning down that signature to a brain region with millimeter precision. There are now several ERP studies showing a common signature (called the ERN, or Error-Related Negativity) for monitoring in speech and non-speech domains (e.g., Ganushchak & Schiller, 2008; Riès, Janssen, Dufau, Alario, & Burle, 2011; see Nozari et al., 2011 for a comprehensive review of this evidence). The ERN is a modality-independent, response-locked ERP component with greater negativity on error than correct trials. A similar component, N2, is observed on high-conflict correct trials, whose origin, similar to the ERN, can be traced back to the Anterior Cingulate Cortex (ACC; e.g., Van Veen & Carter, 2002). These results strongly suggest the presence of a common monitoring mechanism across domains, with a central component, housed in the ACC, which is sensitive to the need for control.

What happens when the need for control is detected? Decades of research on the prefrontal cortex have implicated the VLPFC as being involved in deploying control based on task demands (e.g., Nozari & Thompson-Schill; 2015; Thompson-Schill, D'Esposito, Aguirre, & Farah, 1997). Neuroimaging studies of monitoring and control in language production are sparse, but the results implicate the involvement of the same neural regions observed during monitoring and control of other tasks. For example Gauvin, De Baene, Brass, and Hartsuiker (2016), found activation of the supplementary motor area (pre-SMA), dorsal ACC, and VLPFC during monitoring of a tongue-twister task. Moreover, increased competition by repeated naming of semantically-related pictures has been shown to activate the VLPFC (e.g., Schnur et al., 2009; de Zubicaray, Fraser, Ramajoo, & McMahon, 2017). Similarly, stimulation of the left PFC through transcranial direct current stimulation (tDCS) has been shown to decrease the rate of semantic errors (Nozari, Arnold, & Thompson-Schill, 2014) and phonological errors on words the speakers attend to (Nozari & Thompson-Schill, 2013).

While the neuroimaging and brain stimulation studies provide some evidence for the involvement of the medial and lateral PFC in implementing monitoring and control in language production, this evidence does not imply that the *same population of neurons* in these regions is involved in both linguistic and non-linguistic tasks. For example, it is possible that, at the individual level, slightly different parts of the PFC mediate control in linguistic and non-linguistic tasks (e.g., Fedorenko, Behr, & Kanwisher, 2011), although the significance of this finding must be interpreted in light of the physical constraints of a prefrontal cortex that must accommodate connections from various specialized systems such as language, vision, etc.

In summary, the investigation of common neural correlates for monitoring and control between language production and other domains is relatively new. The evidence so far points to some level of domain-generality, although the exact degree of neuronal overlap within these domain-general regions (e.g., ACC, LPFC) between monitoring and control of language and other domains is not yet known.

4.3 Domain Generality of Functional Adjustment in Control

The evidence for domain-generality in terms of computations and neural substrates has led researchers to push the boundaries on the issue by asking whether the increased need for control in one domain recruits control that can then be applied to another domain. The hypothesis here is that signaling the need for control, regardless of the source of conflict, mobilizes domain-general control resources that can be readily applied to any task in any domain. A small number of studies have reported results that support this prediction. For example Kan, Teubner-Rhodes, Drummey, Nutile, Krupa, and Novick (2013), reported that reading a high-conflict (syntactically-ambiguous) sentence, which signals the need for control to the monitor, led to the recruitment of control that improved performance on high-conflict trials from a word Stroop or a visuospatial task (see also Hsu & Novick, 2016). In contrast to these results, a large body of studies have found no evidence of cross-task transfer of control (e.g., Akçay & Hazeltine, 2011; Egner, Delano & Hirshc, 2007; Kiesel, Kunde, & Hoffmann, 2006; see Egner, 2014 for a review).

We recently tested this prediction directly in language production. Trials from a picture-word interference (PWI) paradigm were interleaved with trials from a visuospatial task (Exp. 1) and a sentence comprehension task (Exp. 2; Freund & Nozari, 2018). All three tasks were designed to contain high-conflict trials (e.g., semantically-related words superimposed on target pictures, arrows with opposing directions, or lexical or syntactic ambiguity, respectively) and low-conflict trials (e.g., target word superimposed on the target picture, same direction arrows, no ambiguity). Since PWI trials alternated with trials from the other tasks, it was possible to assess two effects: *within-task adjustment of control* was assessed by measuring performance on the current trial as a function of a 2-back trial from the *same* task (for each of the PWI, visuospatial, and sentence comprehension tasks), and *cross-task adjustment of control* was assessed by measuring performance on a current trial as a function of 1-back trial from a *different* task (e.g., PWI as a function of visuospatial conflict, etc.).

The results were clear: all three tasks showed robust evidence of within-task adjustment of control, despite being separated by a lag of 4s (Exp. 1) and 8s (Exp. 2) and a trial from a different task to perform in between. On the other hand, there was no sign of performance improvement in any of the three tasks as a function of increased demand for control on the previous trial from a different task. These results speak against a domain-general account in which control is deployed indiscriminately to all systems, and instead indicate that deployment of control is specific to the source of conflict: if conflict is detected in the language production system, control is deployed to that system. In fact, quite possibly, if the need for control is detected in a specific part of the production system (e.g., to resolve lexical competition), then control is deployed to exactly that part of the system (Nozari et al., 2011).

Interestingly, there were suggestions of a *reversed* effect: in some cases, the increased need for control in one task led to the *deterioration* of performance on a high-conflict trial on another task (see Notebaert & Verguts, 2008, for a similar finding). This reversed effect speaks against completely independent control systems. How can we reconcile all these findings? This pattern of results is compatible with a central controller (e.g., part of the LPFC), with at least some level of domain-generality in terms of neurons that implement control, but with specific connections to domain-specific representations. Control is implemented in each task by a special configuration of connections between the central controller and task-specific representations. Critically, this configuration can be adapted in order to facilitate performance the next time the system experiences high conflict in the same task (Freund & Nozari, 2018; Notebaert & Verguts, 2008). Similar to incremental learning processes during lexical selection (Oppenheim et al., 2010) and segmental encoding (Breining et al., 2016, 2018), the connections between the part of the central controller and the domain-specific representations involved in this configuration are strengthened. At the same time, the connections between the same population of neurons in the central controller and other systems are weakened, in order to allocate the limited resources efficiently to the immediate needs. To the extent that the neuronal populations in the central controller overlap in implementing control across tasks, this mechanism of incremental learning facilitates performance on the next trial of the same kind of task, and interferes with the implementation of control on a trial of another task whose connections to the controller have been weakened. Note, however, that in many cases control simply does not transfer from one task to another, without any reliable reversed effect being observed. This finding implies that the overlap between the neuronal populations in the central controller for mediating control in different tasks is limited (see Jiang & Egner, 2014, for neuroimaging evidence supporting this claim).

In summary, increasing the need for control in the language production system does not lead to enhanced implementation of control in other systems and vice versa. This is, however, completely expected from the operation of a limited-capacity central system which is trying to optimize task performance: distribution of limited resources is most effective when control is deployed exactly where it is needed. At the same time, there is some indication that the neuronal populations in the central controller which implement control in different tasks/domains may indeed have a certain degree of overlap.

5. CONCLUSION: HOW SPECIAL IS LANGUAGE PRODUCTION?

We are finally in a position to answer this question: language production is special in so far as its representations are special. I have claimed that even subdomains within production, e.g., lexical items vs. segments, are each special in that they meet Fodor's criteria for eccentricity. What was in question was the assumption that a system with special representations requires specialized operations. Part of the argument for a specialized system was efficiency. Language production is fast, fluent, and relatively error-free, and without the speaker's conscious awareness of the processes that map thoughts onto sounds. So perhaps it is a highly specialized automatic module. The evidence I reviewed in this chapter suggest otherwise. The nature of production processes engenders competition at all levels of production, and empirical evidence suggests that such competition indeed requires control in order to be resolved properly. I thus proposed an alternative: language production is so fast and efficient because of a highly-efficient monitoring-control loop which constantly monitors the need for control and consequently deploys control to the appropriate part of the production system. I then reviewed various accounts of monitoring and argued that a specialized monitoring system such as the perceptual loop system is, in fact, neither the fastest nor the most efficient option.

While the contribution of perceptual loop to detecting errors in overt speech is undeniable, I also discussed two domain-general mechanisms for speech monitoring, both of which are good candidates for a fast and unconscious monitoring process. The first is a forward model which suppresses the perceptual consequences of self-produced actions. This is an excellent candidate for monitoring speech at the articulatory-motor level. Fodor's argument about eccentricity has a critical implication here: while eccentric representations do not necessarily require specialized operations, certain operations do require specific kinds of representations. A monitor that guides motor actions via perceptual consequences of those actions requires separate sets of corresponding motor and perceptual representations. If it is to be applied to subsystems which lack such representations, the process must be redefined.

The second mechanism is a conflict detection mechanism which monitors the difference between the activation of multiple representations in the same part of the system. This is an excellent candidate for monitoring language production, the dynamics of which naturally lead to the activation of multiple representations at various levels. The plausibility of these two mechanisms for speech monitoring (see Hickok, 2012; Nozari et al., 2011; Hanley et al., 2016 for implementations and detailed discussion of the empirical evidence in various populations), together with the empirical evidence for the domain-generality of monitoring and control in language production (e.g., ERP, brain stimulation, and neuroimaging evidence) strongly suggests the involvement of a domain-general process in the regulation of speech production.

In the last section, I cautioned against a sweeping interpretation of this domain-generality. The key is in remembering that these domain-general operations still operate on domain-specific representations. Functionally, this means that a monitoring-control loop for lexical selection consists of domain-specific lexical representations, a domain-general central control system (the same general region(s) in the brain, with some overlap between the subregions that mediate control in specialized domains), and the connections between the two. Signals generated within this loop during monitoring recruit control, which, in neurotypical individuals is commensurate to the need for control. This control, however, does not benefit other systems, as they are simply not part of this particular loop. In fact, in so far as the same neuronal populations in the central controller engage in mediating control across different loops, implementation of control in other loops suffers, just like in the case of overt selective attention. We have shown that in language production, as in other domains, attending to a word increases accuracy on that word at the expense of accuracy on the other words in the sequence (Nozari & Dell, 2012). Our later work showed that stimulation of the central controller (left PFC) exaggerates this effect by widening the gap between the accuracy of attended and unattended items as control increases (Nozari & Thompson-Schill, 2013). Both the conscious attentional effects manipulated in these two studies and the subconscious control effects discussed throughout this chapter follow the same principle: functional-specificity is an inevitable consequence of systems that share a resource-limited domain-general component. This combination of shared computations through (partially shared) machinery, along with specific loops to monitor and regulate domain-specific operations is an optimal solution to guarantee fast, efficient and relatively error-free processing.

To conclude, production processes are monitored and regulated through operations that are very similar to those in other systems, contrary to the notion of special mechanisms for special domains. In that sense, language production is not that special. However, the specificity of linguistic representations to the language domain creates special monitoring and control loops in production that can be selectively trained and selectively damaged. This specificity has a critical clinical implication: the most effective way to train or rehabilitate control in the production system is by using tasks that engage the monitoring and control loop in the context of language production. In this sense, language production is as special as any other cognitive system.

ACKNOWLEDGMENTS

This work was supported in part by the Therapeutic Cognitive Neuroscience Fund at Johns Hopkins University, and in part by the NSF grant NSF-1631993 to Nazbanou Nozari and Akira Omaki. I would like to thank Christopher Hepner, Duane Watson and Kara Federmeier for their feedback on this chapter.

REFERENCES

Akçay, Ç., & Hazeltine, E. (2011). Domain-specific conflict adaptation without feature repetitions. *Psychonomic Bulletin & Review, 18*(3), 505–511.

Baars, B. J., Motley, M. T., & MacKay, D. G. (1975). Output editing for lexical status in artificially elicited slips of the tongue. *Journal of Verbal Learning and Verbal Behavior, 14*(4), 382–391.

Belke, E., Meyer, A. S., & Damian, M. F. (2005). Refractory effects in picture naming as assessed in a semantic blocking paradigm. *The Quarterly Journal of Experimental Psychology Section A, 58*(4), 667–692.

Bock, J. K. (1986). Syntactic persistence in language production. *Cognitive Psychology, 18*(3), 355–387.

Bock, K., & Miller, C. A. (1991). Broken agreement. *Cognitive Psychology, 23*(1), 45–93.

Botvinick, M. M., Braver, T. S., Barch, D. M., Carter, C. S., & Cohen, J. D. (2001). Conflict monitoring and cognitive control. *Psychological Review, 108*(3), 624.

Breining, B., Nozari, N., & Rapp, B. (2016). Does segmental overlap help or hurt? Evidence from blocked cyclic naming in spoken and written production. *Psychonomic Bulletin & Review, 23*(2), 500–506.

Breining, B. L., Nozari, N., & Rapp, B. (2018, July 23). Learning in Complex, Multi-Component Cognitive Systems: Different Learning Challenges Within the Same System. *Journal of Experimental Psychology: Learning, Memory, and Cognition*. Advance online publication. https://doi.org/10.1037/xlm0000630.

Brown-Schmidt, S., & Tanenhaus, M. K. (2008). Real-time investigation of referential domains in unscripted conversation: A targeted language game approach. *Cognitive Science, 32*(4), 643–684.

Chomsky, N. (1975). *Reflections on language*. New York: Pantheon.

De Smedt, K., & Kempen, G. (1987). Incremental sentence production, self-correction and coordination. In *Natural language generation* (pp. 365–376). Dordrecht: Springer.

de Zubicaray, G., Fraser, D., Ramajoo, K., & McMahon, K. (2017). Interference from related actions in spoken word production: Behavioural and fMRI evidence. *Neuropsychologia, 96*, 78–88.

Del Viso, S., Igoa, J. M., & García-Albea, J. E. (1991). On the autonomy of phonological encoding: Evidence from slips of the tongue in Spanish. *Journal of Psycholinguistic Research, 20*(3), 161–185.

Dell, G. S. (1986). A spreading-activation theory of retrieval in sentence production. *Psychological Review, 93*(3), 283.
Dell, G. S., Schwartz, M. F., Martin, N., Saffran, E. M., & Gagnon, D. A. (1997). Lexical access in aphasic and nonaphasic speakers. *Psychological Review, 104*(4), 801.
Dell, G. S., Schwartz, M. F., Nozari, N., Faseyitan, O., & Coslett, H. B. (2013). Voxel-based lesion-parameter mapping: Identifying the neural correlates of a computational model of word production. *Cognition, 128*(3), 380–396.
Egner, T. (2014). Creatures of habit (and control): A multi-level learning perspective on the modulation of congruency effects. *Frontiers in Psychology, 5*, 1247.
Egner, T., Delano, M., & Hirsch, J. (2007). Separate conflict-specific cognitive control mechanisms in the human brain. *NeuroImage, 35*(2), 940–948.
Evans, N., & Levinson, S. C. (2009). The myth of language universals: Language diversity and its importance for cognitive science. *Behavioral and Brain Sciences, 32*(5), 429–448.
Fedorenko, E., Behr, M. K., & Kanwisher, N. (2011). Functional specificity for high-level linguistic processing in the human brain. *Proceedings of the National Academy of Sciences, 108*(39), 16428–16433.
Ferreira, F., & Clifton, C., Jr. (1986). The independence of syntactic processing. *Journal of Memory and Language, 25*(3), 348–368.
Fodor, J. A. (1983). *The modularity of mind*. Cambridge, MA: MIT Press.
Freund, M., Gordon, B., & Nozari, N. (2016). Conflict-based regulation of control in language production. In A. Papafragou, D. Grodner, D. Mirman, & J. C. Trueswell (Eds.), *Proceedings of the 38th Annual Conference of the Cognitive Science Society* (pp. 1625–1630). Austin, TX: Cognitive Science Society.
Freund, M., & Nozari, N. (2018). Is adaptive control in language production mediated by learning? *Cognition, 176*, 107–130.
Ganushchak, L. Y., & Schiller, N. O. (2008). Motivation and semantic context affect brain error-monitoring activity: An event-related brain potentials study. *NeuroImage, 39*(1), 395–405.
Garrett, M. F. (1975). The analysis of sentence production. In G. H. Bower (Ed.), *The psychology of learning and motivation* (pp. 133–175). San Diego, CA: Academic Press.
Gauvin, H. S., De Baene, W., Brass, M., & Hartsuiker, R. J. (2016). Conflict monitoring in speech processing: An fMRI study of error detection in speech production and perception. *NeuroImage, 126*, 96–105.
Gleitman, L. R., January, D., Nappa, R., & Trueswell, J. C. (2007). On the give and take between event apprehension and utterance formulation. *Journal of Memory and Language, 57*(4), 544–569.
Hanley, J. R., Cortis, C., Budd, M. J., & Nozari, N. (2016). Did I say dog or cat? A study of semantic error detection and correction in children. *Journal of Experimental Child Psychology, 142*, 36–47.
Hickok, G. (2012). Computational neuroanatomy of speech production. *Nature Reviews Neuroscience, 13*(2), 135.
Hodges, J. R., Patterson, K., Oxbury, S., & Funnell, E. (1992). Semantic dementia: Progressive fluent aphasia with temporal lobe atrophy. *Brain: A Journal of Neurology, 115*(6), 1783–1806.
Hsu, N. S., & Novick, J. M. (2016). Dynamic engagement of cognitive control modulates recovery from misinterpretation during real-time language processing. *Psychological Science, 27*(4), 572–582.
Jiang, J., & Egner, T. (2014). Using neural pattern classifiers to quantify the modularity of conflict–control mechanisms in the human brain. *Cerebral Cortex, 24*(7), 1793–1805.
Kac, M. (1962). A note on learning signal detection. *IRE Transactions on Information Theory, IT-8*, 126–128.

Kan, I. P., Teubner-Rhodes, S., Drummey, A. B., Nutile, L., Krupa, L., & Novick, J. M. (2013). To adapt or not to adapt: The question of domain-general cognitive control. *Cognition, 129*(3), 637–651.

Kiesel, A., Kunde, W., & Hoffmann, J. (2006). Evidence for task-specific resolution of response conflict. *Psychonomic Bulletin & Review, 13*(5), 800–806.

Laver, J. D. M. (1980). Monitoring systems in the neurolinguistic control of speech production. In V. A. Fromkin (Ed.), *Errors in linguistic performance: Slips of the tongue, ear, pen, and hand* (pp. 287–305). New York: Academic Press.

Levelt, W. J. M. (1983). Monitoring and self-repair in speech. *Cognition, 14*, 41–104.

Levelt, W. J. M. (1989). *Speaking: From intention to articulation*. Cambridge, MA: MIT Press.

Levelt, W. J., Roelofs, A., & Meyer, A. S. (1999). A theory of lexical access in speech production. *The Behavioral and Brain Sciences, 22*(1), 1–38.

Lind, A., Hall, L., Breidegard, B., Balkenius, C., & Johansson, P. (2014). Speakers' acceptance of real-time speech exchange indicates that we use auditory feedback to specify the meaning of what we say. *Psychological Science, 25*(6), 1198–1205.

MacKay, D. G. (1992). Awareness and error detection: New theories and research paradigms. *Consciousness and Cognition: International Journal, 1*(3), 199–225.

Marslen-Wilson, W. D., & Tyler, L. K. (1981). Central processes in speech understanding. *Philosophical Transactions of the Royal Society B, 295*(1077), 317–332.

Meyer, A. S. (1992). Investigation of phonological encoding through speech error analyses: Achievements, limitations, and alternatives. *Cognition, 42*(1), 181–211.

Nooteboom, S. G. (2005). Listening to oneself: Monitoring speech production. In R. J. Hartsuiker, R. Bastiaanse, A. Postma, & F. Wijnen (Eds.), *Phonological encoding and monitoring in normal and pathological speech* (pp. 167–186). Hove: Psychology Press.

Notebaert, W., & Verguts, T. (2008). Cognitive control acts locally. *Cognition, 106*(2), 1071–1080.

Nozari, N. and Hepner, C. (in press). To select or to wait? The importance of criterion setting in debates of competitive lexical selection. *Cognitive Neuropsychology*

Nozari, N. (2018). *The dual origin of semantic errors in access deficit: Activation vs. inhibition deficit* (under review).

Nozari, N., Arnold, J. E., & Thompson-Schill, S. L. (2014). The effects of anodal stimulation of the left prefrontal cortex on sentence production. *Brain Stimulation: Basic, Translational, and Clinical Research in Neuromodulation,* 7(6), 784–792.

Nozari, N., & Dell, G. S. (2012). Feature migration in time: Reflection of selective attention on speech errors. *Journal of Experimental Psychology. Learning, Memory, and Cognition, 38*(4), 1084.

Nozari, N., & Dell, G. S. (2009). More on lexical bias: How efficient can a "lexical editor" be? *Journal of Memory and Language, 60*(2), 291–307.

Nozari, N., Dell, G. S., & Schwartz, M. F. (2011). Is comprehension necessary for error detection? A conflict-based account of monitoring in speech production. *Cognitive Psychology, 63*(1), 1–33.

Nozari, N., Freund, M., Breining, B., Rapp, B., & Gordon, B. (2016). Cognitive control during selection and repair in word production. *Language, Cognition and Neuroscience, 31*(7), 886–903.

Nozari, N., & Novick, J. (2017). Monitoring and control in language production. *Current Directions in Psychological Science, 26*(5), 403–410.

Nozari, N., & Omaki, A. (2018). Syntactic production is not independent of inhibitory control: Evidence from agreement attraction errors. In C. Kalish, M. Rau, J. Zhu, & T. Rogers (Eds.), *Proceedings of the 40th Annual Conference of the Cognitive Science Society* (pp. 824–829). Austin, TX: Cognitive Science Society.

Nozari, N., & Thompson-Schill, S. L. (2013). More attention when speaking: Does it help or does it hurt? *Neuropsychologia, 51*(13), 2770–2780.

Nozari, N., & Thompson-Schill, S. L. (2015). Left ventrolateral prefrontal cortex in processing of words and sentences. In G. Hickok, & S. L. Small (Eds.), *Neurobiology of language* (pp. 569–584). San Diego, CA: Academic Press.

Oppenheim, G. M., Dell, G. S., & Schwartz, M. F. (2010). The dark side of incremental learning: A model of cumulative semantic interference during lexical access in speech production. *Cognition, 114*(2), 227–252.

Pickering, M. J., & Branigan, H. P. (1998). The representation of verbs: Evidence from syntactic priming in language production. *Journal of Memory and language, 39*(4), 633–651.

Pickering, M. J., & Garrod, S. (2013). An integrated theory of language production and comprehension. *Behavioral and Brain Sciences, 36*(4), 329–347.

Pierrehumbert, J. B. (2002). Word-specific phonetics. In C. Gussenhoven, & N. Warner (Eds.), *Laboratory phonology* (p. 101139). Berlin: Mouton de Gruyter.

Pinet, S., & Nozari, N. (2018). "Twisting fingers": the case for interactivity in typed language production. *Psychonomic Bulletin & Review, 25*, 1449–1457.

Pinker, S., & Jackendoff, R. (2005). The faculty of language: what's special about it? *Cognition, 95*(2), 201–236.

Postma, A. (2000). Detection of errors during speech production: A review of speech monitoring models. *Cognition, 77*(2), 97–132.

Postma, A., Kolk, H., & Povel, D.-J. (1990). On the relation among speech errors, disfluencies, and self-repairs. *Language and Speech, 33*(1), 19–29.

Rapp, B., & Goldrick, M. (2000). Discreteness and interactivity in spoken word production. *Psychological Review, 107*(3), 460.

Rayner, K., Carlson, M., & Frazier, L. (1983). The interaction of syntax and semantics during sentence processing: Eye movements in the analysis of semantically biased sentences. *Journal of Verbal Learning and Verbal Behavior, 22*(3), 358–374.

Riès, S., Janssen, N., Dufau, S., Alario, F. X., & Burle, B. (2011). General-purpose monitoring during speech production. *Journal of Cognitive Neuroscience, 23*(6), 1419–1436.

Schlenck, K. J., Huber, W., & Willmes, K. (1987). "Prepairs" and repairs: Different monitoring functions in aphasic language production. *Brain and Language, 30*(2), 226–244.

Schnur, T. T., Schwartz, M. F., Kimberg, D. Y., Hirshorn, E., Coslett, H. B., & Thompson-Schill, S. L. (2009). Localizing interference during naming: Convergent neuroimaging and neuropsychological evidence for the function of Broca's area. *Proceedings of the National Academy of Sciences, 106*(1), 322–327.

Schriefers, H., Meyer, A. S., & Levelt, W. J. (1990). Exploring the time course of lexical access in language production: Picture-word interference studies. *Journal of Memory and Language, 29*(1), 86–102.

Shao, Z., Meyer, A. S., & Roelofs, A. (2013). Selective and nonselective inhibition of competitors in picture naming. *Memory & Cognition, 41*(8), 1200–1211.

Shao, Z., Roelofs, A., & Meyer, A. S. (2012). Sources of individual differences in the speed of naming objects and actions: The contribution of executive control. *The Quarterly Journal of Experimental Psychology: QJEP, 65*(10), 1927–1944.

Tanenhaus, M. K., Spivey-Knowlton, M. J., Eberhard, K. M., & Sedivy, J. C. (1995). Integration of visual and linguistic information in spoken language comprehension. *Science, 268*(5217), 1632–1634.

Thompson-Schill, S. L., D'Esposito, M., Aguirre, G. K., & Farah, M. J. (1997). Role of left inferior prefrontal cortex in retrieval of semantic knowledge: A reevaluation. *Proceedings of the National Academy of Sciences, 94*(26), 14792–14797.

Tian, X., & Poeppel, D. (2010). Mental imagery of speech and movement implicates the dynamics of internal forward models. *Frontiers in Psychology, 1*, 166.

Trude, A., & Nozari, N. (2017). Inhibitory control supports referential context use in language production and comprehension. In G. Gunzelmann, A. Howes, T. Tenbrink, & E. Davelaar (Eds.), *Proceedings of the 39th Annual Conference of the Cognitive Science Society Society* (pp. 1218–1223). Austin, TX: Cognitive Science Society.

Van Veen, V., & Carter, C. S. (2002). The anterior cingulate as a conflict monitor: fMRI and ERP studies. *Physiology & Behavior, 77*(4–5), 477–482.

Wardlow, L. (2013). Individual differences in speakers' perspective taking: The roles of executive control and working memory. *Psychonomic Bulletin & Review, 20*, 766–772.

Warren, R. M. (1970). Perceptual restoration of missing speech sounds. *Science, 167*(3917), 392–393.

Yee, E., & Sedivy, J. C. (2006). Eye movements to pictures reveal transient semantic activation during spoken word recognition. *Journal of Experimental Psychology. Learning, Memory, and Cognition, 32*(1), 1.

Yoon, S. O., Koh, S., & Brown-Schmidt, S. (2012). Influence of perspective and goals on reference production in conversation. *Psychonomic Bulletin & Review, 19*(4), 699–707.

CHAPTER EIGHT

Alternatives and Inferences in the Communication of Meaning

Hannah Rohde*,[1] and Chigusa Kurumada[§]
*Linguistics and English Language, University of Edinburgh, Edinburgh, United Kingdom
[§]Department of Brain and Cognitive Sciences, University of Rochester, Rochester, NY, United States
[1]Corresponding author: Email: hannah.rohde@ed.ac.uk

Contents

1. Two Types of Alternatives 218
2. When Nothing Says Something: Roles of Alternative Forms 223
3. Intonational Phonology: Role of Alternative Meanings 227
4. Coreference: Combining Expectations for Meanings and Forms 235
5. Inferences Based on Alternative Forms and Meanings: A Bayesian Approach 240
6. Where Next? 246
Acknowledgments 252
References 252

Abstract

Communicating meaning is a primary goal of everyday language use. One significant puzzle in the study of linguistic comprehension is that the meaning communicated via language often goes beyond an aggregate of word meaning; comprehenders systematically derive pragmatic—as opposed to lexical or semantic—interpretations by leveraging their extralinguistic contextual knowledge. Key to this process is the insight that comprehenders compare *what was said* against *what could have been said*—the range of alternative messages and alternative utterances—to enrich their inferences about the speaker's intention behind a given choice of the linguistic signal. In this article, we offer a conceptual framework for investigating the roles of alternatives and how they are used in comprehension. The case studies we present span multiple levels, with the common thread that their surface forms fail to uniquely disambiguate the underlying meaning: turn-taking, disfluencies, intonational phonology, and pronoun resolution. We discuss Bayesian statistical inference as a possible computational-level approach to rationally combining assumptions about alternative forms and alternative meanings for pragmatic comprehension.

Psychology of Learning and Motivation, Volume 68
ISSN 0079-7421
https://doi.org/10.1016/bs.plm.2018.08.012

Human language permits speakers and listeners to communicate information about the state of the world, which includes not only what is believed to be true on the part of the speaker but also what is desired or intended as well as counterfactually hypothesized. A fundamental problem in language comprehension arises because speakers do not, and cannot, make explicit every detail of their message. Much is left unsaid, and listeners need to use contextual information (e.g., mutual knowledge, visually available information, prior experiences) to infer the speaker's intended meaning, given what was explicitly said.

Past research in linguistics and psycholinguistics offers the insight that listeners make principled inferences about speakers' intentions based on the assumption that a given linguistic output is selected among possible options warranted in context. Let's take an example (Grice, 1975; Recanati, 1989).

(1) Sam: How many kids do you have?
Nick: Two

In (1), Nick's answer "Two" usually implies that he has no more or fewer than two children, even though his answer is technically true if he has more than two children.[1] This is explained by positing that listeners expect speakers to be generally truthful and cooperative, providing only necessary and sufficient linguistic information given an intended meaning (Grice, 1975). If instead Nick needs to communicate that he has three children, "Three" would have been more informative, better allowing Sam to infer the intended meaning. In other words, the selection of the word "Two" over "None", "One", or "Three" strengthens the interpretation that Nick has *exactly* two children. In fact, such strengthening of meaning via consideration of the context of alternatives occurs regularly. Consider another example.

(2) Neel Kantha – This is a temple which is dedicated to Lord Shiva. Situated at a height of 1700 metres, it takes about 4 hours on foot to reach from Rishikesh. (http://www.himalaya2000.com/)

To many of us, this description implies that walking is the only means to access Neel Kantha. You would find the description to be inaccurate or at

[1] In a semantic analysis, "Nick has three children" entails "Nick has two children" because in all possible worlds in which the former is true, the latter is also true.

least misleading if you later learned that buses and cable cars were also available (even though the 4-hour estimate for walking will remain valid). This is because, as in the case of (1), we expect speakers and writers to be appropriately informative and not to arbitrarily leave out relevant facts. If there were in fact buses and cable cars, the description should have mentioned the relevant alternative options.

What could have been said, as opposed to *what was said* is often called an "alternative". A number of studies have portrayed the ability to derive and enumerate the right set of alternatives as a critical step towards arriving at an intended meaning that is never directly observed (e.g., Degen, 2013; Grice, 1975, 1989; Hirschberg, 1985; Horn, 1972; Jäger & Franke, 2014; Levinson, 2000; Potts, 2005; Sauerland, 2012). A question remains, however, with respect to how various alternatives are evoked and combined to constrain listeners' inferences. In particular, as we discuss below, previous research does not always make a clear distinction between listeners' expectations about two levels of alternatives: a) **alternative linguistic forms** that can be defined at multiple levels of a language's inventory of phonemes, morphemes, lexical items, and syntactic structures (e.g., using the lexical item "Two" instead of "One" or "Three" in (1)); and b) **alternative intended meanings** given a context (e.g., Nick having {ZERO, ONE, TWO, THREE, ...} children). In the case of the Neel Kantha example in (2), how is it that listeners compute from the form of the passage (i.e., the mention of how long the journey would take on foot) that there is only one way to the temple (among the context of possible speaker intentions {FOOTPATH ONLY, FOOTPATH OR BUS, FOOTPATH OR BUS OR CABLE CAR, ...})? Here we argue that, to adequately account for listeners' understanding of speaker intentions, one must avoid conflating these two types of alternatives and instead specify how linguistic comprehension is best understood as a formal inferencing problem.

It turns out to be useful to explain listeners' inferential decoding in terms of their model of a speaker's encoding process. In this way, production and comprehension are inherently linked as two interdependent sides of the communication exchange. This approach contrasts with many models in psycholinguistics, where the tendency is to target one phenomenon or one side of the production/comprehension divide—for example, production models (e.g., Bock & Levelt, 1994) may specify sophisticated speech encoding processes whereby speakers select their messages and choose appropriate linguistic forms, but no claims are made about whether comprehenders capitalize on their knowledge of this system in order to decode the linguistic forms they hear; likewise, comprehension models that rely on competition/integration (MacDonald, 1994; McRae, Spivey-Knowlton, & Tanenhaus, 1998; Spivey & Tanenhaus, 1998) characterize a listener who can integrate a wide variety of cues from context, but the enumeration of that weighted list of cues does not satisfactorily explain how such cues reflect a speaker's choices of particular forms and meanings. In what follows, we elaborate on the notions of alternative forms and meanings, and we show how the two combine in a structured inferential process that serves to link a number of different phenomena.

1. TWO TYPES OF ALTERNATIVES

To convey meaning in a communicative act is to select, as a speaker, an appropriate form to use. The basic assumption common among many theories of language production is that speakers begin this process by formulating their messages (intended meanings) and then selecting linguistic encoding options offered by the grammar of a given language (surface forms; Bock & Levelt, 1994). They assemble conceptual elements (e.g., lemmas) and build structured representations based on syntactic and phonological rules. The possible phonemes, lexical items, and part-of-speech categories are taken to be largely fixed for a given language, and the speaker's choice of an appropriate formulation of their intended meaning is a selection out of the set of available forms. On the other hand, the listeners' job is often thought of as reversing this mapping process: They reconstruct the intended meaning by mapping the observed signal onto increasingly more abstract linguistic representations.

However, the mapping between meanings and forms is not one-to-one; ambiguity is rife, both from the reuse of certain forms within the linguistic system and from the noisy environment in which language is produced and interpreted. What this means is that a listener who encounters an ambiguous form can use what they know of the set of possible alternatives to help recover the speaker's intention. An ambiguous acoustic element can be mapped to a nearby phoneme category; a word is expected to map to one of the available entries in the lexicon; a string of words can be parsed to fit one or more structures that are licensed by the grammar. At the pragmatic level, this context of alternatives has been invoked for analyzing words that participate in scales, along which they are ordered in terms of their evaluative meanings, likelihoods of occurring, and specificities of descriptions (Horn, 1972). It has been hypothesized that listeners recover the intended meaning of a word like "good" by calling to mind a relevant scale (see (3—4)).

(3) Sam: How was your trip to NYC?
Nick: It was good.

(4) BAD < OKAY < GOOD < GREAT < EXCELLENT

A word like "good" is associated with some meaning that is present whenever it is used, namely the meaning NOT BAD and perhaps ABOVE AVERAGE, but it also yields some ambiguity: Its meaning can be associated with any degree of goodness that is not bad (GOOD AND POSSIBLY GREAT AND POSSIBLY EVEN EXCELLENT), or else it can be used to signal a degree of goodness that is not bad but is also no more than good (GOOD BUT NOT GREAT). Given the existence of a scale of alternative meanings like (4) and an ability to reason about cooperative communication (Grice, 1975) (as illustrated in (1) and (2)), a listener who hears something described as "good" is permitted to draw an inference as follows: Had the speaker intended a meaning of GREAT, they would have used that stronger word "great", but they didn't, so the speaker's use of "good" can therefore be enriched to mean GOOD BUT NOT GREAT.

This long-standing approach to the derivation of scalar implicatures has been influential and productive in motivating a great deal of theoretical and empirical work on pragmatic language comprehension. However, it typically sidesteps the questions of where the set of alternative forms comes from and how many there are, problems that have an analogue when one asks what set of candidate meanings a listener must consider. For the interpretation of "good", the two meanings GOOD AND POSSIBLY GREAT and GOOD BUT NOT GREAT are taken to be equally relevant meanings, and it is a computation involving the context of available forms that allows the listener to decide between them.

But not all intended meanings are equally likely. The activation of, and selection among, alternative meanings has been the purview of an increasing number of psycholinguistic studies that try to pin down how listeners use available cues from the context of the utterance to estimate what meaning the speaker likely intends. Candidate lexical items (and the meanings presumably associated with them) are activated based, for example, on the available objects in a visual scene (Altmann & Kamide, 2007; Spivey, Grosjean, & Knoblich, 2005), by a primed concept (Neely, 1977), or more generally by appeal to coherence-driven reasoning and real-world knowledge (Hagoort & van Berkum, 2007; Köhne & Demberg, 2013; Xiang & Kuperberg, 2015). From the perspective of incremental and predictive language processing in general, listeners are posited to assign different weights to candidate lexical items either at the point when they hear a word or even before the word is encountered if they are anticipating how a sentence will continue (Altmann & Kamide, 1999; Brown-Schmidt & Tanenhaus, 2008; Federmeier, 2007; Kuperberg & Jaeger, 2016; Levy, 2008). Much is still unknown about how such parsing and processing behavior might interact with the time-course of derivation of alternatives and pragmatic comprehension based on them.

Another question concerns what representations must be included as members of alternatives. Extant theories often take lexical items as primitives of alternatives. In our daily language use, however, a wider range of elements are often considered and compared against each other. To illustrate, let's think about another example. Imagine Bob and Sarah are sitting together, and Bob says "I love you". Now, in a canonical conversational context, Sarah is obliged to respond, indicating whether she reciprocates the sentiment or not. In other words, the two relevant alternative meanings on the table are LOVE and ¬LOVE ("not" Love). (There are probably more subtle variants in between, but let's stick to those as the two possible states of Sarah's intentions for the purpose of this discussion.)

As for the relevant forms that Sarah's response can take, it is easy for her to choose a response if the intention is LOVE: She can simply reuse Bob's words to utter "I love you too". It is much harder to select an appropriate response if Bob's affection is not reciprocated. If Sarah needs to convey ¬LOVE, there are in fact many things besides "I don't love you" such as "I'm sorry", "well, actually I was meaning to say ...", "it's complicated", etc., some of which could potentially be ambiguous. One can even imagine a situation in which a slight (e.g., less than a second) delay in Sarah's response would already convey some of the answer, via an inference that if someone's intended meaning were LOVE, they would probably be less likely to hesitate in generating their response. If alternatives are defined as possible options that the speaker could have produced, these linguistic and non-linguistic options beyond word choice may need to be considered as relevant alternatives.

A main goal of this paper is to propose a conceptual framework of alternatives that can encompass various aspects of language use that contribute to the pragmatic communication of meaning. Here let us introduce our general approach by elaborating on this example of conveying LOVE and ¬LOVE. Fig. 1 shows a schematic representation of this generative process—one that will be expanded on for the case studies in the upcoming sections.

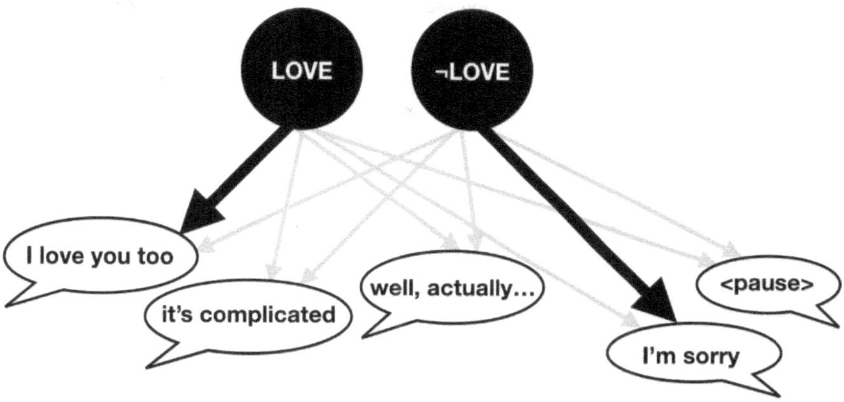

Figure 1 Graphical model of utterance production, showing two underlying meaning states LOVE and ¬LOVE and a sample of possible output forms.

The intended meaning states LOVE and ¬LOVE are shown to be linked to a set of possible surface forms, including the aforementioned hesitation (indicated as <pause>). In the figure, the strength of the form ~ meaning links is indicated by the darkness and width of the arrows, so that, for example, "I love you too" is the most likely utterance given that Sarah's intention is LOVE, whereas "it's complicated" is much less likely. In the model, every form is potentially associated with both underlying meaning states. This allows some probability (albeit near-zero in some cases) to be assigned to the generation of each of the outputs—speakers in LOVE may sometimes fail to articulate that meaning with the best option, just as speakers in ¬LOVE have been known to say all kinds of misleading or awkward things.

Fig. 1 drives home the point that there are two distinct levels of alternatives. The first is concerned with alternative intended **meanings**, represented here with the two black circles. One can think of them as alternatives at the mental state or proposition level, which are abstracted out from any specific linguistic encoding options. The speaker is postulated to be in one of those meaning states amongst the set of alternatives and to then encode that state linguistically. The other level pertains to alternative **forms**, illustrated here as utterances in speech bubbles. Speakers are taken to select linguistic encoding options (e.g., lexical lemmas, grammatical functions, syntactic constructions, phonological forms, phonetic realizations) that best represent their intended meanings. The listener is in turn tasked with leveraging the produced form to make an inference about the intended meaning. In this sense, the process is generative.[2] That is, listeners have a mental model of the process that *generates* the observed linguistic signal. We consider alternative meanings and alternative forms to be variables in this mental model at two different levels: *1) possible meaning options that generate the linguistic signal* and *2) possible signals that can be generated by these meanings.*

This distinction between alternative intended meanings and alternative forms is not always made salient in a discussion about a conventional scale such as GOOD < GREAT < EXCELLENT. This is because the meaning-level alternatives (the concepts of GOOD < GREAT < EXCELLENT) and the linguistic level alternatives (the words "good", "great", "excellent") appear to exhibit close correspondences, at least semantically. However, it is more the norm than the exception in language comprehension that meaning-level alternatives and form-level alternatives yield many-to-many relationships, like the example illustrated in Fig. 1. That is, one meaning can be encoded in multiple alternative forms (including an absence of a form, i.e., pause), and one form can be associated with multiple meanings. The listener therefore needs to navigate the resulting uncertainty to arrive at the most likely intended meaning given an observed form.

In what follows, we explore how the listener may infer a speaker's intended meaning by combining their assumptions about alternative meanings and alternative forms. To do so, we link multiple phenomena not typically analyzed together: filled and unfilled pauses (Section 2), intonational phonology (Section 3), and finally pronoun interpretation (Section 4). Seeing these disparate phenomena as similar requires a new lens, one which formally combines expectations at two different levels of representation. As we mentioned above, much of the previous work that discusses roles of alternatives focuses on those at the level of lexical items (e.g., "good", "great" vs. "excellent", or "some" vs. "all"). By departing from these common examples, we aim to illuminate the utility of postulating two levels of alternatives. Unlike the lexical alternatives that have traditionally been examined, some of the alternative forms we discuss (e.g., silence, disfluency) do not have commonly accepted,

[2] Note that the term 'generative model' has had a long and multi-sided history in the field. We intend it here in the sense of a mental model of the speaker's production process that defines how different possible speaker outputs can be *generated*. Generative Grammar as a conceptual framework was built on similar principles, but the specifics of that framework extend far beyond the underlying sense of 'generative' intended here.

conventionalized, meanings. Rather, listeners actively need to seek possible and likely meanings based on their contextual knowledge and infer how a perceptible form (or a lack thereof) might have been produced (generated) among other alternative forms for a particular intended meaning. Most researchers would acknowledge that context has a role to play, but what is new in this approach is the specification of which contextual features affect which linguistic calculations. Our approach uses an inferential architecture that makes specific predictions about what features can be manipulated with what systematic effects on interpretation. Finally, in Section 5, we lay out a computational-level inference framework for combining the listener's assumptions about alternative meanings and alternative forms.

2. WHEN NOTHING SAYS SOMETHING: ROLES OF ALTERNATIVE FORMS

Let's go back to the LOVE/¬LOVE example discussed above. We mentioned the intuition that a delayed onset of a response can bias the listener's interpretation of the speaker's state to ¬LOVE. How does this interpretation come about?

One important fact about conversation is that contributions from multiple parties are usually very tightly connected, not only in their contents but also in their timing. In a face-to-face conversation between adults, the average gap between two parties' turns is often 200 ms or shorter, with only rare overlap between turns (Casillas, 2013; Levinson & Torreira, 2015; Stivers et al., 2009; de Ruiter, Mitterer, & Enfield, 2006). A deviation from this can be registered as too early or too late, as seen in N400 effects, which are normally associated with contextual expectation violation but disappear with abnormally long pauses (Bögels, Kendrick, & Levinson, 2015). This strong expectation about when to respond creates an arena for an elaborate dance between conversational partners. The speaker must hold the floor while it is still her turn and signal whether and when to end her turn (Sacks, Schegloff, & Jefferson, 1974). The listener, on the other hand, needs to plan her own response so as to jump in as soon a transition of turns is indicated. A skilled speaker can take advantage of this strict temporal coordination to achieve discourse functions. A pregnant pause builds up suspense because the longer-than-expected silence (e.g., "Well, I meant to … tell you … ') forces the listener to wait while feeling compelled to pick up a turn and start talking.

This strong pressure for temporal coordination can turn very subtle signs — including ones produced inadvertently — into alternative forms to communicate pragmatically enriched meanings. Studies on speech disfluency have elegantly explained how this likely occurs. When a speaker experiences difficulty in planning or producing an utterance, her speech may become less fluent. Two main types of disfluencies that have been investigated so far are unfilled pauses (i.e., silence) and filled pauses such as "uh" and "um". Clark and Fox Tree (2002) were among the first to point out that these disfluencies can be produced and employed by the speaker in a systematic manner, just like any other lexical items. In their account, the speaker produces "uh"s and "um"s before items that are difficult to retrieve or produce such as words that are rarely used or those that have not been introduced into discourse yet (see also Brennan & Schober, 2001; Brennan & Williams, 1995; Clark, 1994; Clark & Wasow, 1998; Fox Tree, 1995; Fox Tree & Clark, 1997; Levelt, 1989; Shriberg, 1996; Smith & Clark, 1993). (For other researchers, the production of unfilled pauses is less strategic and need not be based on the speaker's conscious decision to signal their production difficulty e.g., Bock, 1987; Barr, 2001; Bortfield, Leon, Bloom, Schober, & Brennan, 2001; Lake, Humphreys, & Cardy, 2011; Oviatt, 1995).

Listeners, on the other hand, evaluate these disfluencies against their expectations about other (alternative) forms of utterances the speaker could produce. Let's think about the case in which Sarah fails to respond to Bob's declaration of his love within an expected time window (i.e., <pause>). In this case, the delayed onset is compared against possible alternative forms of production such as those in the speech bubbles in Fig. 1. Unlike the case of evaluative adjectives (e.g., "good" vs. "great") that comprise a lexically defined scale, interpretations of a delayed onset do not depend on a clearly ordered set of alternative forms and their associated semantic meanings. Rather, these alternative meanings are socially and situationally defined: Responses are ordered as positive (affirmative, preferred) and negative (contradictory, dispreferred) in terms of the expectation displayed in the initiating action (Kendrick & Torreira, 2015; Levinson, 1983). An optimally-timed response onset is normally associated with a positive response (e.g., LOVE), and an alternative form is associated with a negative response, one that is socially more costly and therefore potentially more difficult to produce (e.g., ¬LOVE). These situationally evoked and constrained ordering of alternatives (both of forms and meanings) are often discussed as ordered on an "ad-hoc" scale as opposed to a lexically defined scale (Stiller, Goodman, & Frank, 2015).

Bob's utterance,"I love you", thus creates an expectation that the LOVE meaning is the probable and expected meaning for Sarah as well, and if Sarah is to convey that meaning, she should be experiencing no difficulties in doing so. A delayed onset contrasts with that scenario, and thereby triggers an inference that Sarah must be conveying an alternative meaning that causes production difficulties. The basic machinery of that inference can be explained in a similar manner to the case of GOOD < GREAT < EXCELLENT. Observing a given form (e.g., "good") raises a question about why the speaker did not produce other options (e.g., "great"), which leads to an inference that the meaning of GREAT was probably not warranted.

Now let's consider the case where Sarah says "well ..." or "uh", which can be taken as her attempt to begin her response (perhaps to avoid inferences associated with a long silence) but to not yet go on record with a committed answer (Fox Tree & Clark, 1997). In a manner similar to what is discussed above for unfilled pauses, these forms are compared against other linguistic items that could be produced in this context. If Sarah were trying to convey the positive, socially more expected meaning (LOVE), there is a form that is readily available (i.e., "I love you, too"). Filled pauses again indicate a deviation from these expected forms, which triggers the inference that Sarah's attempt to articulate her meaning is causing production difficulty, and such difficulty is often attributed to production of linguistic elements that are new, infrequent, or difficult to articulate (e.g., Arnold, Tanenhaus, Altmann, & Fagnano, 2004; Arnold, Hudson Kam, & Tanenhaus, 2007; Barr & Seyfeddinipur, 2010; Loy, Rohde, & Corley, 2017). In this case, it likely leads to an inference that Sarah is in a dispreferred meaning state, i.e., ¬LOVE.

Existing studies on comprehension of disfluencies demonstrate that listeners routinely carry out these inferences in their real-time language comprehension (Arnold et al., 2007, 2004; Bögels et al., 2015; Corley, MacGregor, & Donaldson, 2007; Heller, Arnold, Klein, & Tanenhaus, 2015; Loy et al., 2017). Perception of disfluencies can immediately impact their predictive processing, evoking otherwise less probable or expected elements in a current context. These inferences offer an important insight into theories of alternatives in communication of meanings. That is, alternative meanings and forms are not necessarily static representations stored as part of our linguistic knowledge. Rather, they are emergent constructs in context. Listeners derive expectations about possible and likely meanings according to their assumptions about the context, conversational partners and common ground shared with them (Clark, 1996, 2015; Clark, 2015), prior linguistic experiences, and what is socially and communicatively licensed and preferred. Similarly, the linguistic signal (as well as an absence of it) is evaluated against contextually-constrained expectations about what linguistic encoding options are more or less likely in the context at hand. This resonates with the past 20 years of psycholinguistic research, suggesting that listeners are constantly predicting upcoming input so as to process the linguistic signal despite its fleeting and ambiguous nature (Altmann & Kamide, 1999; Brown-Schmidt & Tanenhaus, 2008; Federmeier, 2007; Kuperberg & Jaeger, 2016; Levy, 2008).

Recent investigations into real-time processing of filled pauses show that listeners are capable of calibrating their expectations at an incredibly fine-grained scale. As we discussed above, an instance of a disfluency such as "uh" or "um" triggers anticipation for a referent that is new in discourse or hard to label for other reasons. However, when listeners are told that the speaker has object agnosia and thus experiences difficulty naming ordinary objects, they do not show a bias toward objects lacking a conventional name when they hear a speaker's disfluency (Arnold et al., 2007). A similar effect of an explicit instruction was observed even with young children, who were told that the speaker is particularly forgetful (Orena & White, 2015). Additionally, listeners show sensitivity to the identity of the speaker in determining which objects have already been referred to, with disfluency biasing toward objects that have not been referred to by the particular speaker (Barr & Seyfeddinipur, 2010; Yoon & Brown-Schmidt, 2016). These results suggest that listeners condition their expectations on factors at the speaker level, fine-tuning their inferences about intended meanings given the specificities of the speaker. This likely includes causal reasoning about why the speaker may be producing disfluencies (Heller et al., 2015). An instance of disfluency produced by a non-native speaker does not trigger an inference about the referent's discourse status or lexical frequency in the same way as when it is produced by a native speaker, as the same assumptions do not hold (Bosker, Quené, Sanders, & de Jong, 2014; see also King, Loy, & Corley, 2018 for other context-driven adjustments given an alternative explanation for observed disfluency). Similar types of speaker-based calibration are possible for pragmatic interpretations of unfilled pauses. For instance, silence of a given duration (e.g., 600 ms) can be considered long for adults but short for young children (Casillas, 2013; Casillas, Bobb, & Clark, 2016; Stivers, Sidnell, & Bergen, 2018) or long in a casual, face-to-face conversation but short in a long-distance phone call. Thus deriving alternative meanings and alternative forms often requires situationally calibrated expectations about interlocutors and their language use, as well as general knowledge about how various meanings are encoded by linguistic alternatives.

3. INTONATIONAL PHONOLOGY: ROLE OF ALTERNATIVE MEANINGS

From the viewpoint of information processing, listeners' expectations for alternative forms play an important role in most domains of speech comprehension. The speech signal is produced and perceived in the presence of noise, and the same physical signal can be mapped onto different linguistic representations (e.g., phonemes) depending on speakers and situations (Liberman, 1960). Listeners therefore need to integrate multiple sources of information to anticipate which forms are more or less likely to be observed in a given context (Cutler, 2015). For instance, perception of a phoneme (e.g., /b/) occurs through integration of information about other, similar-sounding, phonemes (e.g., /p/, /d/, /v/), general knowledge of phonology and phonotactics (e.g., /b/ is more frequent than /v/ in English), distributional patterns of acoustic cues over phoneme categories (e.g., voice onset time), as well as contextual information (e.g., the target sound appears in a word "bathtu__", which makes /b/ more likely than other alternative sounds) (e.g., Clayards, Tanenhaus, Aslin, & Jacobs, 2008; Ganong, 1980; Kleinschmidt & Jaeger, 2015; Kraljic, Brennan, & Samuel, 2008; McMurray & Jongman, 2011; Norris, McQueen, & Cutler, 2003, 2000; Pisoni & Levi, 2007; Samuel, 2001). A number of knowledge sources can thus guide listeners' expectations about possible and likely alternative forms.

Listeners' knowledge and expectations for alternative meanings are often discussed as top-down information that constrain their hypotheses about the speaker's intended forms. The significance of top-down information is best appreciated when alternative forms do not form a closed, easily enumerable class (cf. phonemes). Speech prosody (intonation, in particular) can provide one such case. In English and other languages, stress, accents, and other forms of prosodic emphasis can highlight a linguistic element of importance in semantic and pragmatic communication (e.g., Cruttenden, 1997; Dahan, 2015; Ladd, 2008; Selkirk, 1996). For instance, the utterance "she LOVES you" can be used to convey "she LOVES you, not just LIKES you" while "she loves YOU" can convey "she loves YOU, not SOMEONE ELSE"(Pierrehumbert & Hirschberg, 1990). The speaker can modulate various prosodic cues (e.g., pitch, duration, intensity, loudness) to highlight elements in focus while backgrounding other elements in an utterance. The domain of research that concerns ways in which these different cues are combined to mark different informational statuses is called intonational phonology.

A perennial question in the domain of intonational phonology pertains to mappings between prosodic cues such as perceived pitch contour and intended meanings. Many researchers have largely agreed that speakers make systematic uses of prosodic cues to mark intonational categories, and these categories indicate how a given linguistic element (e.g., a word) should be interpreted in relation to information and assumptions in context (e.g., Beckman & Pierrehumbert, 1986; Calhoun, 2006; Cutler, Dahan, & Donselaar, 1997; Ladd & Morton, 1997; Ladd & Schepman, 2003; Pierrehumbert & Hirschberg, 1990; Watson, Tanenhaus, & Gunlogson, 2008). For instance, the type of emphasis used to convey the meaning [X, not Y] (e.g., LOVE, not LIKE) is often called a contrastive pitch accent and is analyzed separately from an emphasis that can be used for items that are simply new in discourse (Calhoun, 2004; Dilley, Ladd, & Schepman, 2005; Ito & Speer, 2008; Pierrehumbert, 1980; Pierrehumbert & Hirschberg, 1990; Silverman & Pierrehumbert, 1990; Watson et al., 2008). New information is often introduced with a heightened pitch peak aligned with the stressed syllable of a word (e.g., "Who does she love?" ... "she loves BOB") as schematically illustrated on the left in Fig. 2. The annotation conventions of the ToBI system (Silverman et al., 1992) identifies it as an H⋆ (a high tone aligned with a stressed syllable).[3] On the other hand, the contrastive interpretation (e.g., "Does she love Ted?" ... "No, she loves BOB") is often conveyed by a pitch accent that includes a low tone and a high tone aligned with the stressed syllable. It is often annotated as an L + H⋆, as illustrated in the right figure in 2.

Figure 2 New accent vs. contrastive accent (Adapted from Arvaniti, A., & Garding, G (2007). Dialectal variation in the rising accents of American English, *Papers in Laboratory Phonology, 9.*).

[3] We must note that even a cursory glance at the literature makes it clear that these categories are far from self-evident. Some frameworks of intonational phonology reject the assumption that there are categories that can be determined on the basis of acoustic properties of speech e.g., (e.g., Xu, 2005). Substantial work has been done to explore efficacies and limitations of proposing a particular inventory of intonational categories (e.g. Dilley, Breen, Gibson, Bolivar, & Kraemer, 2012, 2006; Wightman, 2002). An exhaustive discussion of such attempts is beyond the scope of the current paper, but we acknowledge the richness of the literature and the fundamental complexity of the topic.

The ease with which a native listener distinguishes these patterns in their acoustic realizations and their associated meanings belies the complexity of the actual processes involved in comprehending intonational meanings. To date, there is little consensus with respect to how listeners distinguish alternative forms, such as those illustrated in Figure (2) as postulated in theories of intonational phonology. A major source of the difficulty is the fact that actual realizations of the forms (i.e., perceptible acoustic cue distributions) are influenced by many other factors (e.g., phonological properties of words, prosodic features of an utterance, speech rate, emotional states of the speaker, individual differences in pitch height and speech rate, dialectal and sociolectal variations), and they are not deterministically mapped onto the theoretically posited intonational categories. Fig. 3 plots the distributions of fundamental frequency (F0) and segment duration of items that are either contrastive or new, produced in a naturalistic task environment (Buxó-Lugo, Toscano, & Watson, 2018). These two cues are chosen because they are often treated as main sources of information used by listeners to distinguish the two meaning categories (Pierrehumbert & Hirschberg, 1990). Actual distributions of these prosodic cues are, however, completely overlapping with each other, making it unlikely that

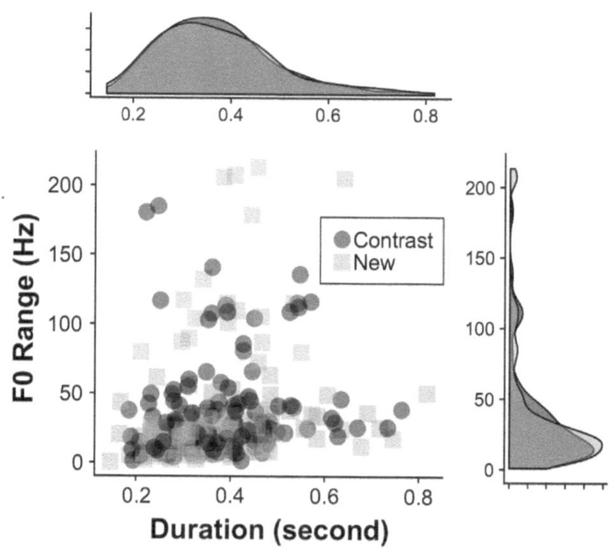

Figure 3 Mean word duration and F0 range of items produced in two different discourse statuses (Contrast (dark gray) vs. New (light gray)), showing large empirical overlap. This data was collected for a recent related study (Buxó-Lugo et al., 2018), re-visualized here courtesy of Andrés Buxó-Lugo.

listeners would arrive at one of these two alternative meanings through the acoustic information alone (without any other normalization schemes).

One potential source of information that can be useful is listeners' expectations about the alternative meanings (Cole, Mo, & Hasegawa-Johnson, 2010; Kurumada, Brown, & Tanenhaus, 2012; Pierrehumbert & Hirschberg, 1990). In a given conversational context, listeners may have a strong expectation for what types of meanings (e.g., new vs. contrast) will likely be conveyed by the speaker. In the example above, the question "Who does she love?" sets up an expectation that a referent introduced in a following utterance will be new information, whereas "she loves BOB" in response to "does she love Ted?" will likely be interpreted as contrastive. That is, listeners can estimate the relative likelihoods with which various alternative meanings may be expressed in a given discourse context. Such expectations for the meanings may, in turn, shape their expectations as to which alternative forms will be used by the speaker.

Kurumada, Brown, Bibyk, and Tanenhaus (2018) tested this prediction, asking how context-driven expectations regarding possible and likely alternative meanings may guide listeners' pragmatic interpretations of speech. The study used the looks-like-an-X construction, as in (5), which can support two interpretations depending on the intonation contour used. Noun-focus intonation puts the emphasis on the final noun (zebra) in (5a) and favors a meaning in which the speaker proffers a guess about the identity of the animal, because the accent placement on the noun works to highlight the relevant set of alternative animals (e.g., "a ZEBRA, not a horse"). On the other hand, contrastive accent on the verb "looks" and the rising intonation at the end of an utterance, as in (5b), can be taken to mean that "it resembles a zebra but it is not one" (Kurumada & Clark, 2016; Kurumada, Brown, Bibyk, Pontillo, & Tanenhaus, 2014). It has been observed that adult speakers often use such constructions to introduce a new, or less familiar, lexical item in child-directed speech (Clark & Wong, 2002). Perhaps, in the case of (5b), the speaker may be looking at an animal with a zebra-like appearance (Fig. 4, right).

(5) **a.** "it looks like a ZEBRA!" [Noun focus, proffered guess meaning = ZEBRA]

b. "it LOOKS like a zebra ..." [Verb focus, correction meaning = ¬ZEBRA]

Figure 4 Visual illustrations of intended alternative meanings of (5): a zebra on the left (intended by (5a) "It looks like a ZEBRA") and an okapi on the right (intended by (5b) "It LOOKS like a zebra ...").

In a series of experiments, Kurumada et al. (under review) manipulated the intonation contour of looks-like-an-X utterances (noun-focus vs. verb-focus) and the discourse context in which those utterances were embedded (bias to GUESS vs. bias to CORRECTION). In one experiment, they manipulated factors that would affect listeners' expectations for alternative meanings only. Participants were randomly assigned to an Expert condition or a Non-expert condition. In the Expert condition, participants were presented with a cover story in which the speaker is a kindergarten teacher talking to children (an expert who is in a position to plausibly correct mistaken evaluations). In the Non-expert condition, the speaker was introduced as a 14 year-old girl talking to her friend in a science museum (a non-expert who is in a position to plausibly be proffering a guess). Each condition was accompanied by pictures and a short text describing the situation, as in (6—7).

(6) Expert condition

Today a kindergarten teacher Ms. Jones had a story time with her students. She brought a book with a lot of pictures of different kinds of objects and animals. The children are very excited about this book. Every time she turns a page, somebody asks, "What's that?"

(7) Non-Expert condition

Today a group of eighth graders went to a natural science museum. Their assignment was to go through different exhibitions and write a journal about what they saw. Catie and Lauren decided to do the assignments together. There were a lot of strange-looking animals and artifacts they had never seen before, and they kept asking each other, "What's that?"

Participants then heard an identical set of 14 instances of "it looks like an X!" vs. "it LOOKS like an x ..." (seven instances each) and provided their interpretation of the speaker's intention by selecting one of the following disambiguated statements regarding the speaker's GUESS, CORRECTION, or uncertainty: a) The speaker is implying "It is an X", b) The speaker is implying "It is not an X", and c) The speaker is implying "I'm not sure". These options were printed in text, in this order, with clickable radio buttons.

As expected, there was a main effect of intonation type such that "it LOOKS like an x ..." yielded significantly more CORRECTION [It is not an X] interpretations. There was no significant effect of the Expert conditions. Crucially, however, the interaction between the intonation type and Expert condition was significant (see Fig. 5). Participants were more likely to interpret the verb-focus intonation as a CORRECTION [It is not an X] when the speaker was presented as an expert (i.e., the kindergarten teacher talking to children), who is likely to know an identity of a referent and correct the listener's assumption ("it LOOKS like a zebra, but it is not."). Remember that the audio stimuli presented to participants in the Expert and in the Non-Expert conditions were identical and hence the differences in the interpretations cannot be attributed to the perception of intonational contours (i.e., alternative forms) per se. It is more likely that the preamble and expectations of the speaker's knowledge states made the meaning of

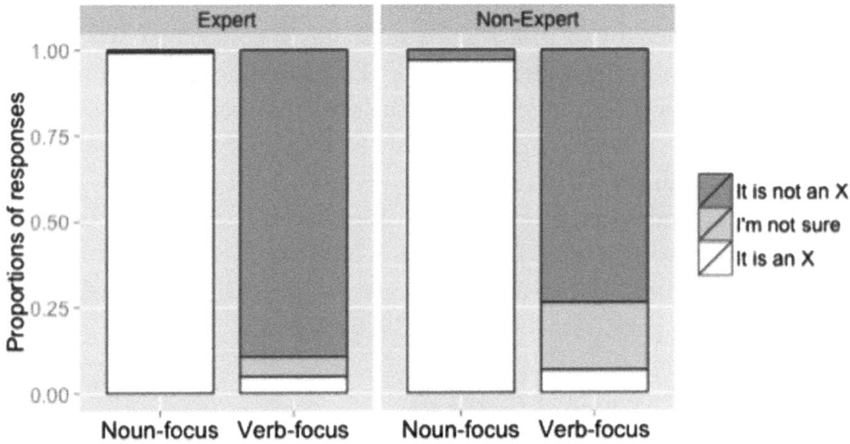

Figure 5 Meaning interpretations across focus and expertise conditions, showing the way context-driven expertise bias modulates the effect of form-driven differences between "It looks like an X" and "It LOOKS like an x" (Adapted from Kurumada, C., Brown, M., Bibyk, S., & Tanenhaus, M.K. (under review). Probabilistic inferences and adaptation in pragmatic interpretation of contrastive prosody.).

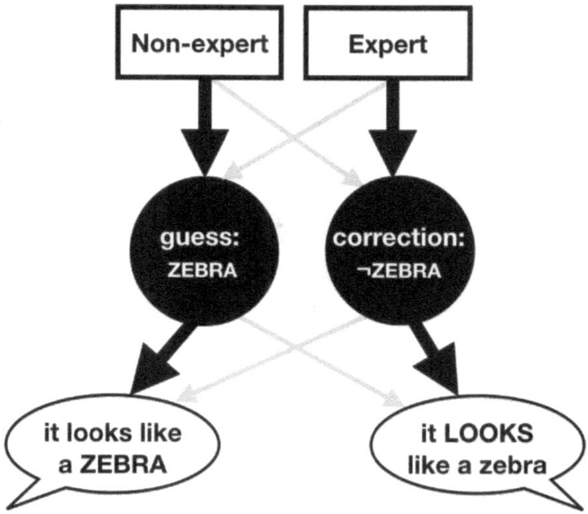

Figure 6 Graphical model of intonation production for "it looks like an X", showing alternative output forms and alternative underlying meaning states with two conditioning contexts.

[it is not an X] more or less likely, which resulted in different interpretations of the physically identical input.

Fig. 6 shows the preferences among possible meanings (conditioned on discourse context) and the preferences among forms (conditioned on intended meaning). Preferences – i.e., favored meanings and their associated forms — are indicated by thicker arrows. These are intended to signal production preferences whereby a teacher is being characterized as more likely to produce the CORRECTION meaning when speaking to students, and a teenager is more likely to produce the GUESS meaning when speaking to a peer. For the choice of forms, we expect the two speakers to possess an equivalent grammar of prosody: A speaker who intends the CORRECTION meaning is posited to favor the "it LOOKS like a zebra" contour, whereas a speaker who intends the GUESS meaning is posited to favor the "it looks like a ZEBRA" tune.[4]

For the listener, the puzzle is how to reverse engineer the likely meaning from the potentially ambiguous tune. The listener encounters, say, an instance of a kindergarten teacher producing "it LOOKS like a zebra".

[4] It is possible that speakers also differ in their preferences for forms in addition to their preferences for meanings. That is, a school teacher, as compared to a teenager, might be more likely to produce the verb-focus contour ("it LOOKS like a zebra") when her intended meaning is CORRECTION.

If the listener's generative model of the speaker's production process looks like Fig. 6, they are aware that this expert speaker is more likely to intend the CORRECTION meaning and that the CORRECTION meaning is more likely to be realized with verb focus (for similar effects of speaker knowledge, see Bergen and Grodner (2012)). In contrast, if the listener believes that a non-expert speaker is less likely to intend the CORRECTION meaning, an outcome of their inference is biased against this meaning even with the same preference at the level of choosing the verb-focus contour given the CORRECTION meaning (see the final column of Fig. 5). The current framework, as graphically illustrated in Fig. 6, can thus capture the long-standing intuition in the field that some of the meanings are *a priori* more or less probable given the listeners' knowledge about the context and that an appropriate integration of such expectations sharpens interpretations of intonation contours (Cole, 2015; Cole et al., 2010; Turnbull, Royer, Ito, Speer, & Turnbull, 2017). Contextually-sensitive weighting of alternative meanings is likely a key to the clear and reliable interpretations listeners can draw from the complex and often ambiguous intonational contours.[5]

Further, when combined with additional computational assumptions, the current approach makes a novel and quantitative prediction as to how these context-based expectations about alternative meanings would influence outcomes of pragmatic interpretations of intonation. Crucially, the current approach posits that listeners' expectations about possible meanings and possible forms are computed as distinct sets of alternatives, and contextual factors can affect them separately. That is, being in a particular context (e.g., listening to a kindergarten teacher who is talking about animals) affects both listeners' expectations about what meanings are more or less likely to be conveyed (alternative meanings) and how each of the meanings might be linguistically encoded (alternative forms). Previous approaches showed in broad strokes that both of these expectations matter in intonational interpretations, but few could articulate a prediction about how they should be integrated with each other

[5] Here we assume at least two types of ambiguity. One pertains to purely perceptual ambiguity resulting from noise in the production and comprehension process. The other pertains to ambiguity in mappings between a perceived intonational contour and a more abstract intonational category: As illustrated in Fig. 3, the same physical contour can be mapped onto two intonational categories (and their associated meanings). In both cases, we believe, listeners may be navigating the ambiguity by adopting systematic inferences along the lines of what we will propose in Section 5. We thank Alice Turk for pointing out these multi-layered sources of ambiguity.

(e.g., Is the relationship additive or interactive?). We come back to this discussion in Section 5 below.

4. COREFERENCE: COMBINING EXPECTATIONS FOR MEANINGS AND FORMS

Similar to the discussions of disfluencies and intonation, the interpretation and production of referring expressions offers a parallel window into the two different types of alternatives discussed in this paper. Let us consider the simple example below.

(8) Do you see the guy sitting by the window? He is Bob's high school friend, Nick.

By *referring expression*, we mean the speaker's choice of a nominal expression to linguistically encode a reference to a particular entity (an object or individual or concept or event). For instance, "the guy (sitting by the window)", "he", "Bob's high school friend", and "Nick" are all referring expressions (alternative forms) that can be used to talk about the same person (an intended meaning). The speaker needs to select an appropriate form depending on linguistic and non-linguistic information available in context to refer to the person. By *coreference*, we mean instances of two or more referring expressions picking out the same entity within or across utterances. For instance, the speaker of (8) needed to choose a referential form at the outset of the second utterance ("Do you see the guy sitting by the window? □ is ...") to clearly indicate the overlapping identity of the person being talked about. Unlike disfluency or intonational phonology, where the alternatives of form can vary continuously, the choices among referring expressions are discrete. A speaker must choose a form out of the (potentially very large) set of possible referring expressions; for our purposes here we will contrast the behavior of pronouns and names (but see Hemforth et al. (2010) for discussion of reference alternatives at the level of syntactic formulation).

Now let us consider how a listener might process various cases of coreference to identify who the speaker intended to refer to. This might appear trivial when looking only at cases such as (8), where there is only one possibility. However, resolving coreference can be a challenge for a listener when there are two possible candidates that cannot be distinguished by gender or animacy. Compare the minimal pair (9).

(9) **a.** John congratulated Bob because he won the race. [he ≈ BOB]
b. John impressed Bob because he won the race. [he ≈ JOHN]

The sentences in (9a-b) are identical except for the verb in the first clause, yet the preferred interpretation of the pronoun (as indicated in brackets) appears to flip from being coreferent with the OBJECT REFERENT BOB in (9a) to the SUBJECT REFERENT JOHN in (9b). The verbs "congratulate" and "impress" are well-studied members of a set of so-called implicit causality (IC) verbs, which are notable for inducing strong biases regarding the likely individual to whom cause can be attributed and thereby influencing the interpretation of ambiguous pronouns in contexts like (9) (Au, 1986; Brown & Fish, 1983; Caramazza, Grober, Garvey, & Yates, 1977; Garvey & Caramazza, 1974; Kehler, Kertz, Rohde, & Elman, 2008; McKoon, Greene, & Ratcliff, 1993).

Subject-biased IC verbs (e.g., "impress", "amaze", "disappoint") describe events in which the cause is typically attributed to the referent in subject position; object-biased IC verbs (e.g., "congratulate", "scold", "thank") typically attribute cause to the object. These are not categorical distinctions; it is possible for the subject to be the causally-implicated referent of an object-biased verb and vice versa. Examples (10a-b) show that the cause of a scolding event can be attributed unambiguously either to the object Bob (who did something undesirable) or to the subject John (for his desire to avoid blame). Examples (10c-d) show that both meanings can be formulated using a pronoun.

(10) John scolded Bob because …
a. … Bob had been boastful.
b. … John was trying to deflect blame from himself.
c. … he had been boastful. [he ≈ BOB]
d. … he was trying to deflect blame from himself. [he ≈ JOHN]

A speaker who intends to describe a situation in which it is Bob's boastfulness that led to the scolding can choose to articulate that meaning with (10a) or (10c); for a situation in which the reason behind the scolding is John's blame deflection, the choices include (10b) and (10d). A listener who encounters a pronoun ("he") and wants to recover the intended meaning can use their generative model of the speaker's production process to weigh the probability of different alternative coreference meanings ({JOHN, BOB})

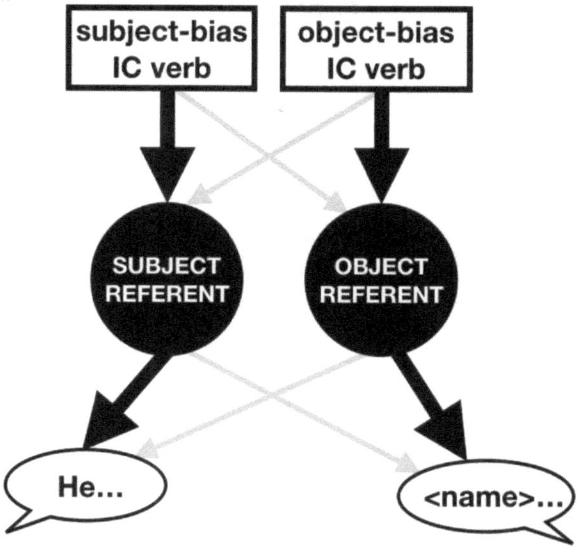

Figure 7 Graphical model of referring-expression production, showing alternative output forms and alternative underlying meaning states with two conditioning contexts.

and the probability of those meanings being articulated with different alternative forms ({"he", "John", "Bob", ...}).

In this way, parallels can be drawn with the looks-like-an-X example from Section 3. The possible meanings and the possible forms are different, but the inferential structure is the same. Fig. 7 shows a model of a speaker's production choices. Again there is a contextual factor (here, verb type) which influences which meaning a speaker is more likely to produce (here, re-mention of the SUBJECT vs. OBJECT referent). The speaker conveys such meanings via a choice among various forms (here, pronominal forms vs. names). In Fig. 7, the thickness of the arrows again indicates production preferences. IC verbs are depicted as influencing the preference over the re-mention of the SUBJECT and OBJECT referents, and the referent's SUBJECT/OBJECT status is depicted as influencing the eventual preferred form of reference.

This latter pronominalization preference is important because it has the power to influence listeners' interpretations even in contexts where a pronoun is fully ambiguous (i.e., "He" where the available referents are JOHN and BOB). The preference for speakers to produce pronouns when re-mentioning a subject referent is often taken to reflect, for English, the presumed topicality of the subject (Givón, 1983) combined with an understanding that continued reference to the same topic is best achieved with a pronoun (Ariel, 1990; Grosz, Joshi, & Weinstein, 1995; Gundel, Hedberg,

& Zacharski, 1993; Lambrecht, 1994). Evidence for this preference comes generally from observations of a "subject preference" for pronoun coreference (Crawley, Stevenson, & Kleinman, 1990; Frederiksen, 1981; Gernsbacher & Hargreaves, 1988; Gernsbacher, Hargreaves, & Beeman, 1989; Järvikivi, van Gompel, Hyönä, & Bertram, 2005), and more specifically from production tasks that measure participants' pronominalization rates (Fukumura & van Gompel, 2010; Rohde & Kehler, 2014). Such production tasks have used IC prompts as in (11) to measure both who participants favor for next mention and what form they prefer to use.

(11) Story-continuation stimuli
  a. [obj-bias IC] John congratulated Bob. ____________________
  b. [subj-bias IC] John impressed Bob. ______________________

For prompts like (11), participants produce more pronouns when the referent is the SUBJECT than when the referent is the OBJECT. Intriguingly, this pattern whereby more pronouns are produced for the subject referent holds even when the subject referent isn't favored for next mention, as is the case with object-biased IC verbs (Fukumura & van Gompel, 2010; Rohde & Kehler, 2014; Kehler & Rohde, 2018; cf. Rosa & Arnold, 2017, for a comparison with transfer-of-possession verbs). In Rohde and Kehler's results for subject-biased IC verbs, subject referents were re-mentioned with pronouns quite often (77.5% of the time, compared to a pronominalization rate of 26.6% when the object referent was re-mentioned). What is interesting is that for object-biased IC verbs, it is still the *subject* referent that is pronominalized at the higher rate (80.8% of the time, compared to a rate of 21.7% for objects). In those cases, subject referents are rarely mentioned in participants' continuations (due to the semantic bias of the object-biased IC verb), but when they are, they are mentioned with a pronominal referring expression.

Based on this, a listener who encounters an ambiguous pronoun is in a position to consider two sources of biases that are at work in the speaker's generative production process. First are the biases about who will be mentioned next (i.e., meaning-driven expectations). Subject-biased IC verbs favor upcoming mention of the SUBJECT; object-biased IC verbs favor the OBJECT. The other bias source pertains to syntactic position. Regardless of the IC verb types, the previous subject is more likely to be mentioned via a pronoun compared to the previous object (i.e., form-driven expectations). The upshot is that comprehenders should use IC cues to inform their

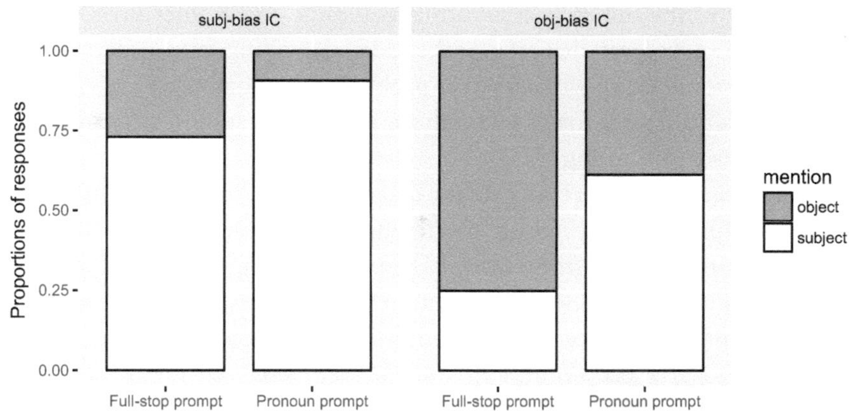

Figure 8 Coreference choices across verb and prompt types, showing the combination of context-driven IC biases and form-driven prompt differences (Adapted from Rohde, H., Kehler, A. (2014). Grammatical and information-structural influences on pronoun production. *Language, Cognition, and Neuroscience, Special Issue on Production of Referring Expressions: Models and Empirical Data,* 912–927.).

meaning-driven expectations about who will be mentioned next before they have encountered any referential form. Then, if they encounter a pronoun, they should update their next-mention expectations in favor of the subject.

This prediction was tested by Rohde and Kehler (2014). They used a story-continuation task that manipulated a contextual cue (subject vs. object biased IC verbs) and the form of the prompt, so that participants saw half the items in a full-stop condition as in (11) and half with a pronoun prompt (as in "John congratulated Bob. He ____________"). The results, shown in Fig. 8, confirm the predicted patterns: The contextual cue guided participants' preferences about which meaning should hold (more continuations that mentioned the subject following subject-biased verbs than object-biased verbs), and the form-based cue encouraged participants to update their preferences (more subject continuations when the prompt contained a pronoun than when it was just a full stop). There was no interaction between verb bias and prompt type.

Note that, for Rohde and Kehler's materials, the pronoun itself is fully ambiguous — if the participant wanted to use the pronoun prompt to corefer with, for example, Bob, the causally implicated referent for the object-biased prompt in (11a), they certainly could (and many of them do). However, the mere presence of a pronoun is enough to yield an increase in subject coreference when compared with the rate of subject coreference in the full-stop condition. As Rohde and Kehler note, the results therefore suggest that comprehenders are taking into account both the context-driven expectations over probable meanings as well as their expectations for different forms given those possible meanings, as captured by the causal structure in Fig. 7.

Why does this matter? The results that Rohde and Kehler (2014) report matter because they distinguish coreference models that rely on a general notion of prominence from models based on the framework advocated here (i.e., differentiating between prominent meanings and preferred forms). A number of coreference models posit a link between a pronoun and the referent who is most prominent, salient, accessible, in-focus, activated, etc. in a context (e.g., Gundel et al.'s (1993), accessibility hierarchy, Arnold's (2001), Expectancy Hypothesis, Grosz et al.'s (1995), Centering Theory), but such models cannot explain the difference shown in Fig. 8 between the full-stop prompt and the pronoun-prompt. Why would participants not simply use the pronoun in the pronoun prompt to refer to the most prominent referent? In the current framework, this difference can be explained by the assumption that the speaker's (or writer's) choice of a referential form out of possible alternatives carries useful information about the intended meaning. Participants' continuations in the full-stop condition reveal which referent they are treating *a priori* as more prominent (a reflection likely of a combination of biases dictated by the IC verb or the prominence of the subject as topic or any number of other factors). In contrast, their continuations in the pronoun-prompt condition reveal their expectations as updated by observing the speaker's choice of a pronoun, which permits them to draw an inference: If the speaker had intended an OBJECT continuation, she would have likely selected an alternative referential form like a name, but she used a pronoun instead and pronouns are commonly used for SUBJECT mentions, so a SUBJECT mention is just that much more likely here. In other words, participants are combining these two sources of biases to reverse-engineer a possible meaning that is most likely to have generated the observed referential form. In the next section we discuss this inference process in detail.

5. INFERENCES BASED ON ALTERNATIVE FORMS AND MEANINGS: A BAYESIAN APPROACH

In the various examples discussed in Sections 1–4, we have consistently identified two levels of alternatives: alternative meanings that speakers may be intending to convey and alternative forms that they can select to convey an intended meaning. An important insight to be reiterated here is that listeners' expectations about these two levels of alternatives can be derived independently. Taking a case of coreference as an example, a listener can *a priori* assign different probability estimates to different meanings (i.e., discourse referents such as BOB, JOHN, TOM, SARAH, ...) in terms of how likely each of them is to be talked about in a current utterance. These estimates are independent of those the listener can derive for different linguistic expressions (e.g., When the intended meaning is BOB, it can be encoded as "Bob", "he", "that guy", "my neighbor"). In this section we discuss how these two levels of probability estimates might be combined to guide the listener's inferences.

Among different approaches, here we focus on those that use Bayesian statistics to tackle this problem. Bayesian statistics are used widely to model subjective probability values (or beliefs) accrued and projected by human agents as opposed to objective probabilities of events (Anderson, 1990). The basic idea is that humans have hypotheses (Hs) about true states of the world that are often not directly observable. These are often called *prior beliefs*. When they observe evidence/data (E), humans evaluate it according to their estimates of the probability of observing E given a particular H. These estimates are called *likelihood estimates*. For example, continuing with the coreference example, listeners can have multiple hypotheses (Hs) about what an intended referent of a pronoun must be (e.g., BOB, JOHN, TOM, SARAH ...). Different Hs are assigned different prior probabilities. When the listener encounters a referential form (e.g., "he"), that constitutes evidence, E. Listeners can derive a likelihood estimate for the piece of evidence given a particular hypothesis (e.g., What's the likelihood of the speaker producing "he" if the intended meaning was BOB?) Using these two parameters, Bayesian inference computes the *posterior probability* of the hypothesis given the evidence.

(12) $$P(H|E) = \frac{P(E|H) \cdot P(H)}{P(E)}$$

In this notation, P stands for probability. P(H), for example, means the probability assigned to a particular hypothesis. P(E|H) is the probably of a piece of evidence given a hypothesis. The right side of the equation represents a product of the likelihood estimate (i.e., P(E|H)) and the prior probability divided by the normalizing constant, P(E). The denominator P(E) is included to make sure that the outcome of the inference is proportional to the probability of observing the given evidence. The left hand side of the equation represents the *posterior probability* of the hypothesis given the data. In the case of the coreference example mentioned above, this represents how likely it is that a given hypothesized meaning (e.g., BOB) is chosen given an observed referential form (e.g., "he"). This Bayesian approach is precisely that outlined in Kehler et al. (2008) for the coreference patterns as presented in Section 4.

A number of recent studies have applied this basic inference model to language comprehension (Phoneme perception and learning: Feldman & Griffiths, 2009; Kleinschmidt & Jaeger, 2015; Norris & McQueen, 2008; Sonderegger & Yu, 2010; Syntactic parsing: Hale, 2006; Levy, 2008; Pragmatic reasoning: Bergen & Grodner, 2012; de Ruiter & Cummins, 2012; Frank & Goodman, 2012; Goodman & Stuhlmüller, 2013; Jurafsky, 2008; Kehler et al., 2008). In the domain of pragmatic inferences and interpretation, one approach has been called "Rational Speech Act (RSA)" model, in which hypothesized meanings can be computed given an instance of observed form (Frank & Goodman, 2012, 2014).[6] It is indeed a powerful reasoning framework that yields a wide range of predictions about how listeners derive their pragmatic interpretations of linguistic input. Let us go back to the earlier example where Bob says "I love you" to Sarah and Sarah responds. In this scenario, Bob can posit alternative meanings, which constitute his hypotheses about Sarah's love (i.e., the distribution $P(H)$). He has a model of language production that assigns different probability values to a particular linguistic form given the hypothesis (i.e., $P(E|H)$), e.g., What would Sarah say if her meaning state was one of reciprocated LOVE? When Bob observes the linguistic evidence (E), he can compute the probability to be assigned to his hypothesis given the observed evidence (i.e., $P(H|E)$).

One major strength of this approach lies in its being a principled, mathematical framework that allows researchers to make quantitative predictions over outcomes of inferences. Specifically, it can capture two key determining components in the communication of meaning: our beliefs about meanings (e.g., what kinds of intentions/meanings are more likely in a given

[6] For those who are interested in this framework, there are a number of informative and accessible papers written from a standpoint of introducing the general reasoning and inference framework to problems related to language processing and acquisition (e.g., Perfors et al., 2011; Sobel, Tenenbaum, Gopnik, 2010; Tenenbaum & Griffiths, 2001).

context) and our model of language (e.g., how a given meaning is mapped onto linguistic encoding options). The Bayesian inference framework provides a systematic approach for how these two knowledge sources might be combined under uncertainty.[7] An advantage of this architecture is that it captures the structured influence of discourse context on pragmatic reasoning about meanings by integrating prior beliefs that are relevant and yet independent of meanings encoded in the linguistic signal. For instance, in the case of the LOVE vs. ¬LOVE example, Bob might *a priori* believe that Sarah is no longer in love with him based on her demeanor or some information he has received from a friend. If the probability of one hypothesis is very high, that prior probability (of, say, ¬LOVE) dominates the outcome of the inference. The prior probability thus represents how likely Bob will infer LOVE or ¬LOVE as an intended meaning in the absence of any linguistic signal from Sarah.

As simple as it may seem, this effect of context on inferred meaning *irrespective of linguistic forms used* is hard to explain in other traditional accounts of linguistic comprehension. This is because many accounts generally assume that listeners first construct a semantic (or logical) meaning of an utterance from semantic and syntactic elements of an utterance. They then enrich it in consideration of relevant components of context. In the current account, the assumption is reversed. Listeners begin to process the linguistic input based on their expectations about possible and likely meanings in an immediate environment as well as on their life-long experiences with linguistic communication. They expect the speaker to produce a linguistic signal that can meaningfully alter (update) the existing hypotheses about intended meanings. As a result, the same physical linguistic evidence can lead to distinct meanings, depending on which meanings were expected. As was illustrated in the example of the assumed level of expertise in the interpretation of "it looks like an X" (Section 3), inferences about the speaker's intended meaning are contingent on a wide variety of contextually-supported assumptions about the speaker (e.g., an expected level of expertise) as well as non-linguistic features of referents (Frank & Goodman, 2012).

[7] We are not proposing that the human brain is necessarily processing information according to the Bayes Rule. Bayesian statistics make no assumptions about memory capacity and metabolic costs for processing information, and often rendered inadequate to capture neural activities. Such an approach, however, offers powerful machinery to make computational level predictions for neural behaviors, and our discussion here also primarily aims to make a computational level prediction for pragmatic inferencing behaviors.

The current approach provides a conceptual framework for understanding how such meaning-related expectations might affect outcomes of pragmatic inferencing.

This way of reasoning itself is, of course, not at all new. Within computational linguistics, Bayesian approaches have long provided the inferential underpinnings for many Natural Language Processing systems ranging from machine translation (Blunsom, Cohn, Dyer, & Osborne, 2009) to word segmentation (Goldwater, Griffiths, & Johnson, 2009) and other tasks (see reviews in Jurafsky & Martin, 2009, and Cohen, 2016). At its core, a Bayesian NLP system builds a generative model of the observed signal, representing the uncertainty about some underlying parameter. For speech recognition, for example, the observed signal is the acoustic signal. The system works by modeling the set of underlying word sequences that could plausibly yield the encountered speech, estimating the prior probability of those sequences of words and the likelihood that such words would be realized with the acoustic forms encountered.

Closer to home, in the domain of pragmatics, a number of observations have been made about the critical role of conceptual statuses of intended meanings under the terms such as *accessibility*, *salience*, *topic/focus-hood*, or *prominence* for coreference (Ariel, 1990; Chafe, 1974; Gundel et al., 1993). For instance, it has been demonstrated that a referent's established salience in a discourse (the probability of a speaker choosing to re-mention it in a subsequent sentence) influences the interpretation of an upcoming pronoun (Arnold, 2001). The current inference framework can correctly predict such "salience" effects by assuming that the semantic factors that influence salience are what lead to a higher prior probability of choosing a given referent for re-mention. It has also been demonstrated that pronominal forms are unique in the frequency of their use when a prominent (subject) entity is being re-mentioned (Gordon, Grosz, & Gilliom, 1993; Grosz et al., 1995). The current framework likewise captures this bias in production choices for prominent subjects. The added value of applying the current framework is in being able to pinpoint which factors influence which choice and to thereby move away from underspecified notions of salience and prominence. The factors that influence a speaker's choice among prominent alternative meanings need not be the same as those that determine prominence for the selection among alternative forms (see Kehler and Rohde (2013) for this argument and the explanation of how salience effects and pronominalization biases can be accounted for within a Bayesian model). Without this model, the appeal to salience and prominence can become circular: The typical story goes that a speaker produces a pronoun to refer to something salient/prominent; and how do we know that the referent is salient/prominent? Well, we know because that referent can be pronominalized! Therefore, in addition to making specific predictions and being quantitative, the current approach avoids this circularity and avoids the simple descriptive enumeration of referent properties that might contribute to salience and prominence.

Another reason that this inferential approach is powerful is that both the prior probabilities and likelihood estimates can be learned and fine-tuned with respect to various conditioning factors. Intuitively, our expectations about possible intended meanings and possible linguistic forms used to convey these meanings vary across contexts. Imagine that Sarah is a generally very expressive and affectionate person, who says "I love you" to all of her friends. In that case, her saying "I love you" might not qualitatively differ from (others') saying "I like you". On the other hand, if Sarah is generally very shy and reserved in the way she expresses her emotions, "I love you" might signal a true declaration of love. In other words, one's expectations for $P(E|H)$ (e.g., What would the speaker say if the intended meaning is LOVE vs. LIKE?) can be conditioned on the identity of the speaker (i.e., $P(E|H$, speaker$)$). As a result, the estimates about $P(H|E)$ will also reflect the speaker, where the same linguistic evidence can support different inferences depending who said it. Conversely, the listeners' estimate about the prior ($P(H)$) may itself vary across listeners since it is not necessarily the case that all individuals across ages and contexts should converge on the same expectations about the messages they hypothesize to be most probable.

The importance of interlocutor identities in linguistic comprehension has been discussed in the long tradition of socio-phonetics (e.g., Drager, 2010b; Hay & Drager, 2007; Labov, 2006). Listeners are sensitive to subtle variations of language use, finding traces of the speaker's linguistic background and social membership. The knowledge about speakers, in turn, changes the way the linguistic signal is processed. It has been shown, for example, that native listeners' perception of vowels (diphthongs) can be influenced by the purported origin of the speaker (Canada vs. Detroit, MI; Niedzielski, 1999). Listeners' perception of /aw/ (e.g., **ou**t) or /ay/ (e.g., n**igh**t) reflects what they expected to hear from a speaker from a particular region. One way to explain this phenomenon is that listeners have different expectations for acoustic values that count as /aw/ and /ay/ for a given dialect, e.g., P(F1 and F2 values | /aw/, Canadian) vs. P(F1 and F2 values | /aw/, Detroiter). The same input then inferentially supports perception of different sound categories depending on these expectations.

More generally, a growing literature targets the extent to which perceived social group membership affects language comprehension, including in phonological and syntactic processing (Drager, 2010a,b; Hanulíková, van Alphen, van Goch, & Weber, 2012; Hay & Drager, 2010; Hay, Warren, & Drager, 2006; Staum Casasanto, 2008; Warren, 2017). In pragmatic interpretation, too, assumptions about the speaker and her knowledge/intention as well as linguistic traits can have systematic effects on outcomes of inferences. Recently, Kurumada, Brown, and Tanenhaus (2017) have provided experimental evidence that native English listeners can adapt their pragmatic interpretations of intonation contours according to distributional characteristics of a given speaker's production. Based on our approach, we can now envision that similar reasoning might apply to investigations of silence and turn-taking. A pause of 600 ms can be perceived as a delay of speech onset, signaling hesitation for a speaker who is generally very eloquent. In other words, P(pause = 600 ms | hesitation) can be really high for this speaker. A pause spanning the same amount of time, however, may not be particularly meaningful for another speaker who is generally careful in choosing words and regularly produces long pauses. Thus the approach described above can pinpoint possible loci of such speaker-based adjustment of inferences, which can facilitate accurate inferences about what the speaker meant to communicate.

6. WHERE NEXT?

As we chronicled in the previous sections, past research has provided a wide variety of observations regarding alternative forms and alternative meanings evoked in communication of meaning. Bayesian statistics offers a principled computational-level account as to how these assumptions and knowledge at different levels of representations might be combined to guide the listener's inferences about communicated meanings. Challenges remain, however, in predicting and simulating how such inferences may be implemented in real-time language comprehension with a rich set of contextual information. How can speakers and listeners converge on a particular set of alternative forms and meanings among an infinite number of possibilities? Indeed, much is still unknown about how probabilistic knowledge about linguistic forms and meanings is stored and represented. Would it be feasible to assume that human language users, with a finite cognitive and computational capacity, are engaging in the inferences outlined in Section 5? How do human infants become proficient in the types of inferences discussed above? Resolving these problems clearly goes beyond the scope of the current work. Here we highlight what we take to be promising avenues for future research in which the insights laid out in the current paper could contribute to theoretical and empirical advances in existing scientific endeavors.

Real-time Processing: As mentioned above, a number of experiments utilize psycholinguistic measurements that are time-locked to the unfolding linguistic signal (e.g., eye-movements, ERP). One of the goals of such studies is to account for the time course of semantic and pragmatic understanding of speech: How do listeners derive contextually-enriched interpretations based on the incoming linguistic signal (Breheny, Ferguson, & Katsos, 2013; Degen & Tanenhaus, 2016; Grodner, Klein, Carbary, & Tanenhaus, 2010; Huang & Snedeker, 2009, 2011; Van Berkum, Brown, & Hagoort, 1999)? Not all of these studies subscribe to the idea that listeners are actively comparing alternative forms and meanings. It is, in fact, actively debated whether listeners' moment-by-moment behavior may be compatible with the type of reverse engineering posited under the Bayesian account (Huang & Snedeker, 2018). Regardless of one's theoretical stance, however, it is important to point out that a typical "Visual World" eye-tracking experiment often presupposes some framework closely related to what we have described here. That is, most eye-tracking studies that use visual referents and linguistic instructions encourage the listener to arrive at alternative (intended) meanings based on visual context as well as to evaluate the incoming linguistic signal with respect to the meaning options.

A classic study conducted by Tanenhaus and his colleagues provides a useful illustration of this (Tanenhaus, Spivey-Knowlton, Eberhard, & Sedivy, 1995). Two types of visual scenes were created with four objects each to modulate expectations for alternative intended meanings. In one, listeners saw four distinct object kinds. In the other, they saw two instances of the same object kind (e.g., two apples). In the latter case, a speaker is presumed to be more likely to intend to distinguish the target apple from the distractor apple, whereas she is less likely to do so when there is only one object that can be referred to as "the apple". In other words, these two scenes altered P(Intended meaning = DISTINGUISHING TWO APPLES). These differing expectations in meanings are associated with expectations about alternative linguistic forms. Listeners have some likelihood estimates for the probability that the speaker will produce a modifier qualifying the target, such as a post-nominal prepositional phrase i.e., P(a post-nominal phrase | Intended meaning = DISTINGUISHING TWO APPLES). According to Bayesian inference, the posterior probability of P(Intended meaning = DISTINGUISHING TWO APPLES | a post-nominal phrase) is higher in the scene with two apples due to the increased prior probability of the meaning.

In the behavioral results, Tanenhaus et al. indeed found that a temporarily ambiguous prepositional phrase "Put the apple **on the towel** ..." is more likely to be interpreted as a post-nominal modifier (i.e., the apple that is on the towel) in the presence of two apples. The same phrase **on the towel** is more readily interpreted as a destination (i.e., Move the apple so it will be on the towel) when there is only one apple in sight. This study, and a number of others following it, illuminate the possibility that the linguistic input is processed according to the rich backdrop of contextually-constrained expectations about possible and likely forms and meanings (e.g., Brown-Schmidt & Tanenhaus, 2008; Federmeier, 2007; Ni, Crain, & Shankweiler, 1996; Papadopoulou & Clahsen, 2006; Sedivy, 2002; Van Berkum et al., 1999).

An outstanding question remains as to whether such an approach can scale to contexts with increased complexity. The majority of visual world eye-tracking studies employ a simple visual context (cf. Brown-Schmidt and Tanenhaus (2008)), and the listener's goal is to resolve the speaker's referential intention. In a more naturalistic context, speakers often intend to convey a multitude of referential, communicative, and social meanings. Identifying a set of alternative meanings must encompass many other sources of information. Additionally, speakers avail themselves of a wide variety of linguistic and non-linguistic means to encode their intended meanings. For instance, a unique reference in the scene with two apples in Tanenhaus et al. (1995) can be just as easily achieved by directing eye-gaze or pointing in the direction of one of the apples (in lieu of a post-nominal modifier). Likewise, skilled listeners take into account contextual expectations and task-relevant affordances of possible referents in a scene (Chambers et al., 2002, 2004). Listeners' successful computation of likelihood estimates for a given linguistic option would therefore be well-placed to take into account a wide variety of options. Furthermore, each speaker may differ in their uses of these different options and combinations of them (Grodner & Sedivy, 2011). Future studies must explore whether and how listeners condition their expectations for alternative forms and meanings on contextual factors including individual differences across speakers.

As for the question of how to scale processing models to the potentially infinite set of alternatives (of form and of intended meaning), this is likewise not a new problem. Just as our use of a Bayesian approach echoes its long-standing utility in computational linguistics, cognitive modeling may benefit from a consideration of the strategies in Natural Language Processing domains where pruning the search space is likewise necessary (e.g., syntactic parsing, Vieira & Eisner, 2017). With pruning strategies, the full combinatorial complexity is reined in by eliminating low-probability options. Such pruning in turn may account for the garden-path processing difficulty that comprehenders experience when they struggle to recover a valid but low-probability parse Jurafsky (1996).

Acquisition of accurate inferencing: The way of reasoning laid out in the current paper presents a significant question about language learning. How do young children acquire the ability to enumerate appropriate alternative forms and meanings, combining them in a principled manner? In fact, pragmatic comprehension of utterances has often been identified as a domain of significant difficulty for young children (e.g., Braine & Rumain, 1981; Chierchia, Crain, Guasti, Gualmini, & Meroni, 2006; Eskritt, Whalen, & Lee, 2008; Gualmini, Crain, Meroni, Chierchia, & Guasti, 2001; Lidz & Musolino, 2002; Noveck, 2000; Papafragou, 2006; Papafragou & Tantalou, 2004), particularly for speaker intentions (e.g., Ackerman, 1981; Ackerman, Szymanski, & Silver, 1990), information status (e.g., Arnold, 2008; Stephens, 2010) and identity of discourse referents (e.g., Surian, 1991). Well after other areas of linguistic abilities become proximal to those of adults, school-aged children continue to fail in pragamtically-enriched interpretations of the linguistic input (e.g., Children older than eight years of age often endorse sentences like "Some elephants have trunks" while adults reject them because "some" implies that there are elephants with no trunk (Noveck, 2000; Smith, 1980)).[8]

[8] This does not necessarily mean that young children are incapable of making pragmatic inferences. To the contrary, there are a number of descriptive studies that observe that children can take into account the speaker's (and the agent's) goals and intentions (Csibra & Gergely, 2009; Meltzoff, 1995; Vouloumanos, Onishi, & Pogue, 2012) and draw contextually-situated meanings (e.g., Akhtar & Tomasello, 1996; Clark & Grossman, 1998; Ochs & Schieffelin, 1979). However, the studies cited here repeatedly found that children's responses in pragmatically-involved language comprehension tasks deviated from those of adults in experimental environments. This might be due to a mismatch between children's construals of the experimental tasks and what the tasks are designed to test. When task demands were alleviated, children can show behaviors similar to adults (Katsos & Bishop, 2011).

Understanding the basic mechanisms of adult language comprehension will facilitate the investigation of how a similar process may develop in children. A particularly intriguing possibility is that the slow and protracted pragmatic development may at least partially reflect children and adults' different assignment of probabilities to alternative forms and meanings due to their diverging linguistic experiences (or even that children are not monitoring for potential alternatives to the same extent as adults; see eye-tracking evidence on awareness of scene ambiguity reported by Rabagliati & Robertson, 2017). If so, outcomes of children's and adults' inferences will necessarily be different even if children are endowed with fully-fledged cognitive and computational abilities.

Some recent results suggest that even young language learners are equipped with the basic computational ability to compare alternative forms and meanings in a rational manner (Endress, 2013; Frank, Goodman, & Tenenbaum, 2009; Perfors, Tenenbaum, Griffiths, & Xu, 2011, 2010). For instance, the manner in which 3- and 4-year-olds form categories reflect inductive biases based on differing prior probabilities of exemplars (Xu & Tenenbaum, 2007). Similar expectations about underlying generative models have been discussed in word learning (Frank et al., 2009), and in syntactic learning (Perfors et al., 2011). Simply put, the generative framework allows learners to use inductive reasoning that supports learning of representations beyond what have been directly observed in the data. Such reasoning is thereby expected to help them effectively bootstrap their linguistic knowledge.

One of the best known examples of such inference may be what is often called Principle of Contrast (Clark, 1987, 1990, 1999, 2009).[9] The basic observation is that children as young as a two years old readily associate an unfamiliar label (e.g., "whisk") with an unfamiliar object in the presence of two objects differing in familiarity (e.g., a spoon and a whisk). This can be understood as a bias stemming from a competition between alternative form-meaning mappings. In this case, for the child, the word "spoon" is likely generated when the intended referential meaning is a spoon

[9] Principle of Contrast is often contrasted with Mutual Exclusivity (Golinkoff, Hirsh-Pasek, Bailey, & Wenger, 1992; Markman, Wasow, & Hansen, 2003). Mutual Exclusivity is considered as a normative bias discussed primarily in the domain of novel noun learning. The bias was proposed to explain children's tendency to maintain one-to-one mapping between a noun and an object. We adopt Principle of Contrast here because we assume that the inference process here involves computation of intended meanings and applies more broadly to lexical classes besides nouns, as proposed by Principle of Contrast.

(i.e., high P("spoon" | SPOON)). This proportionally lowers the probability assigned to the speaker's production of other (alternative) word forms (e.g., low P("whisk" | SPOON)), which contributes to a lower posterior probability of P(SPOON | "whisk") and a higher posterior probability of P(WHISK | "whisk"). In other words, even though the learner does not yet have any precise likelihood estimate of P("whisk" | WHISK), their assumption that words (forms) generally compete for the same meaning can allow them to make a new form-meaning mapping in context. If, on the other hand, learners were simply tracking one-to-one mappings between forms and meanings (without considering them as *alternatives*), such mapping would not be readily predictable.

Extending this approach can facilitate our understanding of how children may expand their ability to extract meaning given the linguistic signal. Here we end with an example in the domain of intonational phonology. Preschoolers are generally known to be insensitive to the types of intonational meanings discussed in Section 3, and the ability develops only gradually through their school years (Cruttenden, 1985; Cutler & Swinney, 1987; Gualmini, Maciukaite, & Crain, 2003; Ito, Bibyk, Wagner, & Speer, 2013; Solan, 1980; Speer & Ito, 2009; Wells, Peppe, & Goulandris, 2004). Kurumada and Clark (2016), however, found that preschoolers are capable of interpreting "It LOOKS like a zebra ..." as ¬ZEBRA as long as they can access an alternative form. In their experiment, four-year-olds were at chance when they only heard the contrast of "It looks like a ZEBRA" (noun-focus) and "It LOOKS like a zebra ..." (verb-focus). However, children from the same age group could reliably derive the ¬ZEBRA meaning when the same speaker produced an alternative form "It is a zebra!" to convey the alternative meaning i.e., ZEBRA. The authors consider that the underlying mechanism guiding the children's inferences can be explained in the same manner as the spoon-whisk example. When the more familiar form "it is a zebra" is used to encode one of the alternative meanings (i.e., ZEBRA), the unfamiliar form "It LOOKS like a zebra ..." becomes more likely to encode an alternative meaning available in context. This reasoning may enable young children to construct the form-meaning mapping even before they become capable of drawing the contrastive interpretation through the intonation pattern (i.e., L + H* on the verb and a rising tone at the end) alone. (For a similar finding in understanding of scalar adjectives, see Barner, Brooks, and Bale (2011).) Comparing alternatives thus allows learners to leverage inferences to make otherwise novel mappings between forms and meanings.

Alternatives provide a powerful framework that makes it possible for listeners and learners to infer what the speaker must have meant by choosing a linguistic form in context.

ACKNOWLEDGMENTS

We thank Andrés Buxó-Lugo for data visualization, and Bethany Gardner and Crystal Lee for feedback during editing. We also thank Eve Clark, Marisa Casillas, Madeleine Long, Kara Federmeier, Alice Turk, and Duane Watson for their feedback on an earlier version of the manuscript. H. Rohde is supported in part by a Leverhulme Trust Prize in Languages and Literatures.

REFERENCES

Ackerman, B. P. (1981). When is a question not answered? the understanding of young children of utterances violating or conforming to the rules of conversational sequencing. *Journal of Experimental Child Psychology, 31*(3), 487–507.

Ackerman, B. P., Szymanski, J., & Silver, D. (1990). Children's use of the common ground in interpreting ambiguous referential utterances. *Developmental Psychology, 26*(2), 234–245.

Akhtar, N., & Tomasello, M. (1996). Two–year–olds learn words for absent objects and actions. *British Journal of Developmental Psychology, 14*(1), 79–93.

Altmann, G. T. M., & Kamide, Y. (1999). Incremental interpretation at verbs: Restricting the domain of subsequent reference. *Cognition, 73*(3), 247–264.

Altmann, G. T. M., & Kamide, Y. (2007). The real-time mediation of visual attention by language and world knowledge: Linking anticipatory (and other) eye movements to linguistic processing. *Journal of Memory and Language, 57*, 502–518.

Anderson, J. R. (1990). *The adaptive character of thought.* Hillsdale, NJ: Erlbaum.

Ariel, M. (1990). *Accessing noun phrase antecedents.* Routledge.

Arnold, J. E. (2001). The effects of thematic roles on pronoun use and frequency of reference. *Discourse Processes, 31*, 137–162.

Arnold, J. E. (2008). The bacon not the bacon: How children and adults understand accented and unaccented noun phrases. *Cognition, 108*(1), 69–99.

Arnold, J. E., Hudson Kam, C., & Tanenhaus, M. (2007). If you say thee uh- you're describing something hard: The on-line attribution of disfluency during reference comprehension. *Journal of Experimental Psychology: Learning, Memory, and Cognition, 33*, 914–930.

Arnold, J. E., Tanenhaus, M. K., Altmann, R., & Fagnano, M. (2004). The old and thee, uh, new. *Psychological Science*, 578–582.

Arvaniti, A., & Garding, G. (2007). Dialectal variation in the rising accents of American English. *Papers in Laboratory Phonology, 9*, 547–576.

Au, T. K. (1986). A verb is worth a thousand words: The causes and consequences of interpersonal events implicit in language. *Journal of Memory and Language, 25*, 104–122.

Barner, D., Brooks, N., & Bale, A. (2011). Accessing the unsaid: The role of scalar alternatives in children's pragmatic inference. *Cognition, 118*(1), 84–93.

Barr, D. J. (2001). Trouble in mind: Paralinguistic indices of effort and uncertainty in communication. In S. Santi, I. Guaïtella, C. Cave, & G. Konopczynski (Eds.), *Oralité et gestualité, communication multimodale, intéraction* (pp. 597–600). Paris: L'Harmattan.

Barr, D. J., & Seyfeddinipur, M. (2010). The role of fillers in listener attributions for speaker disfluency. *Language & Cognitive Processes, 25*(4), 441–455.

Beckman, M. E., & Pierrehumbert, J. B. (1986). *Intonation structure in Japanese and English.*

Bergen, L., & Grodner, D. (2012). Speaker knowledge influences the comprehension of pragmatic inferences. *Journal of Experimental Psychology-learning Memory and Cognition, 38*(5), 1450–1460.
Blunsom, P., Cohn, T., Dyer, C., & Osborne, M. (2009). A gibbs sampler for phrasal synchronous grammar induction. In *Proceedings of the joint conference of the 47th annual meeting of the ACL and the 4th international joint conference on natural language processing of the AFNLP* (pp. 782–790).
Bock, J. K. (1987). An effect of the accessibility of word forms on sentence structures. *Journal of Memory and Language, 26*, 119–137.
Bock, K., & Levelt, W. J. M. (1994). Language production: Grammatical encoding. In M. Gernsbacher (Ed.), *Handbook of psycholinguistics* (pp. 945–984). London: Academic Press.
Bögels, S., Kendrick, K. H., & Levinson, S. C. (2015). Never say no… How the brain interprets the pregnant pause in conversation. *PLoS One, 1–15.*
Bortfield, H., Leon, S. D., Bloom, J. E., Schober, M. F., & Brennan, S. E. (2001). Disfluency rates in conversation: Effects of age, relationship, topic, role, and gender. *Language and Speech, 32*, 229–259.
Bosker, H. R., Quené, H., Sanders, T., & de Jong, N. H. (2014). Native 'um's elicit prediction of low-frequency referents, but non-native 'um's do not. *Journal of Memory and Language, 75*, 104–116.
Braine, M. D., & Rumain, B. (1981). Development of comprehension of "or": Evidence for a sequence of competencies. *Journal of Experimental Child Psychology, 31*(1), 46–70.
Breheny, R. R., Ferguson, H. H. J., & Katsos, N. (2013). Investigating the timecourse of accessing conversational implicatures during incremental sentence interpretation. *Language & Cognitive Processes, 28*(4), 443–467.
Brennan, S. E., & Schober, M. F. (2001). How listeners compensate for disfluencies in spontaneous speech. *Journal of Memory and Language, 44*, 274–296.
Brennan, S. E., & Williams, M. (1995). The feeling of another's knowing: Prosody and filled pauses as cues to listeners about the metacognitive states of speakers. *Journal of Memory and Language, 34*, 383–398.
Brown-Schmidt, S., & Tanenhaus, M. K. (2008). Real-time investigation of referential domains in unscripted conversation: A targeted language game approach. *Cognitive Science, 32*, 643–684.
Brown, R., & Fish, D. (1983). The psychological causality implicit in language. *Cognition, 14*, 237–273.
Buxó-Lugo, A., Toscano, J. C., & Watson, D. G. (2018). Effects of participant engagement on prosodic prominence. *Discourse Processes, 55*(3), 305–323.
Calhoun, S. (2004). Phonetic dimensions of intonational categories: The case of L+H* and H*. *Proceedings of Speech Prosody, 2004*, 103–106.
Calhoun, S. (2006). *Information structure and the prosodic structure of English: A probabilistic relationship* (Ph.D. thesis). University of Edinburgh.
Caramazza, A., Grober, E., Garvey, C., & Yates, J. (1977). Comprehension of anaphoric pronouns. *Journal of Verbal Learning and Verbal Behaviour, 16*, 601–609.
Casillas, M. (2013). *Learning to take turns on time: Perception and production processes involved in keeping inter-turn gaps short* (Ph.D. thesis). Stanford University.
Casillas, M., Bobb, S. C., & Clark, E. V. (2016). Turn-taking, timing, and planning in early language acquisition. *Journal of Child Language, 43*(6), 1310–1337.
Chafe, W. (1974). Language and consciousness. *Language, 50*, 111–133.
Chambers, C. G., Tanenhaus, M. K., Eberhard, K. M., Filip, H., & Carlson, G. N. (July 2002). Circumscribing referential domains during real-time language comprehension. *Journal of Memory and Language, 47*(1), 30–49.

Chambers, C. G., Tanenhaus, M. K., & Magnuson, J. S. (2004). Actions and affordances in syntactic ambiguity resolution. *Journal of Experimental Psychology: Learning, Memory, and Cognition, 30*(3), 687–696.
Chierchia, G., Crain, S., Guasti, M. T., Gualmini, A., & Meroni, L. (2006). The acquisition of disjunction: Evidence for a grammatical view of scalar implicatures. In A. H.-J. Do (Ed.), *BUCLD 25 proceedings* (pp. 157–168). Cascadilla Press.
Clark, E. V. (1987). The principle of contrast: A constraint on language acquisition. In B. MacWhinney (Ed.), *Mechanisms of language acquisition* (pp. 1–33). Erlbaum.
Clark, E. V. (1990). On the pragmatics of contrast. *Journal of Child Language, 17*(2), 417–431.
Clark, H. H. (1994). Managing problems in speaking. *Speech Communication, 15*, 243–250.
Clark, H. H. (1996). *Using language*. Cambridge: Cambridge University Press.
Clark, E. V. (1999). Acquisition in the course of conversation. *Studies in the Linguistic Sciences, 29*(2), 1–18.
Clark, E. V. (2009). *First language acquisition* (vol. 2). Cambridge University Press.
Clark, E. V. (2015). Common ground. In B. MacWhinney, & W. O'Grady (Eds.), *The handbook of language emergence* (pp. 328–353). London: Wiley-Blackwell.
Clark, H. H., & Fox Tree, J. E. (2002). Using uh and um in spontaneous speaking. *Cognition, 84*, 73–111.
Clark, E. V., & Grossman, J. B. (1998). Pragmatic directions and children's word learning. *Journal of Child Language, 25*(1), 1–18.
Clark, H. H., & Wasow, T. (1998). Repeating words in spontaneous speech. *Cognitive Psychology, 37*, 201–242.
Clark, E. V., & Wong, A. (2002). Pragmatic directions about language use: Words and word meanings. *Language in Society, 31*, 181–212.
Clayards, M., Tanenhaus, M. K., Aslin, R. N., & Jacobs, R. A. (2008). Perception of speech reflects optimal use of probabilistic speech cues. *Cognition, 108*(3), 804–809.
Cohen, S. B. (2016). *Bayesian analysis in natural language processing. Synthesis lectures on human language technologies*. Morgan and Claypool.
Cole, J. (2015). Prosody in context: A review. *Language, Cognition and Neuroscience, 30*(1–2), 1–31.
Cole, J., Mo, Y., & Hasegawa-Johnson, M. (2010). Signal-based and expectation-based factors in the perception of prosodic prominence. *Laboratory Phonology, 1*(2).
Corley, M., MacGregor, L. J., & Donaldson, D. I. (2007). It's the way that you, er, say it: Hesitations in speech affect language comprehension. *Cognition, 105*, 658–668.
Crawley, R., Stevenson, R., & Kleinman, D. (1990). The use of heuristic strategies in the interpretation of pronoun. *Journal of Psycholinguistic Research, 4*, 245–264.
Cruttenden, A. (1985). Intonation comprehension in ten-year-olds. *Journal of Child Language, 12*, 643–661 (ST – Intonation comprehension in ten–year).
Cruttenden, A. (1997). *Intonation*. Cambridge University Press.
Csibra, G., & Gergely, G. (2009). Natural pedagogy. *Trends in Cognitive Sciences, 13*(4), 148–153.
Cutler, A. (2015). *Native listening: Language experience and the recognition of spoken words*.
Cutler, A., Dahan, D., & Donselaar, W. (1997). Prosody in the comprehension of spoken language: A literature review. *Language and Speech, 40*(2), 141–201.
Cutler, A., & Swinney, D. A. (1987). Prosody and the development of comprehension. *Journal of Child Language, 14*(01), 145–167.
Dahan, D. (2015). Prosody and language comprehension. *Wiley Interdisciplinary Reviews: Cognitive Science, 6*(5), 441–452.
de Ruiter, J. P., Mitterer, H., & Enfield, N. J. (2006). Projecting the end of a speaker's turn: A cognitive cornerstone of conversation. *Language, 82*(3), 515–535.
de Ruiter, J. P., & Cummins, C. (2012). A model of intentional communication: AIRBUS (Asymmetric Intention Recognition with Bayesian Updating of Signals). In S. Brown-Schmidt, J. Ginzburg, & S. Larsson (Eds.), *Proceedings of SemDial (SeineDial), 16th Workshop on the Semantics and Pragmatics of Dialogue* (pp. 149–150).

Degen, J. (2013). *Alternatives in pragmatic reasoning* (Ph.D. thesis). University of Rochester.
Degen, J., & Tanenhaus, M. K. (January 2016). Availability of alternatives and the processing of scalar implicatures: A visual world eye-tracking study. *Cognitive Science, 40*(1), 172–201.
Dilley, L., Breen, M., Gibson, E., Bolivar, M., & Kraemer, J. (2006). A comparison of inter-transcriber reliability for two systems of prosodic annotation: Rap (rhythm and pitch) and tobi (tones and break indices). In *Proceedings of the international conference on spoken language processing. Pittsburgh, PA.*
Dilley, L., Breen, M., Gibson, E., Bolivar, M., & Kraemer, J. (2012). A comparison of inter-transcriber reliability for two systems of prosodic annotation: RaP (rhythm and pitch) and ToBI (tones and break indices). In *Corpus linguistics and linguistic theory* (vol. 8, pp. 277–312).
Dilley, L. C., Ladd, D. R., & Schepman, A. (2005). Alignment of L and H in bitonal pitch accents: Testing two hypotheses. *Journal of Phonetics, 33*(1), 115–119.
Drager, K. (2010a). Sensitivity to grammatical and sociophonetic variability in perception. *Laboratory Phonology, 1*(1), 473–480.
Drager, K. (July 2010b). Sociophonetic variation in speech perception. *Language and Linguistics Compass, 4*(7), 473–480. https://doi.org/10.1111/j.1749-818X.2010.00210.x.
Endress, A. D. (2013). Bayesian learning and the psychology of rule induction. *Cognition, 127*(2), 159–176.
Eskritt, M., Whalen, J., & Lee, K. (2008). Preschoolers can recognize violations of the gricean maxims. *British Journal of Developmental Psychology, 26*(3), 435–443.
Federmeier, K. D. (2007). Thinking ahead: The role and roots of prediction in language comprehension. *Psychophysiology, 44*, 491–505.
Feldman, N. H., & Griffiths, T. L. (2009). A rational account of the perceptual magnet effect. *Proceedings of the 29th Annual Conference of the Cognitive Science Society, 1*, 6.
Fox Tree, J. E. (1995). The effects of false starts and repetitions on the processing of subsequent words in spontaneous speech. *Journal of Memory and Language, 34*, 709–738.
Fox Tree, J. E., & Clark, H. H. (1997). Pronouncing "the" as "thee" to signal problems in speaking. *Cognition, 62*, 151–167.
Frank, M. C., & Goodman, N. D. (2012). Predicting pragmatic reasoning in language games. *Science, 336*, 998.
Frank, M. C., & Goodman, N. D. (2014). Inferring word meanings by assuming that speakers are informative. *Cognitive Psychology, 75*, 80–96. https://doi.org/10.1016/j.cogpsych.2014.08.002.
Frank, M. C., Goodman, N. D., & Tenenbaum, J. B. (2009). Using speakers' referential intentions to model early cross-situational word learning. *Psychological Science, 20*(5), 578–585.
Frederiksen, J. (1981). Understanding anaphora: Rules used by readers in assigning pronominal referents. *Discourse Processes, 4*, 323–347.
Fukumura, K., & van Gompel, P. G. (2010). Choosing anaphoric expressions: Do people take into account likelihood of reference? *Journal of Memory and Language, 62*, 52–66.
Ganong, W. F. (1980). Phonetic categorization in auditory word perception. *Journal of Experimental Psychology: Human Perception and Performance, 6*, 110–125.
Garvey, C., & Caramazza, A. (1974). Implicit causality in verbs. *Linguistic Inquiry, 5*, 459–464.
Gernsbacher, M. A., & Hargreaves, D. J. (1988). Accessing sentence participants: The advantage of first mention. *Journal of Memory and Language, 27*, 699–717.
Gernsbacher, M. A., Hargreaves, D. J., & Beeman, M. (1989). Building and accessing clausal representations: The advantage of the first mention versus the advantage of clause recency. *Journal of Memory and Language, 28*, 735–755.
Givón, T. (1983). Topic continuity in discourse: An introduction. In T. Givón (Ed.), *Topic continuity in discourse: A quantitative, cross-language study* (pp. 1–42). Amsterdam: John Benjamins.

Goldwater, S., Griffiths, T., & Johnson, M. (2009). A bayesian framework for word segmentation: Exploring the effects of context. *Cognition, 112*, 21–54.
Golinkoff, R. M., Hirsh-Pasek, K., Bailey, L. M., & Wenger, N. R. (1992). Young children and adults use lexical principles to learn new nouns. *Developmental Psychology, 28*(1), 99–108.
Goodman, N. D., & Stuhlmüller, A. (2013). Knowledge and implicature: Modeling language understanding as social cognition. *Topics in Cognitive Science.*
Gordon, P. C., Grosz, B. J., & Gilliom, L. A. (1993). Pronouns, names, and the centering of attention in discourse. *Cognitive Science, 17*, 311–347.
Grice, P. (1975). Logic and conversation. In P. Cole, & J. Morgan (Eds.), *Syntax and semantics* (vol. 3, pp. 41–58). New York: Academic Press.
Grice, H. P. (1989). *Studies in the way of words* (vol. 65). Harvard University Press.
Grodner, D. J., Klein, N. M., Carbary, K. M., & Tanenhaus, M. K. (2010). Some, and possibly all, scalar inferences are not delayed: Evidence for immediate pragmatic enrichment. *Cognition, 116*(1), 42–43.
Grodner, D. J., & Sedivy, J. C. (2011). The effect of speaker-specific information on pragmatic inferences. In N. Pearlmutter, & E. Gibson (Eds.), *The processing and acquisition of reference* (pp. 239–272). MIT Press.
Grosz, B. J., Joshi, A. K., & Weinstein, S. (1995). Centering: A framework for modeling the local coherence of discourse. *Computational Linguistics, 21*, 203–225.
Gualmini, A., Crain, S., Meroni, L., Chierchia, G., & Guasti, M. T. (2001). At the semantics/pragmatics interface in child language. In *Proceedings of SALT XI. Ithaca, NY* (pp. 231–247).
Gualmini, A., Maciukaite, S., & Crain, S. (2003). Children's insensitivity with contrastive stress with only. In *Proceedings of the 25th Penn linguistics colloquium* (vol. 9, pp. 87–110). University of Pennsylvania.
Gundel, J. K., Hedberg, N., & Zacharski, R. (1993). Cognitive status and the form of referring expressions in discourse. *Language, 69*(2), 274–307.
Hagoort, P., & van Berkum, J. (May 2007). Beyond the sentence given. *Philosophical Transactions of the Royal Society of London. Series B, Biological Sciences, 362*(1481), 801–811.
Hale, J. (2006). Uncertainty about the rest of the sentence. *Cognitive Science, 30*(4), 609–642.
Hanulíková, A., van Alphen, P. M., van Goch, M. M., & Weber, A. (2012). When one person's mistake is another's standard usage: The effect of foreign accent on syntactic processing. *Journal of Cognitive Neuroscience, 24*(4), 878–887.
Hay, J., & Drager, K. (2010). Stuffed toys and speech perception. *Linguistics, 48*(4), 865–892.
Hay, J., & Drager, K. (September 2007). Sociophonetics. *Annual Review of Anthropology, 36*(1), 89–103. https://doi.org/10.1146/annurev.anthro.34.081804.120633.
Hay, J., Warren, P., & Drager, K. (2006). Factors influencing speech perception in the context of a merger-in-progress. *Journal of Phonetics, 34*(4), 458–484.
Heller, D., Arnold, J. E., Klein, N., & Tanenhaus, M. K. (April 2015). Inferring difficulty: Flexibility in the real-time processing of disfluency. *Language and Speech, 58*(2), 190–203.
Hemforth, B., Konieczny, L., Scheepers, C., Colonna, S., Schimke, S., Baumann, P., & Pynte, J. (2010). Language specific preferences in anaphor resolution: Exposure or gricean maxims? In S. Ohlsson, & R. Catarambone (Eds.), *Proceedings of the 32nd annual conference of the cognitive science society* (pp. 2218–2223).
Hirschberg, J. (1985). *A theory of scalar implicature* (Ph.D. thesis). Philadelphia: University of Pennsylvania.
Horn, L. R. (1972). *On the semantic properties of logical operators in English* (Ph.D. thesis). UCLA.
Huang, Y. T., & Snedeker, J. (2009). Online interpretation of scalar quantifiers: Insight into the semantics-pragmatics interface. *Cognitive Psychology, 58*, 376–415.
Huang, Y. T., & Snedeker, J. (2011). Logic and conversation revisited: Evidence for a division between semantic and pragmatic content in real-time language comprehension. *Language & Cognitive Processes, 26*, 1161–1172 (933044199).

Huang, Y. T., & Snedeker, J. (2018). Some inferences still take time: Prosody, predictability, and the speed of scalar implicatures. *Cognitive Psychology, 102*, 105–126.
Ito, K., Bibyk, S. A., Wagner, L., & Speer, S. R. (2013). Interpretation of contrastive pitch accent in six- to eleven-year-Old English-speaking children (and adults). *Journal of Child Language, 1*, 27.
Ito, K., & Speer, S. R. (2008). Anticipatory effects of intonation: Eye movements during instructed visual search. *Journal of Memory and Language, 58*(2), 541–573.
Jäger, G., & Franke, M. (2014). Pragmatic back-and-forth reasoning. In S. P. Reda (Ed.), *Pragmatics, semantics and the case of scalar implicature* (pp. 170–200). Houndmills, Basingstoke, Hampshire: Palgrave Macmillan.
Järvikivi, J., van Gompel, R., Hyönä, J., & Bertram, R. (2005). Ambiguous pronoun resolution: Contrasting the first mention and subject preference accounts. *Psychological Science, 16*, 260–264.
Jurafsky, D. (1996). A probabilistic model of lexical and syntactic access and disambiguation. *Cognitive Science, 20*(2), 137–194.
Jurafsky, D. (2008). *Pragmatics and computational linguistics*. Wiley-Blackwell.
Jurafsky, D., & Martin, J. H. (2009). *Speech and language processing: An introduction to natural language processing, speech recognition, and computational linguistics*. Prentice-Hall.
Katsos, N., & Bishop, D. V. M. (2011). Pragmatic tolerance: Implications for the acquisition of informativeness and implicature. *Cognition, 120*, 67–81.
Kehler, A., Kertz, L., Rohde, H., & Elman, J. L. (2008). Coherence and coreference revisited. *Journal of Semantics, 25*(1), 1–44.
Kehler, A., & Rohde, H. (2013). A probabilistic reconciliation of coherence-driven and centering-driven theories of pronoun interpretation. *Theoretical Linguistics, 39*, 1–37.
Kehler, A., & Rohde, H. (2018). Prominence and coherence in a Bayesian theory of pronoun interpretation. *Journal of Pragmatics: Special Issue on Prominence in Pragmatics*.
Kendrick, K. H., & Torreira, F. (2015). The timing and construction of preference: A quantitative study. *Discourse Processes, 52*(4), 255–289.
King, J. P. J., Loy, J. E., & Corley, M. (2018). Contextual effects on online pragmatic inferences of deception. *Discourse Processes, 55*(2), 123–135.
Kleinschmidt, D. F., & Jaeger, T. F. (April 2015). Robust speech perception: Recognize the familiar, generalize to the similar, and adapt to the novel. *Psychological Review, 122*(2), 148–203.
Köhne, J., & Demberg, V. (2013). The time-course of processing discourse connectives. In *Proceedings of the 35th annual meeting of the cognitive science society* (pp. 2760–2765).
Kraljic, T., Brennan, S. E., & Samuel, A. G. (2008). Accommodating variation: Dialects, idiolects, and speech processing. *Cognition, 107*(1), 54–81.
Kuperberg, G. R., & Jaeger, T. F. (2016). What do we mean by prediction in language comprehension? *Language, Cognition and Neuroscience, 31*(1), 32–59.
Kurumada, C., Brown, M., Bibyk, S., Pontillo, D. F., & Tanenhaus, M. K. (2014). Is it or isn't it: Listeners make rapid use of prosody to infer speaker meanings. *Cognition, 133*(2), 335–342.
Kurumada, C., Brown, M., Bibyk, S., & Tanenhaus, M. K. (2018). *Probabilistic inferences and adaptation in pragmatic interpretation of contrastive prosody* (under review).
Kurumada, C., Brown, M., & Tanenhaus, M. K. (2012). Prosody and pragmatic inference: It looks like speech adaptation. In N. Miyake, D. Peebles, & R. P. Cooper (Eds.), *Proceedings of the 34th annual conference of the cognitive science society* (pp. 647–652).
Kurumada, C., Brown, M., & Tanenhaus, M. K. (2017). Effects of distributional information on categorization of prosodic contours. *Psychological Bulletin and Review*.
Kurumada, C., & Clark, E. (2016). Pragmatic inferences in context: Learning to interpret contrastive prosody. *Journal of Child Language*.
Labov, W. (2006). A sociolinguistic perspective on sociophonetic research. *Journal of Phonetics, 34*(4), 500–515.

Ladd, D. R. (2008). *Intonational phonology* (2nd ed.). Cambridge University Press.
Ladd, D. R., & Morton, R. (1997). The perception of intonational emphasis: Continuous or categorical? *Journal of Phonetics, 25*(3), 313–342.
Ladd, D. R., & Schepman, A. (2003). Sagging transitions between high pitch accents in English: Experimental evidence. *Journal of Phonetics, 31*(1), 81–112.
Lake, J. K., Humphreys, K. R., & Cardy, S. (2011). Listener vs. speaker-oriented aspects of speech: Studying the disfluencies of individuals with autism spectrum disorders. *Psychonomic Bulletin & Review, 18*, 135–140.
Lambrecht, K. (1994). *Information structure and sentence form: Topic, focus, and the mental representations of discourse referents*. Cambridge: Cambridge University Press.
Levelt, W. J. M. (1989). *Speaking: From intention to articulation*. Cambridge, MA: MIT Press.
Levinson, S. C. (1983). *Pragmatics (Cambridge textbooks in linguistics)*. Cambridge, UK: Cambridge University Press.
Levinson, S. (2000). *Presumptive meanings*. The MIT Press.
Levinson, S. C., & Torreira, F. (2015). Timing in turn-taking and its implications for processing models of language. *Frontiers in Psychology, 6*, 731.
Levy, R. (2008). Expectation-based syntactic comprehension. *Cognition, 106*(3), 1126–1177.
Liberman, M. (1960). Some acoustic correlates of word stress in {A}merican {E}nglish. *Journal of the Acoustical Society of America, 32*, 451–454.
Lidz, J., & Musolino, J. (2002). Children's command of quantification. *Cognition, 84*(2), 113–154.
Loy, J., Rohde, H., & Corley, M. (2017). Effects of disfluency in online interpretation of deception. *Cognitive Science, 41*, 1434–1456.
MacDonald, M. C. (1994). Probabilistic constraints and syntactic ambiguity resolution. *Language & Cognitive Processes, 9*(2), 157–201.
Markman, E. M., Wasow, J. L., & Hansen, M. B. (2003). Use of the mutual exclusivity assumption by young word learners. *Cognitive Psychology, 47*(3), 241–275.
McKoon, G., Greene, S., & Ratcliff, R. (1993). Discourse models, pronoun resolution, and the implicit causality of verbs. *Journal of Experimental Psychology, 19*, 1040–1052.
McMurray, B., & Jongman, A. (2011). What information is necessary for speech categorization? Harnessing variability in the speech signal by integrating cues computed relative to expectations. *Psychological Review, 118*(2), 219–246.
McRae, K., Spivey-Knowlton, M. J., & Tanenhaus, M. K. (1998). Modeling the influence of thematic fit (and other constraints) in on-line sentence comprehension. *Journal of Memory and Language, 38*(3), 283–312.
Meltzoff, A. N. (1995). Understanding the intentions of others: Re-enactment of intended acts by 18-month-old children. *Developmental Psychology, 31*(5), 838.
Neely, J. (1977). Semantic priming and retrieval from lexical memory: Roles of inhibitionless spreading activation and limited capacity attention. *Journal of Experimental Psychology: General, 106*, 226–254.
Ni, W., Crain, S., & Shankweiler, D. (1996). Sidestepping garden paths: Assessing the contribu- tions of syntax, semantics and plausibility in resolving ambiguities. *Language & Cognitive Processes, 11*(3), 283–334.
Niedzielski, N. (1999). The effect of social information on the perception of sociolinguistic variables. *Journal of Language and Social Psychology, 18*(1), 62–85.
Norris, D., Cutler, A., & McQueen, J. M. (2000). The optimal architecture for simulating spoken-word recognition. In *Cognitive science in Australia, 2000: Proceedings of the fifth biennial conference of the Australasian Cognitive Science Society*. Adelaide: Causal Productions.
Norris, D., & McQueen, J. (2008). Shortlist b: A bayesian model of continuous speech recognition. *Psychological Review, 115*(2), 357–395.
Norris, D., McQueen, J. M., & Cutler, A. (2003). Perceptual learning in speech. *Cognitive Psychology, 47*(2), 204–238.

Noveck, I. A. (2000). When children are more logical than adults: Experimental investigations of scalar implicature. *Cognition, 78*(2), 165–188.
Ochs, E., & Schieffelin, B. (1979). *Developmental pragmatics*. Academic Press.
Orena, A. J., & White, K. S. (2015). I forget what That's called! Children's online processing of disfluencies depends on speaker knowledge. *Child Development, 86*(6), 1701–1709.
Oviatt, S. (1995). Predicting spoken disfluencies during human-computer interaction. *Computer Speech & Language, 9*, 19–35.
Papadopoulou, D., & Clahsen, H. (2006). Ambiguity resolution in sentence processing: The role of lexical and contextual information. *Journal of Linguistics, 42*, 109–138.
Papafragou, A. (2006). From scalar semantics to implicature: Children's interpretation of aspectuals. *Journal of Child Language, 33*(4), 721–757.
Papafragou, A., & Tantalou, N. (2004). *Children's computation of implicatures*.
Perfors, A., Tenenbaum, J. B., Griffiths, T. L., & Xu, F. (2011). A tutorial introduction to Bayesian models of cognitive development. *Cognition, 120*(3), 302–321.
Perfors, A., Tenenbaum, J. B., & Wonnacott, E. (2010). Variability, negative evidence, and the acquisition of verb argument constructions. *Journal of Child Language, 37*(3), 607–642.
Pierrehumbert, J. (1980). *The phonology and phonetics of English intonation* (Ph.D. thesis). MIT.
Pierrehumbert, J., & Hirschberg, J. (1990). The meaning of intonational contours in the interpretation of discourse. In P. R. Cohen, J. Morgan, & M. E. Pollack (Eds.), *Intentions in communication* (pp. 271–311). MIT Press.
Pisoni, D. B., & Levi, S. V. (2007). Representations and representational specificity in speech perception and spoken word recognition. *The Oxford Handbook of Psycholinguistics*, 3–18.
Potts, C. (2005). *The logic of conventional implicatures*. Oxford University Press UK.
Rabagliati, H., & Robertson, A. (2017). How do children learn to avoid referential ambiguity? insights from eyetracking. *Journal of Memory and Language, 94*, 15–27.
Recanati, F. (1989). The pragmatics of what is said. *Mind & Language, 4*(4), 295–329.
Rohde, H., & Kehler, A. (2014). Grammatical and information-structural influences on pronoun production. *Language, Cognition, and Neuroscience, Special Issue on Production of Referring Expressions: Models and Empirical Data*, 912–927.
Rosa, E. C., & Arnold, J. E. (2017). Predictability affects production: Thematic roles can affect reference form selection. *Journal of Memory and Language, 94*, 43–60.
Sacks, H., Schegloff, E. A., & Jefferson, G. (1974). A simplest systematics for the organization of turn-taking for conversation. *Language, 50*(4), 696–735.
Samuel, A. G. (2001). Knowing a word affects the fundamental perception of the sounds within it. *Psychological Science, 12*(4), 348–351.
Sauerland, U. (2012). The computation of scalar implicatures: Pragmatic, lexical or grammatical? *Language and Linguistics Compass, 6*(1), 36–49.
Sedivy, J. C. (2002). Invoking discourse-based contrast sets and resolving syntactic ambiguities. *Journal of Memory and Language, 46*, 341–370.
Selkirk, E. (1996). Sentence prosody: Intonation, stress and phrasing. In J. A. Goldsmith (Ed.), *The handbook of phonological theory* (pp. 550–569). Blackwell Publishing.
Shriberg, E. (1996). Disfluencies in switchboard. In *Proceedings international conference on spoken language processing (ICSLP)* (pp. 11–14). Addendum.
Silverman, K., Beckman, M., Pitrelli, J., Ostendorf, M., Wightman, C., Price, P., … Hirschberg, J. (1992). ToBI: A standard for labeling English prosody. In *International conf. On spoken language processing* (vol. 2, pp. 867–870). Banff: International Conf. on Spoken Language Processing.
Silverman, K., & Pierrehumbert, J. (1990). The timing of prenuclear high accents in English. In *Papers in laboratory phonology I* (pp. 72–106). Cambridge University Press.
Smith, C. L. (1980). Quantifiers and question answering in young children. *Journal of Experimental Child Psychology, 30*(2), 191–205. http://www.sciencedirect.com/science/article/pii/0022096580900570.

Smith, V. L., & Clark, H. H. (1993). On the course of answering questions. *Journal of Memory and Language, 32*, 25–38.

Sobel, D., Tenenbaum, J., & Gopnik, A. (2010). Children's causal inferences from indirect evidence: Backwards blocking and Bayesian reasoning in preschoolers. *Cognitive Science, 28*(3), 303–333.

Solan, L. (1980). Contrastive stress and children's interpretation of pronouns. *Journal of Speech & Hearing Research, 23*(3), 688–698.

Sonderegger, M., & Yu, A. C. L. (2010). A rational account of perceptual compensation for coarticu- lation. In S. Ohlsson, & R. Catrambone (Eds.), *Proceedings of the 32nd annual conference of the cognitive science society* (pp. 375–380).

Speer, S. R., & Ito, K. (2009). Prosody in first language acquisition – acquiring intonation as a tool to organize information in conversation. *Language and Linguistics Compass, 3*(1), 90–110.

Spivey, M. J., Grosjean, M., & Knoblich, G. (2005). Continuous attraction toward phonological competitors. *Proceedings of the National Academy of Sciences, 102*(29), 10393–10398.

Spivey, M. J., & Tanenhaus, M. K. (1998). Syntactic ambiguity resolution in discourse: Modeling the effects of referential content and lexical frequency. *Journal of Experimental Psychology: Learning, Memory, and Cognition, 24*(6), 1521–1543.

Staum Casasanto, L. (2008). Does social information influence sentence processing?. In *Proceedings of the 30th annual meeting of the cognitive science society* (pp. 799–804).

Stephens, N. (2010). *Given-before-new: The effects of discourse on argument structure in early child language* (Ph.D. thesis). Stanford University.

Stiller, A. J., Goodman, N. D., & Frank, M. C. (2015). Ad-hoc implicature in preschool children. *Language Learning and Development, 11*(2), 176–190.

Stivers, T., Enfield, N. J., Brown, P., Englert, C., Hayashi, M., Heinemann, T., ... Levinson, S. C. (June 2009). Universals and cultural variation in turn-taking in conversation. *Proceedings of the National Academy of Sciences, 106*(26), 10587–10592.

Stivers, T., Sidnell, J., & Bergen, C. (2018). Children's responses to questions in peer interaction: A window into the ontogenesis of interactional competence. *Journal of Pragmatics, 124*, 14–30.

Surian, L. (1991). Do children exploit the maxim of antecedent in order to interpret ambiguous descriptions? *Journal of Child Language, 18*(2), 451–457.

Tanenhaus, M., Spivey-Knowlton, M., Eberhard, K., & Sedivy, J. (1995). Integration of visual and linguistic information in spoken language comprehension. *Science, 268*(5217), 1632–1634.

Tenenbaum, J. B., & Griffiths, T. L. (2001). Generalization, similarity, and Bayesian inference. *The Behavioral and brain sciences, 24*, 629–640. discussion 652–791.

Turnbull, R., Royer, A. J., Ito, K., Speer, S. R., & Turnbull, R. (2017). Prominence perception is dependent on phonology, semantics, and awareness of discourse. *Language, Cognition and Neuroscience, 0*(0), 1–17.

Van Berkum, J. J. A., Brown, C. M., & Hagoort, P. (1999). Early referential context effects in sentence processing: Evidence from event-related brain potentials. *Journal of Memory and Language, 41*, 147–182.

Vieira, T., & Eisner, J. (Augest 2017). Learning to prune: Exploring the frontier of fast and accurate parsing. *Transactions of the Association for Computational Linguistics (TACL), 5*, 263–278.

Vouloumanos, A., Onishi, K. H., & Pogue, A. (2012). Twelve-month-old infants recognize that speech can communicate unobservable intentions. *Proceedings of the National Academy of Sciences, 109*(32), 12933–12937.

Warren, P. (2017). The interpretation of prosodic variability in the context of accompanying sociophonetic cues. *Laboratory Phonology, 8*(1), 1–21.

Watson, D., Tanenhaus, M. K., & Gunlogson, C. (2008). Interpreting pitch accents in online comprehension: H* vs. L+H*. *Cognitive Science, 32*(7), 1232–1244.
Wells, B., Peppe, S., & Goulandris, N. (2004). Intonation development from five to thirteen. *Journal of Child Language, 31*(04), 749–778.
Wightman, C. W. (2002). Tobi or not tobi?. In *Proceedings of Speech Prosody. Aix-en-Provence, France* (pp. 25–29).
Xiang, M., & Kuperberg, G. (2015). Reversing expectations during discourse comprehension. *Language, Cognition and Neuroscience, 30*(6), 648–672.
Xu, Y. (2005). Speech melody as articulatorily implemented communicative functions. *Speech Communication*, 220–251.
Xu, F., & Tenenbaum, J. B. (2007). Word learning as Bayesian inference. *Psychological Review, 114*(2), 245–272.
Yoon, S. O., & Brown-Schmidt, S. (2016). *Influence of the historical discourse record on language processing in dialogue.* Discourse Processes.

CHAPTER NINE

Reading Ahead by Hedging Our Bets on Seeing the Future: Eye Tracking and Electrophysiology Evidence for Parafoveal Lexical Processing and Saccadic Control by Partial Word Recognition

Elizabeth R. Schotter[1]
Department of Psychology, University of South Florida, Tampa, FL, United States
[1]Corresponding author. E-mail: eschotter@usf.edu

Contents

1. The Reading Speed Paradox 265
2. *Hedging Our Bets*: Coordinating Eye Movements and Language Processing 266
3. *Seeing the Future*: Obtaining Parafoveal Preview of Upcoming Words 267
4. *Reading Ahead*: Using Parafoveal Preview to Initiate Word Recognition 269
   4.1 Traditional Accounts of Parafoveal Preview: *Trans-saccadic Integration* 270
   4.2 Another Account of Parafoveal Preview: *Forced Fixations* 271
   4.3 A Hybrid Account of Parafoveal Preview Effects 274
5. Prior Debates About Parafoveal Processing: *Serial Versus Parallel Lexical Processing* 276
   5.1 Semantic Preview Benefit 277
   5.2 Parafoveal-on-Foveal Effects 278
6. Evidence for Parafoveal Processing From Electrophysiology Studies 280
   6.1 Tests of the Assumptions of Saccadic Control Models of Reading With ERPs 280
   6.2 RSVP With Flanker Paradigm 281
   6.3 Fixation Related Potentials (FRPs) Paradigm 285
7. Modeling Parafoveal Preview Effects 288
   7.1 Modeling Forced Fixations 288
   7.2 Modeling Trans-saccadic Integration 290
8. Summary 291
References 292

Psychology of Learning and Motivation, Volume 68
ISSN 0079-7421
https://doi.org/10.1016/bs.plm.2018.08.011

Abstract

How can cognition influence reading speed (i.e., the timing of saccades—eye fixations) when language processing (i.e., word recognition) seems so much slower? This paper suggests that the answer to this question requires two assumptions: (1) the triggering of saccades is based on partial, rather than complete, word recognition and (2) readers preview words in parafoveal vision prior to fixating them, allowing for a head-start on linguistic processing. I propose a novel theoretical framework about the use of parafoveal preview in reading (i.e., a hybrid account of forced fixations and trans-saccadic integration) and provide evidence for that framework from eye tracking studies, electrophysiology studies, and computational models.

... the relative position of the visual fixation pause in the total complex can at present be estimated only roughly as somewhere in the middle of the [reading] process, between the pre-fixational vision and the utterance, or appreciation.

Raymond Dodge (1907, p. 60)

More succinctly, we do not wait until we look at a word to start reading it, and we are not done by the time we decide to leave. Dodge and his contemporaries had access to only rudimentary empirical methods to study reading, but made incredible theoretical breakthroughs in explaining the process in detail. This paper covers the insights we have gained some 110 years later using advanced empirical methods like high-resolution eye tracking, electroencephalography, and combined recordings of the two. What humbles the modern cognitive scientist is that such advanced equipment only serves to affirm the conclusions those researchers surmised long ago using only logic and simple observation.

Skilled reading progresses at a remarkable speed (i.e., 200–400 words per minute) despite how many complex cognitive processes are involved (e.g., word recognition, sentence parsing, discourse representation, eye-movement control, etc.; Huey, 1908; Rayner, 1998; Rayner, Schotter, Masson, Potter, & Treiman, 2016). This surprising efficiency has led some researchers to insinuate that there is not enough time for language processing to influence moment-to-moment processes in reading (how long the eyes linger before moving—i.e., *fixation durations*; McConkie & Yang, 2003; Vitu, 2003). Despite substantial empirical and computational evidence suggesting that timing of eye movements (i.e., *saccades*, which intervene between fixations) is directly controlled by ongoing language processing (Reingold, Reichle, Glaholt, & Sheridan, 2012), we are still posed with an apparent paradox. How can fixation durations only last, on average, 200 milliseconds when electrophysiology studies suggest that neurocognitive responses to a word's meaning (i.e., the *N400 component*, which peaks around 400 ms; see Kutas & Federmeier, 2011) occur well after that (Rayner & Clifton, 2009)? In this paper, I propose that there are two aspects of the reading process that might help unravel this apparent paradox: (1) *hedging our bets* on word recognition—the triggering of saccades based on partial, rather than complete, word recognition and (2) *seeing the future* words in the sentence—the ability to preview words in parafoveal vision prior to fixating them during natural reading. I argue that these two properties of the reading system contribute to a specialized cognitive architecture that allows for *reading ahead*—allocating attention to words before our eyes reach them. This architecture causes saccade behavior during reading to be composed of several underlying types, which are sometimes determined by a parafoveal preview of a word, and sometimes a combination of both the parafoveal preview and the subsequently foveated information obtained from the word.

1. THE READING SPEED PARADOX

Natural reading is not merely serial identification of fixated words, but rather a highly coordinated process in which prior context, foveal information, and parafoveal preview interact to guide the eyes' pursuit of information from the text (i.e., saccades). If we were to speculate about the natural reading process (in which entire sentences or paragraphs are available at all times and readers are able to fixate the words for however long, and in whatever order they please) based solely on results from single word or Rapid Serial Visual Presentation (RSVP) studies, we might arrive at erroneous conclusions about the speed with which readers can identify words and progress through text. For example, *lexical decision latencies* (the time required to judge whether a letter string is a word or nonword and press a corresponding button) last approximately 600 ms. If we were to subtract the time required for other processes (e.g., button press: ~400 ms; Ratcliff, Gomez & McKoon, 2004), we are left with an estimate for the time needed for word recognition around 200 ms. Because fixation durations last approximately 250 ms, and saccade programming must be initiated approximately 125 ms before then (Becker & Jürgens, 1979; Rayner, Slowiczek, Clifton, & Bertera, 1983; Rayner, 1998, 2009), reading behavior can only be explained by overlapping processes (e.g., word processing, saccade programming, etc.) that depend on parafoveal preview to maintain efficiency (Reichle & Reingold, 2013; Reingold et al., 2012). Moreover, the timing of the N400 suggests that substantial linguistic information is processed by the brain after a saccade would have moved the eyes away from the word, and well after that saccade plan was initiated.

How can we reconcile these seemingly incompatible timelines? Obviously, one answer to this question is that these measurements (i.e., button press reaction time the lexical decision task, saccade latencies/fixation durations in reading, and N400 peak latencies in ERPs) index different aspects of word recognition. For example, the lexical decision task only requires that the person know whether the stimulus is a word or not to do the task, whereas during reading in eye tracking and ERP experiments the person, presumably, needs to also know *which* word it is. In addition, ERP and eye tracking studies, which presumably measure the same process (i.e., language comprehension), do so in very different ways; ERP studies focus on *time-fixed processes* (i.e., those that occur with a consistent timing in response to a particular event—the onset of the presentation of the word) whereas eye tracking studies focus on *time-variable processes* (i.e., those that cause a saccade to leave a word sooner versus later, or cause the eyes to move forward versus backward). When focusing on the measurement advantage of one of these methodologies we may obscure the strength of the other, losing the precision to detect its signal. In this paper, however, I focus on two other explanations: saccades during reading are initiated by hedging our bets on word recognition and, words are preprocessed before they are fixated during parafoveal preview.

2. *HEDGING OUR BETS*: COORDINATING EYE MOVEMENTS AND LANGUAGE PROCESSING

A simple way to solve this apparent paradox is already a fundamental aspect of the architecture of the reading system, as explained by some current models of saccadic control in reading: word recognition processes initiate saccade plans forward from words, but do so based on partial, rather than complete word recognition. For example, *direct control models* assume that saccades are controlled by ongoing cognitive processing (e.g., Reichle, Pollatsek, Fisher, & Rayner, 1998), which contrasts with *indirect control models* (e.g., McConkie & Yang, 2003) that assume cognition does not, or does only rarely, intervene with a random or autonomous saccadic control mechanism (see Reingold et al., 2012). The clearest evidence for direct control comes from the fact that fixation durations last longer on words that are more difficult to process cognitively or linguistically (e.g., words that are uncommon or unexpected; Rayner, 1998, 2009). This suggests that saccades can be initiated sooner when word recognition is in the process of completing faster. In fact, divergence points from a survival analysis of fixation duration distributions show that the effect of word frequency is observed as early as 145 ms after fixation onset, early enough to affect the majority of fixations (i.e., 91%: Reingold et al., 2012).

Some of these direct control models (e.g., E-Z Reader: Reichle et al., 1998; Reichle, 2011) posit that word identification involves *two successive stages*, (1) an initial assessment of familiarity after which saccades are triggered, and (2) subsequent retrieval of meaning, pronunciation, etc (cf. SWIFT: Engbert, Nuthmann, Richter, & Kliegl, 2005; Schad & Engbert, 2012). Because the completion of the first stage of word recognition triggers saccade planning, this architecture implies that readers plan saccades forward from a word *before they have completely identified it*, and consequently before accessing its semantics (i.e., processes that generate the N400). This can explain why readers will skip over a word that makes no sense in the sentence context, provided that it is short and familiar (e.g., in "She was sure she would the all the tests." the underlined word *the* is skipped more often than the correct word, *ace*, and as often as the subsequent felicitous occurrence of *the*; Angele & Rayner, 2013; see also Abbott, Angele, Ahn, & Rayner, 2015; Angele, Laishley, Rayner, & Liversedge, 2014).

Importantly, in order for this aspect of the reading architecture to completely solve the timing issue, readers would have to plan saccades extremely early during fixations. For example, if a fixation were to last 200 ms and saccade plans take approximately 125 ms to execute (Becker & Jürgens, 1979), the plan would have had to have been initiated within the first 75 ms. Given that it takes approximately 60 ms for information about the word to transfer from the retina to the brain (see Reichle & Reingold, 2013), this means that only 15 ms worth of linguistic information would feed into the decision to make a saccade. This seems implausible and too risky to be efficient; therefore, there must be some other source of linguistic information that readers use to determine when they have enough information about a word to make a saccade forward.

3. *SEEING THE FUTURE*: OBTAINING PARAFOVEAL PREVIEW OF UPCOMING WORDS

Reading speed also derives partly from the opportunity to obtain information from a word before directly viewing it, while the word is in parafoveal vision (i.e., during *parafoveal preview*; see Schotter, Angele, & Rayner, 2012

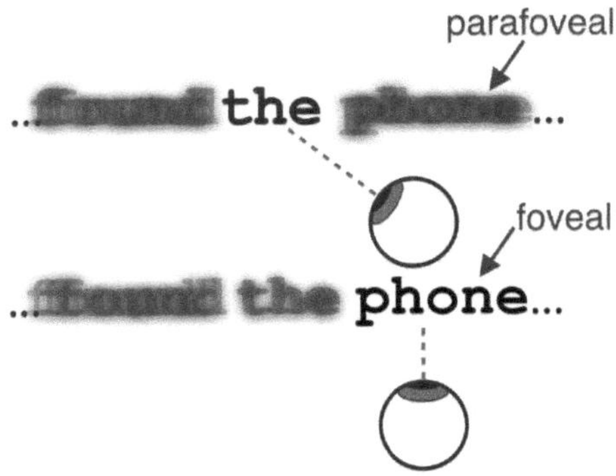

Figure 1 Schematic of the visual fidelity of a word (*phone*) when viewed in different areas of the visual field (i.e., as a parafoveal versus a foveal word) on two successive fixations.

for a review). Parafoveal vision provides poorer perception of the text than direct, *foveal vision* (Fig. 1); thus, any linguistic processing of the word that occurs during parafoveal preview is less precise than processing during foveal perception. Although foveal information enters the processing stream with higher perceptual fidelity, parafoveal information enters the processing stream sooner and therefore has potential to influence reading directly. In the same study using divergence point estimates in a survival analysis, mentioned above Reingold et al. (2012) used a gaze-contingent boundary paradigm, to determine how early the effect of word frequency is observed when the preview is available compared to when it is denied by a parafoveal nonword mask (i.e., when word processing cannot start until direct fixation). That analysis revealed that the effect of word frequency was detectable about 111 ms earlier when preview was available, suggesting that readers had done a substantial amount of word processing based on the parafoveal preview.

As I discuss in detail below, although it is clear that parafoveal preview allows the reading system to gain a head-start on word processing, it is not entirely clear how such preprocessing is subsequently utilized. For example, is parafoveal information used to facilitate processing on the subsequent fixation (i.e., for *trans-saccadic integration* with foveal information: Cutter, Drieghe, & Liversedge, 2015; Pollatsek, Lesch, Morris, & Rayner, 1992; Rayner, 1975) or independently for saccade planning (i.e., *skipping*: Rayner, 2009 or *forced fixations*: Schotter & Leinenger, 2016) and subsequent word recognition? I argue that both processes operate, and the likelihood of each scenario depends on the ease of lexically processing the parafoveal preview.

4. *READING AHEAD*: USING PARAFOVEAL PREVIEW TO INITIATE WORD RECOGNITION

Studies that investigate parafoveal processing in reading use a *gaze-contingent display change (boundary) paradigm* (Rayner, 1975), in which a parafoveal preview stimulus changes to a different *foveal target* word during the saccade to the target (readers are not consciously aware of this change if it occurs during the saccade, due to saccadic suppression; Matin, 1974). Reading time (e.g., fixation duration on the target) is generally compared between conditions in which the preview is available (i.e., *valid*) compared to when it is masked with a different stimulus (i.e., is *invalid*) and, on average, fixation durations are shorter by about 20–50 ms for valid compared to invalid previews (see Rayner, 2009; Reingold et al., 2012; Schotter et al. 2012; Vasilev & Angele, 2017). However, as discussed below, some recent findings using a different experimental design suggest that the notion of "valid" versus "invalid" preview with respect to the target stimulus may not capture the complete nature of parafoveal preview effects (Fig. 2).

The findings from boundary paradigm studies have broader implications for reading because they may explain why readers sometimes misinterpret words that do not change (i.e., because they look like another, more common word—a *higher frequency neighbor*—that would also fit into the sentence). For example, readers sometimes initially misinterpret the word *birth* in the sentence "Mary knew that giving *birch* trees to the park would beautify it." Upon reading the rest of the sentence they would become confused and reread to confirm the correct word meaning (i.e., *birch*; Slattery, 2009; see also Gregg & Inhoff, 2016; Johnson, 2009). Thus, although I

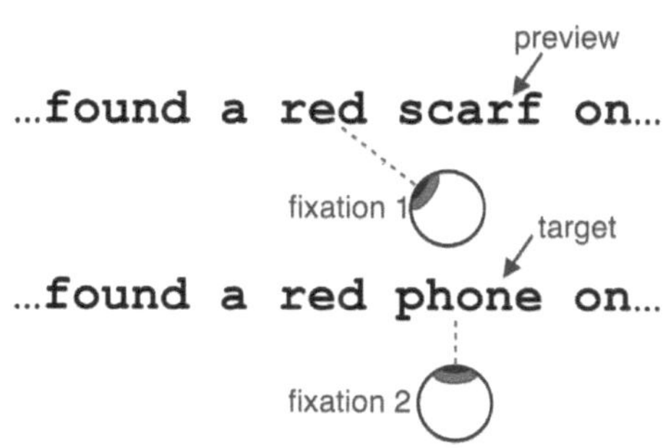

Figure 2 A schematic of the gaze-contingent boundary paradigm (Rayner, 1975). The identity of a particular stimulus changes between two successive fixations—on the first fixation, the preview stimulus (*scarf*) is presented and on the second fixation the target stimulus (*phone*) is presented.

suggested that parafoveal preview and saccade planning based on partial word recognition contribute to reading efficiency, there is also potential for these mechanisms to cause inefficiencies if readers incorrectly presume a word's identity based on the parafoveal preview and can only properly identify it by rereading.

4.1 Traditional Accounts of Parafoveal Preview: *Trans-saccadic Integration*

If parafoveal information is not used to trigger saccades, and a different word is ultimately fixated (i.e., after a display change), what happens to the word recognition process? Traditionally, the preview "validity" effects described above have been explained by trans-saccadic linguistic integration accounts, which posit that parafoveal preview information is merged or compared with foveal target information once directly fixated. On these accounts, processing is easier when preview and target are similar and therefore more easily integrated; consequently, fixation durations are shorter, leading to a *preview benefit* (e.g., Cutter et al., 2015; Pollatsek et al., 1992; Rayner, 1975). An alternative way to conceptualize this effect is that processing is more difficult when preview and target are dissimilar and less easily integrated; consequently, fixation durations are longer, leading to a *preview cost* (e.g., Kliegl, Hohenstein & McDonald, 2013; Marx, Hawelka, Schuster, & Hutzler, 2015). On these accounts, parafoveal information is accessed, but foveal information is required to modify or interact with the information that was pre-activated from the parafovea in order to trigger saccades. This idea was noted by early reading researchers,

> *As in subsequent fixation the peripherally seen word comes to the area of clear vision I conjecture that the inhibitory function of clear perception becomes more prominent, shutting out of the competition all of the residua aroused by the more general peripheral stimulation except those further stimulated by the new, more definite details.*
>
> ***Dodge (1907, p. 57)***

An alternative explanation for these effects is an explanation by misidentification, akin to the misidentification effects described via the *birch/birth* example above (e.g., Slattery, 2009; see also Gregg & Inhoff, 2016; Johnson, 2009). Parafoveal information may initiate the recognition process, but due to poor parafoveal perception, perceptual units within the word are imprecisely activated. Because sublexical representations of the preview and target are similar, these imprecise activations initiate activation of the target word, even though it is not yet presented. This implies that parafoveal preview activates a set of lexical items, rather than a single one.

Most studies that investigated and found evidence for trans-saccadic integration have used nonwords as invalid previews, which are on their own not recognizable and would not lead to the completion of the first stage of word recognition (i.e., saccade triggering). These "invalid" previews *should* impose a cost on reading efficiency, but it is unclear why: possibly due to a difference from the target (i.e., trans-saccadic integration failure) or possibly due to a lack of recognizability.

4.2 Another Account of Parafoveal Preview: *Forced Fixations*

Although trans-saccadic integration has successfully accounted for the majority of findings in gaze-contingent studies (see Cutter et al., 2015; Schotter et al., 2012; Vasilev & Angele, 2017), not all findings are compatible with this idea; a recent set of work that has investigated the effect of preview recognizability (i.e., independent of a relationship to the target) found data that cannot be explained by trans-saccadic integration. Schotter and Leinenger (2016) used the boundary paradigm with unrelated high and low frequency target words (e.g., "The boy found a red *phone/scarf* on his way to school.") and previews that were either identical to the target (e.g., *phone* was a preview for *phone* or *scarf* was a preview for *scarf*) or the other member of the pair (i.e., the unrelated higher- or lower-frequency word, *phone* was a preview for *scarf* or *scarf* was a preview for *phone*, respectively). In the high frequency target condition, the unrelated preview led to longer fixation durations on the target than the identical preview—a *standard preview benefit*. However, in the low frequency target condition, the unrelated preview led to shorter fixation durations on the target than the identical preview—a *reversed preview benefit* (Fig. 3).

Because trans-saccadic integration accounts assume that target processing is better following similar compared to dissimilar previews, they cannot explain this reversed preview benefit, and likewise cannot explain *preview plausibility benefits*, where readers spend less time on a target when a completely unrelated preview was plausible compared to implausible (e.g., Schotter & Jia, 2016; Veldre & Andrews, 2016a). In contrast, an alternative account can explain these effects because it posits that the durations of (at least some) fixations are determined by information obtained during parafoveal preview, regardless of the subsequently fixated target information. On this account, previewing a word that is easy to process (i.e., plausible

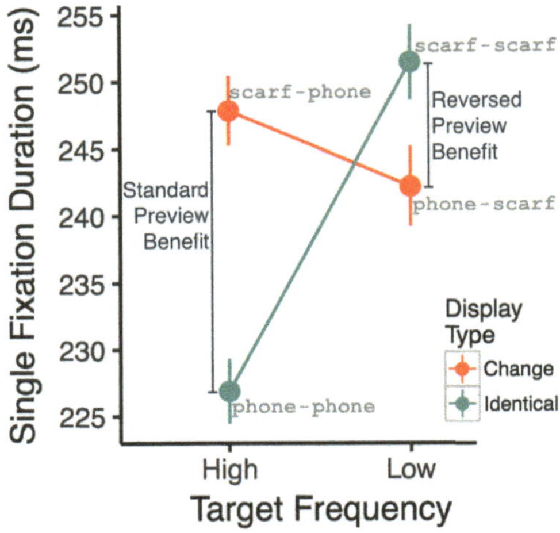

Figure 3 Diagram of the design and results from Schotter and Leinenger (2016, Experiment 1). Single fixation duration on the target word as a function of target frequency, preview frequency, and display type. Example preview-target stimulus pairs are shown next to each condition mean and the comparisons used to show standard preview benefit and reversed preview benefit are noted. Error bars represent $\pm$ 1 SEM. *Figure adapted from Schotter, E. R., & Leinenger, M. (2016). Reversed preview benefit effects: Forced fixations emphasize the importance of parafoveal vision for efficient reading.* Journal of Experimental Psychology: Human Perception and Performance, 42, *2039–2067.*

and/or high frequency) can lead the reading system to not only skip over it (Rayner, 2009), but also to make a *forced fixation* on it (Schotter & Leinenger, 2016). In fact, prior modeling work suggests that parafoveal information can sometimes be processed to such a degree that it reaches the second stage of word identification based on the preview (i.e., semantic processing is initiated and saccades away from the word are initiated: Schotter, Reichle, & Rayner, 2014). Again, this idea is not new,

> *Unquestionably, the peripheral vision is sometimes satisfactory enough in itself without demanding clearer vision. Such is undoubtedly the case in reading for zones bordering on the macula [i.e., the parafovea].*
>
> ***Dodge (1907, pp. 58–59)***

Forced fixations are short single fixations on the word before moving forward, even when the subsequently fixated target word is completely unrelated to the preview. More precisely, forced fixations are cases in which the system would otherwise skip the word—because it seemed easy to recognize—but the skipping decision was made too late, at a point when the saccade toward the to-be-skipped word could not be cancelled (see Becker & Jürgens, 1979; Rayner et al., 1983). During forced fixations, the reader should be insensitive to properties of the fixated word because attention has already shifted to the subsequent one (Morrison, 1984). Schotter, Leinenger, and von der Malsburg (2018) tested this prediction, by explicitly probing the reader's encoding of the text with two-alternative forced choice questions with the preview and target word as response options (see also Schotter & Jia, 2016). Readers reported reading the target word the majority of the time, except when they skipped over it or fixated it for less than 100 ms and did not reread (Fig. 4). If the reader had reread the target word, they were extremely likely to report the target (i.e., rather than the preview), regardless of initial skipping or fixation duration because the re-encounter with the target word overrode the initial representation (Booth & Weger, 2013; Schotter, Tran, & Rayner, 2014).

As Schotter, Leinenger, et al. (2018) noted, during forced fixations on the target it appears as if readers did not encode the word they fixated - they *fixated it with their eyes but skipped it with their minds*. Importantly, in contrast to indirect control theories of reading (e.g., McConkie & Yang, 2003; Vitu, 2003), the likelihood and duration of forced fixations are determined by lexical properties, specifically the ease of lexically processing

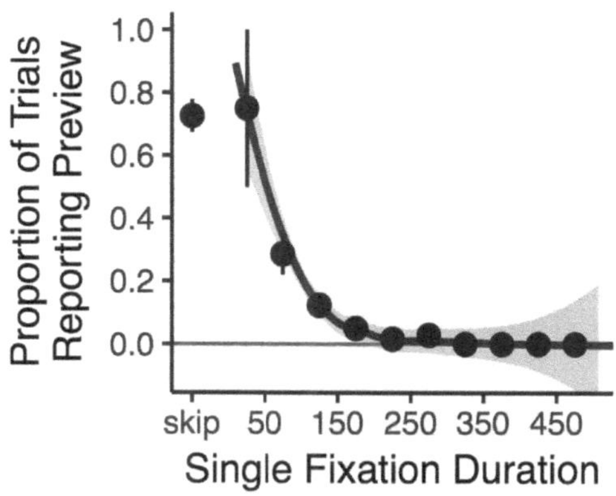

Figure 4 Probability of reporting the preview as a function of fixation behavior (skipping or 50-ms bins of single fixation duration) as a function of which word was reported in response to the probe question for trials in which the target was not reread. The blue line represents a loess smoothed fit line for non-binned data and the grey envelope represents a 95% CI, error bars on binned data represent ± 1 SEM. *Figure adapted from Schotter, E. R., Leinenger, M., & von der Malsburg, T. (2018a). When your mind skips what youreyes fixate: How forced fixations lead to comprehension illusions in reading.* PsychonomicBulletin & Review, *online first.*

the parafoveal preview. Forced fixations are more likely when the parafoveal preview is easy to process (e.g., high frequency and/or plausible) than when it is not, and the saccade should be initiated sooner with increasing ease of recognition.

4.3 A Hybrid Account of Parafoveal Preview Effects

As mentioned above, neither account is sufficient to explain the variety of parafoveal preview effects reported in the literature. Only trans-saccadic integration can account for standard preview benefit effects observed for parafoveal nonword previews that are orthographically similar (Rayner, 1975) or phonologically similar (Miellet & Sparrow, 2004) to the target; because nonword previews are not easy to recognize they would not lead to skips or forced fixations. Thus, preview benefits from nonwords that are linguistically related to the target suggest that sub-lexical information is obtained from the preview and, in some way, is used to facilitate processing once the target is fixated, perhaps via trans-saccadic linguistic integration or perhaps because the nonword preview was mistaken for the target word, due to their similarity. In contrast, only forced fixations can account for fixations that are relatively short on the target (i.e., those that would otherwise be skips) when the preview and target are substantially different. Because trans-saccadic integration relies on similarity, these types of trials should lead to particularly long fixations. However, when the preview is a plausible higher frequency word that is unrelated to the target, as in the *phone-scarf* example above (Schotter & Leinenger, 2016), fixation durations are shorter than when the preview and target are identical. Thus, there are multiple ways in which a parafoveal preview affects saccade triggering in reading, either independent of the target (i.e., via forced fixations) or in combination with it (i.e., via trans-saccadic integration). Importantly, these two mechanisms regard the relationship between word identification and saccadic control, and do not require differing underlying word recognition processes—the difference is a matter of speed or efficiency.

Using a quantile regression analysis, Schotter and Leinenger (2016) demonstrated that fixations of different durations show different effects of the preview, target, or comparison thereof (Fig. 5). Short fixations showed an influence of preview word frequency but not target word frequency whereas long fixations showed an influence of target word frequency but not preview word frequency. Lastly, intermediate quantiles showed an interaction (i.e., an effect of display change). These data suggest that, even

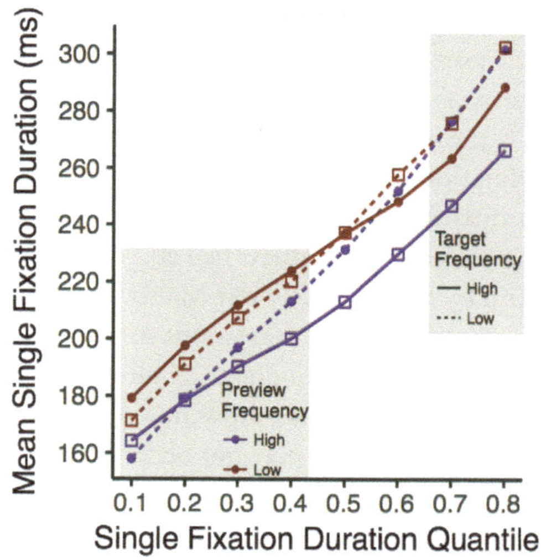

Figure 5 Diagram of the results of the quantile regression from Schotter and Leinenger (2016). Mean single fixation duration as a function of fixation duration quantile, preview frequency, and target frequency. *Figure adapted from Schotter, E. R., & Leinenger, M. (2016). Reversed preview benefit effects: Forced fixations emphasize the importance of parafoveal vision for efficient reading.* Journal of Experimental Psychology: Human Perception and Performance, 42, *2039–2067.*

within a single distribution of fixations obtained within a single study, there is evidence for different populations of fixations that are influenced by different linguistic properties.

This variety of preview effects may be explained by a hybrid mechanism of saccade triggering (Reingold et al., 2012) that includes both forced fixations and trans-saccadic integration (Schotter & Leinenger, 2016). On this hybrid account, forced fixations reflect one saccade triggering mechanism that is engaged when the preview is easy to process, whereas preview costs reflect another mechanism that intervenes when trans-saccadic integration failure occurs (i.e., a lack of similarity between the preview and target). Phenomenologically, forced fixations are short fixations followed by progressive saccades whereas fixations that are influenced by trans-saccadic integration failure are relatively longer and/or followed by refixations or regressions. Because forced fixations are relatively short, there is little to no time for the foveal target information to enter the system (e.g., due to the retina-brain lag; Reichle & Reingold, 2013) by the time the progressive saccade leaving that word is executed, and therefore little influence of the target on the duration or likelihood of these progressive saccades.

If trans-saccadic integration fails, it may only affect subsequent fixations and saccades (i.e., on the post-target word). When forced fixations do not occur (i.e., when the preview is difficult to process), fixations on the target are longer and there is more opportunity to obtain foveal information from it; these fixation durations are influenced exclusively by the frequency of the target (longest quantiles), or by the lack of similarity between the preview and target (intermediate quantiles: Fig. 5).

Subsequent work by Schotter, von der Malsburg, and Leinenger, (2018) found a differential time course of regressions in response to trans-saccadic integration based on preview frequency. Their data suggest that it may not be the case that all short fixations show no effect of trans-saccadic integration failure, but rather the effect may not occur early enough to influence fixation durations on, or saccades out of, the target word. To test whether trans-saccadic integration (or failure thereof) depends on the likelihood of forced fixations (i.e., preview frequency), they manipulated a *probe region* at the end of the sentence that caused one, neither, or both words to become implausible, and inserted a *buffer region* between it and the preview/target region—in which both words remained plausible—to distinguish regressions due to trans-saccadic integration failure from those due to implausibility. Regressions due to trans-saccadic integration failure occurred earlier (i.e., from the target region) when the preview was low frequency, and later (i.e., from the buffer region) when the preview was high frequency. These findings align with the quantile regression analysis reported by Schotter and Leinenger (2016) (see also Fig. 5, above) showing a different time course of effects that depends on preview word properties (i.e., frequency). In contrast, regressions out of the probe region (for trials in which there were not already regressions out of prior regions—due to trans-saccadic integration failure) were only influenced by the plausibility of the target word, not properties of the preview or the display change. This finding aligns with the analysis reported by Schotter et al. (2018) showing that, most of the time, the completion of word recognition is primarily based on information obtained during direct fixation on a word.

5. PRIOR DEBATES ABOUT PARAFOVEAL PROCESSING: *SERIAL VERSUS PARALLEL LEXICAL PROCESSING*

The above hybrid account of parafoveal processing is a relatively new idea (although, as noted, it echoes conjectures made by researchers over a century ago), which considers parafoveal preview in a different way than much of the research that preceded it. The majority of prior research focused on the implications of parafoveal processing for models of saccadic control in reading, particularly the theoretical debate between models that assume that word identification occurs serially across words (e.g., *E-Z Reader*: Reichle et al., 1998; see Reichle, Liversedge, Pollatsek, & Rayner, 2009) or in parallel (e.g., *SWIFT*: Engbert et al., 2005; Schad & Engbert, 2012). These debates have primarily focused on two issues: (1) whether readers obtain semantic information during parafoveal preview (i.e., *semantic preview benefit*) and (2) whether lexical processing of the parafoveal word affects fixation durations on the preceding word (i.e., *parafoveal-on-foveal effects*). These are important theoretical questions, but it may be that the debates, as currently characterized, are ill-formed by setting up alternatives that are too strongly dichotomized. In fact, the framework I have laid out above suggests that these two issues may not tease the models apart because an overlap in linguistic processing of the upcoming word and the fixation on the current word is possible within the architecture of a serial lexical processing model. These two issues have captured the attention and focus of electrophysiology studies of parafoveal preview (see electrophysiological evidence sections below), but in doing so have limited the potential for those studies to fully explore the relationship between cognitive-linguistic processing and saccadic control in reading.

5.1 Semantic Preview Benefit

Because semantic processing of words is presumed to occur after processing of other properties (e.g., orthography and phonology), researchers have argued that evidence for semantic processing during parafoveal preview would be more compatible with parallel lexical processing models (e.g., Engbert et al., 2005; Reilly & Radach, 2006) than serial lexical processing models (e.g., Reichle et al., 1998). Early studies did not find evidence for semantic preview benefit by comparing preview words that were semantically related and unrelated to the subsequently fixated target word (Altarriba, Kambe, Pollatsek, & Rayner, 2001; Rayner, Balota, & Pollatsek, 1986; Rayner, Schotter, & Drieghe, 2014). However, subsequent studies did find evidence for semantic preview benefit using similar experimental designs (Hohenstein & Kliegl, 2014; Hohenstein, Laubrock, & Kliegl, 2010; Yan, Richter, Shu, & Kliegl, 2009; Yan, Zhou, Shu, & Kliegl, 2012; Yang, 2013; Yang, Wang, Tong, & Rayner, 2012). Initially, these differences were explained cross-linguistically; for example, semantic preview benefit might be more likely in orthographically shallow languages like German, or languages for which there is a more direct connection between orthography and semantics like Chinese, than it is in an orthographically deep alphabetic language like English (Laubrock & Hohenstein, 2012; Schotter et al., 2012). However, subsequent work revealed that semantic preview benefit occurs even in English (e.g., for synonymous previews; Schotter, 2013) and the presence depends on whether the initial letter is capitalized (i.e., draws attention to the preview: Rayner & Schotter, 2014), the sentence is constraining (Schotter, Lee, Reiderman, & Rayner, 2015), and, most importantly, whether the preview is plausible in the sentence context (Schotter & Jia, 2016; Veldre & Andrews, 2016a, 2016b; cf. Yang, Wang, Slattery, & Rayner, 2014 for similar effects in Chinese). This latter finding aligns with the idea of forced fixations, described above; to the extent that the parafoveal preview is easy to process (i.e., plausible), it can lead to shorter fixation durations, potentially irrespective of a relationship to the target. As discussed above, this latter explanation is not necessarily incompatible with a serial account because simulations of the data from Schotter (2013) showed that the E-Z Reader model had reached the presumed semantic activation (i.e., second) stage of word recognition based on the parafoveal preview, and consequently had initiated saccade programming (Schotter et al., 2014).

5.2 Parafoveal-on-Foveal Effects

While the above literature focuses on whether semantic information can be accessed from the parafovea to influence subsequent fixations on the target, the core of the theoretical debate surrounding parafoveal-on-foveal (PoF) effects regards whether lexical (and semantic) information accessed in the parafovea influences reading of the previously fixated word (see Drieghe, 2011; Hyönä, 2011; Schotter et al., 2012). Models that assume words are processed serially would not predict lexical properties of the upcoming word to affect saccade decisions for the preceding word, but parallel models should, in theory. Note that parallel models do not explicitly have a mechanism to implement PoF effects; for example, the most prominent of these models (i.e., the SWIFT model; Engbert et al., 2005) only includes a mechanism for *foveal* influences on saccade timing, not *parafoveal* influences (cf. Schad & Engbert, 2012).

Some evidence for PoF effects comes from non-reading tasks. For example, some researchers suggest PoF effects occur when participants make lexical or semantic decisions about a centrally presented word with related or unrelated flankers (Bradshaw, 1974), but these effects did not hold when fixation position was monitored to ensure that subjects did not move their eyes to the flanker words (Inhoff, 1982; Inhoff & Rayner, 1980).

In a reading context, most positive evidence for lexical PoF effects comes from corpus analyses (Kennedy & Pynte, 2005; Kennedy, 1998; Kliegl, 2007; Kliegl, Nuthmann, & Engbert, 2006), which lack tight experimental control that allows for causative inference. In contrast, experimentally controlled studies have not found evidence for lexical PoF effects (Angele & Rayner, 2011; Angele, Slattery, Yang, Kliegl, & Rayner, 2008; Carpenter & Just, 1983; Henderson & Ferreira, 1993; Inhoff, Starr, & Shindler, 2000; Rayner, Juhasz, & Brown, 2007; Staub, Rayner, Pollatsek, Hyona, and Majewski, 2007). Importantly, in a study that combined a corpus analysis and an experimental manipulation of word frequency and of access to the word in parafoveal vision via a moving parafoveal mask, small lexical PoF effects (i.e., 1 ms effects) were observed both when the parafoveal preview was visible and when it was masked (Angele et al., 2015). These data suggest that evidence for lexical PoF effects observed in corpus analyses should be judged with skepticism, because they may not be due to parafoveal lexical information influencing incoming saccade decisions, but rather something about the structure of the sentence (i.e., a property of the currently fixated word is correlated with the frequency of the upcoming word).

It is important to note about PoF effects that the debate regarding the models centers on whether parafoveal processing of the word affects *fixation durations* on the prior word. Some ERP studies, which are reviewed below, expand the definition of these effects beyond fixation durations and interpret ERP components relating to the parafoveal word as PoF effects. However, such evidence of parafoveal processing is not necessarily incompatible with a serial attention shift model (i.e., if it occurs after the lexical processing for the fixated word; see Schotter et al., 2014) and therefore should be considered a parafoveal preview effect rather than a PoF effect. More specifically, those effects reflect the type of processing implied by forced fixations, rather than trans-saccadic integration. That is, parafoveal processing progresses so far as to cause skipping or pre-initiated saccades away from the parafoveal word, and sometimes to higher level word processing (e.g., the kind reflected in the N400).

6. EVIDENCE FOR PARAFOVEAL PROCESSING FROM ELECTROPHYSIOLOGY STUDIES

As mentioned at the beginning of this article, there is an apparent paradox posed by comparisons between eye tracking and ERP studies with respect to the timing of word recognition. Lexical information about words influences eye movements as early as the decision to skip them (i.e., *before* they are fixated), but also affects ERP components that peak sometimes as late as 400 ms after the word is presented. Rayner and Clifton (2009) suggested that this discrepancy may be explained by methodological differences, for example (1) the presentation rate of the RSVP stream in ERP research tends to be much slower than the timing of natural reading, (2) RSVP studies traditionally have presented only one word at a time, eliminating the possibility of parafoveal preview, and (3) ERP research has focused on the N400 component, but there may be earlier, smaller components that reflect effects analogous to those observed in the eye movement record. As will be clear from much of the review that follows, I doubt the first explanation, but the combination of the second and third is quite similar to the hypothesis proposed above - mainly that single word RSVP studies have not allowed for parafoveal preview, and saccade latencies and ERPs (or at least the N400) measure different aspects of the reading process.

6.1 Tests of the Assumptions of Saccadic Control Models of Reading With ERPs

Sereno, Rayner, and Posner (1998) compared the timelines of the effects of lexical properties (i.e., word frequency, and orthographic regularity) using the same stimuli in both a reading task with eye tracking and a lexical decision task using ERPs (see also Dambacher & Kliegl, 2007). They found an early effect of word frequency in the ERP record and proposed a timeline of lexical access during reading that overlapped with (but was not complete by) the time that the saccade program was initiated (see also Sereno & Rayner, 2003). This conclusion is based on deduction and comparisons between independent measurements of eye movements and ERPs, but has also been tested directly by measuring ERPs in the time window leading up to a saccade (Reichle, Tokowicz, Liu, & Perfetti, 2011). Reichle et al.'s participants made lexical decisions to simultaneously presented but spatially separated pairs of letter strings, one at fixation and one in the periphery, such that they had to fixate each word sequentially and make lexical decisions to each one. To investigate the neurocognitive processes that led to the decision to move the eyes, they time-locked the ERP waveform to the onset of the saccade away from the first word and examined the ERP components that occurred prior to it. They included a frequency manipulation to see whether it had an influence on such neurocognitive processes and found an effect of word frequency as early as 125–150 ms before saccade onset, which is consistent with saccade planning time estimates (e.g., Becker & Jürgens, 1979; Rayner et al., 1983) and with the assumptions of a model in which there are two stages of word recognition and the completion of the first one initiates saccade programs (e.g., the E-Z Reader model).

A recent review of ERP evidence of the neurophysiological constraints on the eye-mind link (i.e., the coordination of cognitive processing and saccadic control) suggested that "the temporal constraints are too severe to permit direct lexical control of eye movements without a significant amount of parafoveal processing" (Reichle & Reingold, 2013, p. 1). However, as mentioned by Rayner and Clifton (2009), most ERP studies have not allowed parafoveal preview due to the use of single word RSVP. Since Rayner and Clifton's paper, several new studies have been published that address these limitations, and start to bridge the gap between reading research using eye movements and ERPs. In general, these studies take two approaches: (1) use an *RSVP with flankers paradigm* to measure ERPs in the canonical way (i.e., using RSVP to ensure tight experimental control) and provide parafoveal preview via parafoveal flankers or (2) use a *Fixation Related Potentials (FRPs) paradigm* to measure ERPs during natural reading with saccades and correct for artifacts in the EEG record afterward (Fig. 6). These studies have revealed evidence of parafoveal lexical processing, but as mentioned above, it is unclear whether such processing occurs during or after lexical processing of the fixated word, and therefore whether these data are compatible or incompatible with a serial model of lexical identification. Moreover, many of these studies have not used a reading for comprehension task, so it may be unclear whether the results generalize to reading scenarios. Nonetheless, these studies provide important preliminary data for future work that may directly test these theories.

6.2 RSVP With Flanker Paradigm

Barber, Doñamayor, Kutas, and Münte (2010) tested whether the brain registers semantic information from parafoveal words using a modification to the traditional ERP paradigm. German readers read sentences in which

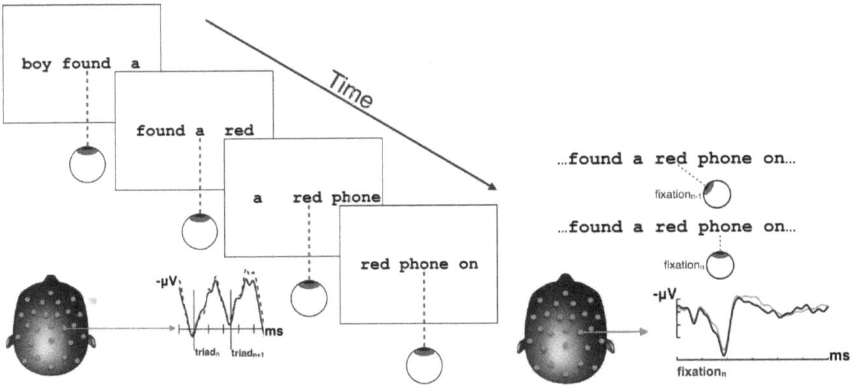

Figure 6 Schematic diagram of the RSVP with flankers paradigm (left) and the Fixation Related Potentials paradigm (right).

the words were presented via RSVP in triads such that the centrally fixated word was flanked two degrees (i.e., in the parafovea) bilaterally by the preceding and succeeding word. On the critical trials, the rightward parafoveal word was either the appropriate word in the sentence, or was replaced with a semantically incongruent word. They found that the incongruence elicited an N400 effect, suggesting that the readers had begun to process the word's meaning prior to it being directly fixated. Barber, Ben-Zvi, Bentin, and Kutas (2011) conducted a similar paradigm in which the flanker strings were pseudowords except for the third triad of each sentence, which was flanked either by two pseudowords, or by a pseudoword/word combination in which the word randomly appeared on the left or right, and was either semantically congruent or incongruent with the sentence. They compared native readers of English and Hebrew, reading in their respective languages, to assess whether any parafoveal effects they observe were related to canonical reading direction. In studies of eye movements in reading, these two languages lead to different asymmetries of the allocation of parafoveal attention: parafoveal processing extends further to the right in English (and other languages that are read from left-to-right) and extends further to the left in Hebrew and Arabic, which are read from right-to-left (Pollatsek, Bolozky, Well, & Rayner, 1981; see also; Jordan et al., 2014). In the ERP study, parafoveally presented incongruent words elicited larger P2 amplitudes than congruent words and this effect was observed only when the word was in the rightward parafovea for English readers and only when the word was presented in the leftward parafovea for Hebrew readers (Barber, Ben-Zvi, Bentin, & Kutas, 2011). Thus, these data suggest that

ERP evidence for parafoveal processing in this paradigm is related to the parafoveal allocation of attention that is generally part of reading.

In subsequent work Barber, Meij, and Kutas (2013) tested whether the degree to which readers access semantic information parafoveally depends on the sentence context. Akin to the idea that foveal load decreases the amount of parafoveal preview (i.e., because processing of the currently fixated word is more attention-consuming; Henderson & Ferreira, 1990), they found larger responses to parafoveally presented incongruous words when the fixated word was predictable, thereby allowing more cognitive resources to be allocated parafoveally. Similar effects were observed by Payne, Stites, and Federmeier (2016); the brain's response to the orthographic legality of the upcoming parafoveal word was modulated by the semantic predictability of the fixated word. Zhang, Li, Wang, and Wang (2015) replicated these findings and found that the effect was observed even when the word that created the congruity was read earlier in the sentence, suggesting that the effects were due to sentence-level integration processes, rather than word-level semantic priming. Thus, these studies suggest that parafoveal processing indicates semantic processing of upcoming words and is modulated by the difficulty of foveal processing. These findings alone do not distinguish between serial and parallel lexical processing models in reading without knowing more about the time course of the parafoveal processing relative to the foveal processing. If foveal load affects parafoveal processing by delaying the time at which it can start, even a serial model can account for foveal load effects (Schotter et al., 2014).

Stites, Payne, and Federmeier (2017) found that readers' brain responses demonstrated graded responses to the parafoveal words' semantic fit into the sentence (i.e., not only an effect of congruous versus incongruous words, but an effect of plausibility or *degree* of expectedness). In contrast, responses to foveal words showed only an effect of anomaly, suggesting that they had done a substantial amount of word processing from the parafovea and therefore needed to allocate less cognitive effort to unexpected, but not anomalous, words when they subsequently fixated them. Simultaneous work showed that the processing of words across the visual field differs for older readers: they showed both a graded effect of semantic fit in response to words viewed both parafoveally and foveally (Payne & Federmeier, 2017), suggesting that older readers process words parafoveally, but perhaps do not complete the processing then and therefore require more foveal processing than younger adults. These data may be compatible with the idea of trans-saccadic integration, but because there were no manipulations of "preview validity" (i.e., the identity of the words did not change between when they were presented parafoveally and foveally) such conclusions are premature.

Only a few studies have addressed the question of whether parafoveal information is integrated with subsequent foveal information. Barber et al. (2013) also manipulated the congruity of the word when it was presented in foveal vision, which allows for two important comparisons. First, they were able to directly compare foveal and parafoveal responses to semantic incongruity, which were clearly larger for foveally presented words (i.e., there was a greater amplitude difference between the congruent and incongruent conditions). Second, they were able to test whether similarity between the parafoveal and foveal stimulus had an effect on the brain's response, and it seems as if it did not. Barber et al. (2013) suggested that this lack of an effect is difficult to explain, but as discussed earlier, trans-saccadic integration is not the sole mechanism for parafoveal preview effects. Subsequent work by Li, Niefind, Wang, Sommer, and Dimigen (2015) on Chinese readers did find an effect of similarity between the parafoveal and foveal stimulus, in that valid previews elicited a more positive N1 component once the word was view foveally than did invalid previews. Clearly more work is needed to determine the conditions under which such trans-saccadic integration can be observed. However, we may not expect all findings in the eye movement literature to have analogous findings in the ERP literature because many aspects of the reading architecture may change during RSVP.

The previously reviewed studies suggest that parafoveal lexical/semantic processing does occur, which is an important demonstration that this paradigm works, but does not address the theoretical debates, or the new theoretical framework I laid out above. One question about the generalizability of these studies is whether, during natural reading, the foveal words in these triads *would have been fixated*. Words tend to be skipped when they are predictable (see Staub, 2015), short and/or high frequency (see Drieghe, Brysbaert, Desmet, & De Baecke, 2004); in the RSVP with flankers paradigm, forcing a reader to fixate a word they might otherwise not have, or for longer than they normally would have if they had fixated it, might encourage parafoveal processing and therefore over-estimate the degree of it relative to natural reading.

Together, these studies align well with much of the research reviewed above: readers are able to access semantic information from upcoming words when reading for comprehension, especially when the word is upcoming (i.e., to the right for readers of German and English, or to the left for readers of Hebrew) and when the fixated word easy to process (i.e., highly expected). As mentioned, it is unclear whether the tight experimental control over the presentation of the words caused more parafoveal processing than readers might otherwise take advantage of (i.e., they may have already moved their eyes to fixate the word before the transition to the next word triad).

6.3 Fixation Related Potentials (FRPs) Paradigm

The theoretical framework I laid out above specifically regards how parafoveal linguistic processing relates to saccadic control during reading. One of the limitations to the RSVP with flankers paradigm is that participants are required to hold fixation, so any aspect of the reading process that relates to attention shifting to the upcoming word prior to the saccade may be diminished or eliminated. In fact Kornrumpf, Niefind, Sommer, and Dimigen (2016) found that electrophysiological measures of parafoveal processing (e.g., an effect they termed the N1 preview effect) were smaller in the RSVP with flankers paradigm than in FRP paradigm when participants were "reading" word lists. Similar findings were reported by Niefind and Dimigen (2016), who replicated the study (i.e., compared the FRP and RSVP with flankers paradigms when subjects were "reading" word lists) and also incorporated a preview validity manipulation. They orthogonally manipulated preview word frequency and validity much like Schotter and Leinenger (2016) (see also Risse & Kliegl, 2014); they found effects of parafoveal word frequency during the fixation on the previous word in the FRP paradigm, but not the ERP with flankers paradigm, perhaps because the participants were less likely to shift attention to the parafovea when they were not making saccades. These findings also contrast with the parafoveal N400 effects in the RSVP with flankers paradigm studies reported above, perhaps because "reading" word lists reduces lexical parafoveal processing; this could be because the stimuli reduced the benefit of prior semantic/syntactic context, or because the task introduced cognitive load associated with doing something other than understanding (i.e., making a decision about whether one of the words in the list was an animal). In fact Kornrumpf et al. (2016) reported that foveal load reduced parafoveal effects in both eye movements and electrophysiological measures.

Baccino and Manunta (2005), who introduced the method of time-locking brain activity and saccades (i.e., FRPs), investigated PoF effects (i.e., effects of parafoveal word properties during fixation on the previous word) while participants made semantic decisions about word pairs that they needed to make saccades between. They found little evidence for semantic PoF effects, but did find effects of parafoveal word status (i.e., whether the parafoveal stimulus was a word or nonword). As noted above, such sublexical PoF effects do not distinguish between serial and parallel models of word identification. Moreover, they only investigated early FRP components (i.e., those within the time window of the fixation) so it is unclear whether there was any evidence for semantic processing, as other studies have shown that parafoveal semantic effects are observed in the N400 component (see review, above). López-Peréz, Dampuré, Hernández-Cabrera, and Barber (2016) replicated Baccino and Manunta (2005) but used a larger time window (i.e., included the N400) and incorporated a display change to investigate preview validity effects. They found a parafoveal N400 effect in the FRPs, but no PoF effect in the eye movements. Such differences in these findings suggest that parafoveal N400 effects should not be considered PoF effects, but rather parafoveal preview effects (see discussion in the PoF effects section, above).

Dimigen, Kliegl, and Sommer (2012) used the boundary paradigm and measured FRPs while subjects "read" word lists to investigate trans-saccadic integration. They reported a *preview positivity* that peaked approximately 200–280 ms after fixation on the target word that only differentiated the identical condition from display change conditions but did not differentiate semantically related from unrelated conditions. This preview positivity might indicate trans-saccadic integration (failure), and the timeline is within a reasonable fixation duration for it to cause increased fixation durations or regressions observed in the eye movement record. This contrasts with a later timing of preview validity effects reported by López-Peréz et al. (2016), which may have been due to a task difference (i.e., two words versus longer lists of words) or a difference in stimuli. Given how few studies have been conducted on preview validity effects in FRPs, more work needs to be done to determine which components are most reflective of trans-saccadic integration processes.

All the studies on FRPs reported above have not used a reading for comprehension task, but rather a task in which subjects are required to make saccades between words and make some sort of decision. Such a task has methodological advantages in that it makes it more likely that subjects will fixate each word, and will be likely to do so only once, which reduces data loss (e.g., due to skipping) and makes data processing and analysis more akin to traditional ERP studies. However, such tasks are not completely analogous to reading for comprehension and it is not entirely clear how such a task change affects the underlying cognitive processes involved. It is clear that task changes not only affect saccadic control (e.g., proofreading leads to less skipping and longer fixations than reading), but also cognitive processing associated with word identification (e.g., word frequency, and sometimes predictability, effects are larger in proofreading than in reading for comprehension; Kaakinen & Hyona, 2010; Schotter, Bicknell, et al., 2014). The only study, published to date, which investigated parafoveal processing using FRPs in a naturalistic reading task was reported by Kretzschmar, Bornkessel-Schlesewsky, and Schlesewsky (2009). They had subjects read sentences that were manipulated such that a particular word was highly expected (i.e., "The opposite of black is ____") and either presented the predicted word (i.e., *white*), a related word (i.e., *yellow*) or unrelated word (i.e., *nice*). Note, however, that the word in the last condition is not only unrelated, but also implausible on its own, so it is unclear whether the effects they observed were due to relatedness or plausibility (see discussion of forced fixations, above). They found a foveal N400 effect (i.e., once the word was fixated) that differentiated the predicted from either unpredicted word, and a parafoveal N400 effect (i.e., during the last fixation prior to landing on the target) that differentiated the unrelated from both the expected and the related word. The differential pattern of findings between the parafoveal effect and the foveal effect suggests that readers had accessed the word's meaning prior to fixating it, but the nature of that process was different than the semantic processing they had done during direct fixation; for example, they suggested an initial process based on semantic spreading activation and then a second process based on pure predictability (see Kretzschmar et al., 2009). Alternatively, these data may reflect a singular process of word recognition in which readers home in on a particular word (i.e., narrow a set of lexical candidates), and do so using a range of information (e.g., prior linguistic context, parafoveal preview, etc.).

It is exciting that a growing number of studies are co-registering eye movements and EEG, and these studies have the potential to help shed light on questions about the extent, nature, and timing of parafoveal lexical processing in reading. At the moment, these studies only provide preliminary evidence that parafoveal lexical processing occurs, even to the point of semantic processing of the parafoveal word, but much more work needs to be done to determine (1) whether this processing occurs after lexical processing of the fixated word, or in parallel with it, and (2) how such processing interacts with saccade planning (e.g., word skipping, forced fixations, trans-saccadic integration failure that may lead to regressions, etc.). As more of these co-registration studies are planned, it is important to connect their design with what we know about the timing of lexical processing and saccadic control, and the constraints they mutually impose on each other (i.e., the assumptions of computational models of oculomotor control and cognitive processing in reading). To this end, the final section of this paper covers current issues in modeling parafoveal preview effects in reading.

7. MODELING PARAFOVEAL PREVIEW EFFECTS

Because the effects of parafoveal preview have important implications for models of saccadic control in reading, it is important to understand how such models would be able to explain the types of effects summarized here. Few simulations have actually addressed parafoveal processing directly, so many of the explanations that follow are merely conjecture—the actual modeling work still needs to be done. Furthermore, these modeling efforts have not yet incorporated any data from the RSVP with flankers paradigm or the FRP paradigm.

7.1 Modeling Forced Fixations

As mentioned above, current models of oculomotor control in reading posit that the decision to start planning a saccade forward from a word is made before complete recognition of it. For example, the E-Z Reader model (Reichle et al., 1998) posits two stages of word identification; the first stage is a *familiarity check* (L_1), which initiates both saccade programming and a second stage of *lexical access* (L_2), in which word identification is completed. Given this architecture, forced fixations can already be explained within the E-Z Reader model (Reichle et al., 1998) and potentially any model that has two-stage lexical processing and two-stage saccade program assumptions (e.g., Engbert et al., 2005). In fact, the forced fixations account that Schotter and Leinenger (2016) proposed was based on the modeling approach of Schotter et al. (2014), who used the E-Z Reader model to simulate data from Schotter (2013) and to estimate how far into lexical processing the model has progressed based on the preview. They found that there were times in which the model reached the second stage of word identification (i.e., L_2, which is initiated by the completion of L_1, which also initiates saccade programming).

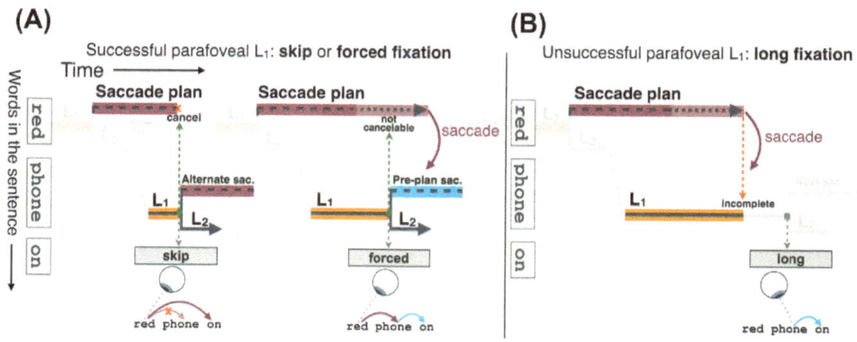

Figure 7 Schematic of how the architecture of the E-Z Reader model can lead to three different types of fixation/saccade behaviors.

Within the architecture of the E-Z Reader model, fixation behavior on the upcoming word depends on the timing of identification of the upcoming word relative to the timing of saccade planning toward that word (Fig. 7). In addition to the two stages of word identification mentioned above (represented with solid horizontal lines), E-Z Reader posits two stages of saccade planning (represented with dashed horizontal lines); during the first stage (M_1: dark purple bar), the current saccade plan can be canceled but during the second stage (M_2: light purple bar) the current saccade cannot be cancelled. If L_1 (orange bar) for the upcoming word completes during the M_1 stage of the saccade program toward it, that saccade would be cancelled and replaced with a skip (Fig. 7A, left side: lower dark purple bar). However, if L_1 for the upcoming word completes during the M_2 stage, the saccade toward that word could not be cancelled, and the system would instead pre-initiate the subsequent saccade program forward from the upcoming word (Fig. 7A, right side: lower blue bar) because saccades can be programmed in parallel (Becker & Jürgens, 1979; Morrison, 1984; Rayner et al., 1983). The pre-initiation of saccade programs, due to an inability to execute a skip, leads to the intervening fixation on that word being relatively short—these are forced fixations (Fig. 7A, right side). When skips or forced fixations do not happen, the word will be fixated for a relatively longer amount of time, and foveal information is needed in order to initiate the progressive saccade (Fig. 7B). If the display had changed in these cases, the system must deal with the fact that the new information obtained in the fovea differs from the information that has been previously obtained from the parafovea and may be subject to trans-saccadic integration failure.

7.2 Modeling Trans-saccadic Integration

The issue of modeling trans-saccadic integration is more complicated than the currently implemented forced fixations scenario. To adequately model how preview and target information are adjudicated when both are represented, one needs a detailed account of the process of word identification in the context of saccadic control in reading that incorporates parafoveal preview. Extant modeling work has not yet done this, and instead has focused on explaining standard preview effects with two approaches that make different assumptions. One approach assumes that lexical processing does not start until fixation on the target (e.g., Pollatsek, Reichle, & Rayner, 2006; Sheridan & Reichle, 2016), analogous to the idea that there is only an influence of the target word. The other approach assumes that lexical processing resets after a display change (e.g., Risse, Hohenstein, Kliegl, & Engbert, 2014), analogous to the idea of trans-saccadic integration failure.

It may be the case that delaying lexical processing is what happens when the parafoveal preview is completely unrecognizable (i.e., an orthographically illegal nonword or an x mask), but such a process could not explain why nonword previews that are orthographically or phonologically related to the target provide a benefit relative to unrelated previews (i.e., trans-saccadic integration). Likewise, completely resetting lexical processing after a display change would not be able to account for these effects, either. Therefore, any model simulations that set out to explain standard preview benefit effects (i.e., those that align with the idea of trans-saccadic integration) must, to some degree, take into account how the words are represented *as they are being processed*. That is, presumably these standard preview benefit effects are due to the "resetting" of processing being less severe or costly when the preview and target are similar compared to when they are ilar. This may mean that sublexical components are activated in the process of word identification and are used for trans-saccadic integration. Alternatively, these sub-lexical units may be used to narrow down a set of lexical candidates—to the extent that the nonword preview leads the reader toward word forms like the target word, that word will be processed more easily if it is subsequently encountered. Teasing apart these alternatives is an exciting area for future research because adequately simulating these effects might give us a better idea of the ongoing process of word representation during reading.

8. SUMMARY

In this paper, I reviewed eye tracking and electrophysiological evidence suggesting that readers start to read a word before looking at it (i.e., obtain parafoveal preview), and discussed different theories about how that preview affects the reading process. I suggested that the key to understanding reading efficiency is understanding that, although the durations of fixations are determined by language processing, readers plan to move their eyes forward from a word before they have completely identified it (i.e., hedge their bets on the success of word recognition). When the word is processed in parafoveal vision and is (or rather, seems) easy to identify, the reader may make saccade decisions (i.e., skip over it or make a forced fixation on it) and potentially not even register what the word was when they did fixate it. When this does not happen, the parafoveal information may be compared or integrated with subsequently obtained foveal information (i.e., via trans-saccadic integration). I suggest that, together, these two processes contribute to a hybrid mechanism of saccadic control.

Many open questions remain about the role of parafoveal processing in reading, and we should not forget to return to the theories and explanations that have been proposed over a century ago, as many of them are still supported by work that is being conducted using methodologies those researchers likely had never imagined were possible. As Dodge notes,

> *The adequacy of a fixation is a relative matter, depending ... on central conditions, and in part on a number of peripheral circumstances ... A complete analysis of the relative influence of these various factors would be a most desirable piece of experimental work for which considerable material is already at hand.*
>
> **Dodge (1907, p. 27)**

To answer all of the questions about parafoveal processing in reading would be an amazing scientific feat; there has already been a great amount of work aimed at addressing these questions by various researchers, across many fields. But, as should be clear from the fact that we have been investigating these questions for over a century, in order to gain more traction on these issues the somewhat insular fields eye movements in reading and electrophysiology of language comprehension should engage in more collaboration and cross-pollination of ideas. For example, what kind of information is used in the decision to trigger saccades, if it is not complete word recognition? Does this architecture suggest that pre-saccade plan and post-saccade plan word recognition are two distinct processes that are influenced by different linguistic properties or cognitive processes, or that there is it a "threshold" in a singular process that creates the distinction? If so, is this threshold something that is fixed for a reader, develops with experience, or is flexible from moment to moment? Currently, models of saccadic control contain parameters that account for three benchmark effects on fixation durations (e.g., word length, frequency, and predictability) but we currently do not have a detailed account of the *process* of word identification during reading.

Electrophysiology studies may help shed light on the word identification in reading, but there are few extant studies using either the RSVP with flankers paradigm or the FRP paradigm, and many used tasks that only *approximate* reading. Furthermore, when taking advantage of the strengths of one methodology, we often lose the rich information (i.e., the signal) afforded by the other. Therefore, more work needs to be done to detail how ERP components index different aspects of parafoveal processing (e.g., which component relates to parafoveal processing of frequency, sensibility, etc., and which component relates to trans-saccadic integration). The framework I have laid out here may help with this endeavor, by allowing researchers to group trials by saccade behavior (e.g., skip, forced fixation, long fixation, etc.) and investigate the neurocognitive components that should, in theory, reflect similar processing (e.g., word recognition based on the preview, or trans-saccadic integration). With respect to the idea of trans-saccadic integration, it is not clear whether it is one process (e.g., a holistic comparison between the preview and target representations), a constellation of processes (e.g., orthographic comparisons, phonological comparisons, semantic comparisons, etc.), or not an integration process at all (i.e., the consequence of the narrowing of lexical candidates based on sublexical and contextual information).

REFERENCES

Abbott, M. J., Angele, B., Ahn, Y. D., & Rayner, K. (2015). Skipping syntactically illegal the previews: The role of predictability. Journal of Experimental Psychology: *Learning, Memory, and Cognition, 41*, 1703–1714.

Altarriba, J., Kambe, G., Pollatsek, A., & Rayner, K. (2001). Semantic codes are not used in integrating information across eye fixations in reading: Evidence from fluent Spanish-English bilinguals. *Perception & Psychophysics, 63*(5), 875–890.

Angele, B., & Rayner, K. (2011). Parafoveal processing of word n + 2 during reading: Do the preceding words matter? *Journal of Experimental Psychology. Human Perception and Performance, 37*, 1210–1220.

Angele, B., & Rayner, K. (2013). Processing the in the parafovea: Are articles skipped automatically? *Journal of Experimental Psychology: Learning, Memory, and Cognition, 39*, 649–662.

Angele, B., Laishley, A. E., Rayner, K., & Liversedge, S. P. (2014). The effect of high-and low-frequency previews and sentential fit on word skipping during reading. *Journal of Experimental Psychology: Learning, Memory, and Cognition, 40*, 1181–1203.

Angele, B., Schotter, E. R., Slattery, T. J., Tenenbaum, T. L., Bicknell, K., & Rayner, K. (2015). Do successor effects in reading reflect lexical parafoveal processing? Evidence from corpus-based and experimental eye movement data. *Journal of Memory and Language, 79*, 76–96.

Angele, B., Slattery, T. J., Yang, J., Kliegl, R., & Rayner, K. (2008). Parafoveal processing in reading: Manipulating n + 1 and n + 2 previews simultaneously. *Visual Cognition, 16*, 697–707.

Baccino, T., & Manunta, Y. (2005). Eye-fixation-related potentials: Insight into parafoveal processing. *Journal of Psychophysiology, 19*(3), 204.

Barber, H. A., Ben-Zvi, S., Bentin, S., & Kutas, M. (2011). Parafoveal perception during sentence reading? An ERP paradigm using rapid serial visual presentation (RSVP) with flankers. *Psychophysiology, 48*(4), 523–531.

Barber, H. A., Doñamayor, N., Kutas, M., & Münte, T. (2010). Parafoveal N400 effect during sentence reading. *Neuroscience Letters, 479*(2), 152–156.

Barber, H. A., Meij, M., & Kutas, M. (2013). An electrophysiological analysis of contextual and temporal constraints on parafoveal word processing. *Psychophysiology, 50*(1), 48–59.

Becker, W., & Jürgens, R. (1979). An analysis of the saccadic system by means of double step stimuli. *Vision Research, 19*, 967–983.

Booth, R. W., & Weger, U. W. (2013). The function of regressions in reading: Backward eye movements allow rereading. *Memory & Cognition, 41*, 82–97.

Bradshaw, J. L. (1974). Peripherally presented and unreported words may bias the perceived meaning of a centrally fixated homograph. *Journal of Experimental Psychology, 103*, 1200–1202.

Carpenter, P. A., & Just, M. A. (1983). What your eyes do while your mind is reading. In K. Rayner (Ed.), *Eye movements in reading: Perceptual and language processes* (pp. 275–307). Hillsdale: Academic Press.

Cutter, M. G., Drieghe, D., & Liversedge, S. P. (2015). How is information integrated across fixations in reading. In A. Pollatsek, & R. Treiman (Eds.), *Oxford handbook of reading* (pp. 245–260). Oxford: Oxford University Press.

Dambacher, M., & Kliegl, R. (2007). Synchronizing timelines: Relations between fixation durations and N400 amplitudes during sentence reading. *Brain Research, 1155*, 147–162.

Dimigen, O., Kliegl, R., & Sommer, W. (2012). Trans-saccadic parafoveal preview benefits in fluent reading: A study with fixation-related brain potentials. *Neuroimage, 62*(1), 381–393.

Dodge (1907).

Drieghe, D. (2011). Parafoveal-on-foveal effects on eye movements during reading. In S. P. Liversedge, I. D. Gilchrist, & S. Everling (Eds.), *Oxford library of psychology. The Oxford handbook of eye movements* (pp. 839–855). New York, NY, US: Oxford University Press.

Drieghe, D., Brysbaert, M., Desmet, T., & De Baecke, C. (2004). Word skipping in reading: On the interplay of linguistic and visual factors. *European Journal of Cognitive Psychology, 16*, 79–103.

Engbert, R., Nuthmann, A., Richter, E. M., & Kliegl, R. (2005). SWIFT: A dynamical model of saccade generation during reading. *Psychological Review, 112*(4), 777.

Gregg, J., & Inhoff, A. W. (2016). Misperception of orthographic neighbors during silent and oral reading. *Journal of Experimental Psychology: Human Perception and Performance, 42*, 799–820.

Henderson, J. M., & Ferreira, F. (1990). Effects of foveal processing difficulty on the perceptual span in reading: Implications for attention and eye movement control. *Journal of Experimental Psychology. Learning, Memory, and Cognition, 16*, 417–429.

Henderson, J. M., & Ferreira, F. (1993). Eye movement control during reading: Fixation measures reflect foveal but not parafoveal processing difficulty. *Canadian Journal of Experimental Psychology, 47*, 201–221.

Hohenstein, S., & Kliegl, R. (2014). Semantic preview benefit during reading. *Journal of Experimental Psychology. Learning, Memory, and Cognition, 40*, 166–190.

Hohenstein, S., Laubrock, J., & Kliegl, R. (2010). Semantic preview benefit in eye movements during reading: A parafoveal fast-priming study. *Journal of Experimental Psychology. Learning, Memory, and Cognition, 36*(5), 1150.

Huey (1908).

Hyönä, J. (2011). Foveal and parafoveal processing during reading. In S. P. Liversedge, I. D. Gilchrist, & S. Everling (Eds.), *Oxford library of psychology. The Oxford handbook of eye movements* (pp. 819–838). New York, NY, US: Oxford University Press.

Inhoff, A. W. (1982). Parafoveal word perception: A further case against semantic preprocessing. *Journal of Experimental Psychology. Human Perception and Performance, 8*, 137–145.

Inhoff, A. W., & Rayner, K. (1980). Parafoveal word perception: A case against semantic preprocessing. *Perception & Psychophysics, 27*, 457–464.

Inhoff, A. W., Starr, M., & Shindler, K. L. (2000). Is the processing of words during eye fixations in reading strictly serial? *Perception & Psychophysics, 62*, 1474–1484.

Johnson, R. L. (2009). The quiet clam is quite calm: Transposed-letter neighborhood effects on eye movements during reading. *Journal of Experimental Psychology: Learning, Memory, and Cognition, 35*, 943–969.

Jordan, T. R., Almabruk, A. A., Gadalla, E. A., McGowan, V. A., White, S. J., Abedipour, L., & Paterson, K. B. (2014). Reading direction and the central perceptual span: Evidence from Arabic and English. *Psychonomic Bulletin & Review, 21*, 505–511.

Kaakinen, J. K., & Hyönä, J. (2010). Task effects on eye movements during reading. *Journal of Experimental Psychology: Learning, Memory, and Cognition, 36*, 1561–1566.

Kennedy, A. (1998). The influence of parafoveal words on foveal inspection time: Evidence for a processing trade-off. In G. Underwood (Ed.), *Eye guidance in reading and scene perception* (pp. 149–179). Oxford: Elsevier.

Kennedy, A., & Pynte, J. (2005). Parafoveal-on-foveal effects in normal reading. *Vision Research, 45*, 153–168.

Kliegl, R. (2007). Toward a perceptual-span theory of distributed processing in reading: A reply to Rayner, Pollatsek, Drieghe, Slattery, and Reichle (2007). *Journal of Experimental Psychology. General, 136*, 530–537.

Kliegl, R., Hohenstein, S., Yan, M., & McDonald, S. A. (2013). How preview space/time translates into preview cost/benefit for fixation durations during reading. *The Quarterly Journal of Experimental Psychology: QJEP, 66*, 581–600.

Kliegl, R., Nuthmann, A., & Engbert, R. (2006). Tracking the mind during reading: The influence of past, present, and future words on fixation durations. *Journal of Experimental Psychology. General, 135*, 12–35.

Kornrumpf, B., Niefind, F., Sommer, W., & Dimigen, O. (2016). Neural correlates of word recognition: A systematic comparison of natural reading and rapid serial visual presentation. *Journal of Cognitive Neuroscience, 28*(9), 1374–1391.

Kretzschmar, F., Bornkessel-Schlesewsky, I., & Schlesewsky, M. (2009). Parafoveal versus foveal N400s dissociate spreading activation from contextual fit. *NeuroReport, 20*, 1613–1618.

Kutas, M., & Federmeier, K. D. (2011). Thirty years and counting: Finding meaning in the N400 component of the event-related brain potential (ERP). *Annual Review of Psychology, 62*, 621–647.

Laubrock, J., & Hohenstein, S. (2012). Orthographic consistency and parafoveal preview benefit: A resource-sharing account of language differences in processing of phonological and semantic codes. *Behavioral and Brain Sciences, 35*(5), 292–293.

Li, N., Niefind, F., Wang, S., Sommer, W., & Dimigen, O. (2015). Parafoveal processing in reading Chinese sentences: Evidence from event-related brain potentials. *Psychophysiology, 52*(10), 1361–1374.

López-Peréz, P. J., Dampuré, J., Hernández-Cabrera, J. A., & Barber, H. A. (2016). Semantic parafoveal-on-foveal effects and preview benefits in reading: Evidence from Fixation Related Potentials. *Brain and Language, 162*, 29–34.

Marx, C., Hawelka, S., Schuster, S., & Hutzler, F. (2015). An incremental boundary study on parafoveal preprocessing in children reading aloud: Parafoveal masks overestimate the preview benefit. *Journal of Cognitive Psychology, 27*, 549–561.

Matin, E. (1974). Saccadic suppression: A review and an analysis. *Psychological Bulletin, 81*, 899–917.

McConkie, G. W., & Yang, S.-N. (2003). How cognition affects eye movements during reading. In J. Hyönä, R. Radach, & H. Deubel (Eds.), *The Mind's eye: Cognitive and applied aspects of eye movement research* (pp. 413–427). Amsterdam: Elsevier.

Miellet, S., & Sparrow, L. (2004). Phonological codes are assembled before word fixation: Evidence from boundary paradigm in sentence reading. *Brain and Language, 90*, 299–310.

Morrison, R. E. (1984). Manipulation of stimulus onset delay in reading: Evidence for parallel programming of saccades. *Journal of Experimental Psychology. Human Perception and Performance, 10*, 667–682.

Niefind, F., & Dimigen, O. (2016). Dissociating parafoveal preview benefit and parafovea-on-fovea effects during reading: A combined eye tracking and EEG study. *Psychophysiology, 53*(12), 1784–1798.

Payne, B. R., & Federmeier, K. D. (2017). Pace yourself: Intraindividual variability in context use revealed by self-paced event-related brain potentials. *Journal of Cognitive Neuroscience, 29*(5), 837–854.

Payne, B. R., Stites, M. C., & Federmeier, K. D. (2016). Out of the corner of my eye: Foveal semantic load modulates parafoveal processing in reading. *Journal of Experimental Psychology. Human Perception and Performance, 42*(11), 1839.

Pollatsek, A., Bolozky, S., Well, A. D., & Rayner, K. (1981). Asymmetries in the perceptual span for Israeli readers. *Brain and Language, 14*, 174–180.

Pollatsek, A., Lesch, M., Morris, R. K., & Rayner, K. (1992). Phonological codes are used in integrating information across saccades in word identification and reading. *Journal of Experimental Psychology. Human Perception and Performance, 18*, 148–162.

Pollatsek, A., Reichle, E. D., & Rayner, K. (2006). Tests of the EZ Reader model: Exploring the interface between cognition and eye-movement control. *Cognitive Psychology, 52*, 1–56.

Ratcliff, Gomez, & McKoon (2004).

Rayner, K. (1975). The perceptual span and peripheral cues in reading. *Cognitive Psychology, 7*, 65–81.

Rayner, K. (1998). Eye movements in reading and information processing: 20 years of research. *Psychological Bulletin, 124*, 372–422.

Rayner, K. (2009). Eye movements and attention in reading, scene perception, and visual search. *The Quarterly Journal of Experimental Psychology: QJEP, 62*, 1457–1506.

Rayner, K., Balota, D. A., & Pollatsek, A. (1986). Against parafoveal semantic preprocessing during eye fixations in reading. *Canadian Journal of Psychology/Revue canadienne de psychologie, 40*(4), 473.

Rayner, K., & Clifton, C. (2009). Language processing in reading and speech perception is fast and incremental: Implications for event-related potential research. *Biological Psychology, 80*(1), 4–9.

Rayner, K., Juhasz, B. J., & Brown, S. J. (2007). Do readers obtain preview benefit from word n + 2? A test of serial attention shift versus distributed lexical processing models of eye movement control in reading. *Journal of Experimental Psychology. Human Perception and Performance, 33*, 230–245.

Rayner, K., Slowiaczek, M. L., Clifton, C., & Bertera, J. H. (1983). Latency of sequential eye movements: implications for reading. *Journal of Experimental Psychology: Human Perception and Performance, 9*, 912–922.

Rayner, K., & Schotter, E. R. (2014). Semantic preview benefit in reading English: The effect of initial letter capitalization. *Journal of Experimental Psychology. Human Perception and Performance, 40*(4), 1617.

Rayner, K., Schotter, E. R., & Dreighe, D. (2014). Lack of semantic parafoveal preview benefit in reading revisited. *Psychonomic Bulletin & Review, 21*, 1067–1072.

Rayner, K., Schotter, E. R., Masson, M. E. J., Potter, M. C., & Treiman, R. (2016). So much to read, so little time: How do we read, and can speed reading help? *Psychological Science in the Public Interest, 17*, 4–34.

Reichle (2011).

Reichle, E. D., Liversedge, S. P., Pollatsek, A., & Rayner, K. (2009). Encoding multiple words simultaneously in reading is implausible. *Trends in Cognitive Sciences, 13*, 115–119.

Reichle, E. D., Pollatsek, A., Fisher, D. L., & Rayner, K. (1998). Toward a model of eye movement control in reading. *Psychological Review, 105*, 125–157.

Reichle, E. D., & Reingold, E. M. (2013). Neurophysiological constraints on the eye-mind link. *Frontiers in Human Neuroscience, 7*, 1–6.

Reichle, E. D., Tokowicz, N., Liu, Y., & Perfetti, C. A. (2011). Testing an assumption of the E-Z Reader model of eye-movement control during reading: Using event-related potentials to examine the familiarity check. *Psychophysiology, 48*, 993–1003.

Reilly, R. G., & Radach, R. (2006). Some empirical tests of an interactive activation model of eye movement control in reading. *Cognitive Systems Research, 7*, 34–55.

Reingold, E. M., Reichle, E. D., Glaholt, M. G., & Sheridan, H. (2012). Direct lexical control of eye movements in reading: Evidence from a survival analysis of fixation durations. *Cognitive Psychology, 65*, 177–206.

Risse, S., Hohenstein, S., Kliegl, R., & Engbert, R. (2014). A theoretical analysis of the perceptual span based on SWIFT simulations of the n + 2 boundary paradigm. *Visual Cognition, 22*(3–4), 283–308.

Risse, S., & Kliegl, R. (2014). Dissociating preview validity and preview difficulty in parafoveal processing of word n + 1 during reading. *Journal of Experimental Psychology. Human Perception and Performance, 40*(2), 653.

Schad, D. J., & Engbert, R. (2012). The zoom lens of attention: Simulating shuffled versus normal text reading using the SWIFT model. *Visual Cognition, 20*(4–5), 391–421.

Schotter, E. R. (2013). Synonyms provide semantic preview benefit in English. *Journal of Memory and Language, 69*, 619–633.

Schotter, E. R., Angele, B., & Rayner, K. (2012). Parafoveal processing in reading. *Attention, Perception, & Psychophysics, 74*, 5–35.

Schotter, E. R., & Jia, A. (2016). Semantic and plausibility preview benefit effects in English: Evidence from eye movements. *Journal of Experimental Psychology. Learning, Memory, and Cognition, 42*, 1839–1866.

Schotter, E. R., Lee, M., Reiderman, M., & Rayner, K. (2015). The effect of contextual constraint on parafoveal processing in reading. *Journal of Memory and Language, 83*, 118–139.

Schotter, E. R., & Leinenger, M. (2016). Reversed preview benefit effects: Forced fixations emphasize the importance of parafoveal vision for efficient reading. *Journal of Experimental Psychology. Human Perception and Performance, 42*, 2039–2067.

Schotter, E. R., Leinenger, M., & von der Malsburg, T. (2018). When your mind skips what your eyes fixate: How forced fixations lead to comprehension illusions in reading. *Psychonomic Bulletin & Review* (in press).

Schotter, E. R., Reichle, E. D., & Rayner, K. (2014). Rethinking parafoveal processing in reading: Serial attention models can account for semantic preview benefit and n+2 preview effects. *Visual Cognition, 22*, 309–333.

Schotter, E. R., von der Malsburg, T., & Leinenger, M. (2018). Forced fixations, trans-saccadic integration, and word recognition: Evidence for a hybrid mechanism of saccade triggering in reading. *Journal of Experimental Psychology Learning Memory and Cognition* (in press).

Sereno, S. C., & Rayner, K. (2003). Measuring word recognition in reading: Eye movements and event-related potentials. *Trends in Cognitive Sciences, 7*(11), 489–493.

Sereno, S. C., Rayner, K., & Posner, M. I. (1998). Establishing a time-line of word recognition: Evidence from eye movements and event-related potentials. *Neuroreport, 9*(10), 2195–2200.

Sheridan, H., & Reichle, E. D. (2016). An analysis of the time course of lexical processing during reading. *Cognitive Science, 40*(3), 522–553.

Slattery, T. J. (2009). Word misperception, the neighbor frequency effect, and the role of sentence context: Evidence from eye movements. *Journal of Experimental Psychology. Human Perception and Performance, 35*, 1969–1975.

Staub, A., Rayner, K., Pollatsek, A., Hyönä, J., & Majewski, H. (2007). The time course of plausibility effects on eye movements in reading: Evidence from noun-noun compounds. *Journal of Experimental Psychology. Learning, Memory, and Cognition, 33*, 1162–1169.

Staub, A. (2015). The effect of lexical predictability on eye movements in reading: Critical review and theoretical interpretation. *Language and Linguistics Compass, 9*, 311–327.

Stites, M. C., Payne, B. R., & Federmeier, K. D. (2017). Getting ahead of yourself: Parafoveal word expectancy modulates the N400 during sentence reading. *Cognitive, Affective & Behavioral Neuroscience, 17*(3), 475–490.

Vasilev, M. R., & Angele, B. (2017). Parafoveal preview effects from word N + 1 and word N + 2 during reading: A critical review of Bayesian meta-analysis. *Psychonomic Bulletin & Review, 24*, 666–689.

Veldre, A., & Andrews, S. (2016a). Is semantic preview benefit due to relatedness or plausibility? *Journal of Experimental Psychology. Human Perception and Performance, 42*, 939–952.

Veldre, A., & Andrews, S. (2016b). Parafoveal preview effects depend on both preview plausibility and target predictability. *The Quarterly Journal of Experimental Psychology: QJEP*, 1–12.

Vitu, F. (2003). The basic assumptions of E-Z Reader are not well-founded. *Behavioral and Brain Sciences, 26*, 506–507.

Yang, J. (2013). Preview effects of plausibility and character order in reading Chinese transposed words: Evidence from eye movements. *Journal of Research in Reading, 36*(S1).

Yang, J., Li, N., Wang, S., Slattery, T. J., & Rayner, K. (2014). Encoding the target or the plausible preview word? The nature of the plausibility preview benefit in reading Chinese. *Visual Cognition, 22*, 193–213.

Yang, J., Wang, S., Tong, X., & Rayner, K. (2012). Semantic and plausibility effects on preview benefit during eye fixations in Chinese reading. *Reading and Writing, 25*, 1031–1052.
Yan, M., Richter, E., Shu, H., & Kliegl, R. (2009). Readers of Chinese extract semantic information from parafoveal words. *Psychonomic Bulletin & Review, 16*, 561–566.
Yan, M., Zhou, W., Shu, H., & Kliegl, R. (2012). Lexical and sublexical semantic preview benefits in Chinese reading. *Journal of Experimental Psychology. Learning, Memory, and Cognition, 38*(4), 1069.
Zhang, W., Li, N., Wang, X., & Wang, S. (2015). Integration of sentence-level semantic information in parafovea: Evidence from the RSVP-flanker paradigm. *PLoS One, 10*(9), e0139016.

CHAPTER TEN

Individual Differences in the Real-Time Neural Dynamics of Language Comprehension

Darren Tanner[1], Maria Goldshtein and Benjamin Weissman
Department of Linguistics, University of Illinois at Urbana-Champaign, Champaign, IL, United States
[1]Corresponding author: E-mail: dstanner@gmail.com

Contents

1. Introduction 300
2. Language-Related ERP Components 305
3. Some Quantitative Individual Differences 307
4. Qualitative Individual Differences in Non-native and Bilingual Morphosyntactic Processing 310
5. Monolingual Morphosyntactic, Semantic, and Thematic Processing and Some Psycholinguistic Conundrums 314
6. Higher-Order Inferencing and Multi-Modal Communication 319
7. Conclusions and Challenges for the Future 327
References 329

Abstract

Recordings of event-related brain potentials (ERPs) provide a rich source of information about the cognitive systems supporting real-time language use. However, the interpretation of ERPs can be complicated by individual differences that aren't reflected in traditional analyses or visualizations. This is problematic, as failure to recognize important and systematic individual differences has in some cases led to inappropriate interpretations of ERP effects, with neurocognitive models of language comprehension sometimes being built on these inappropriate interpretations. In this chapter we review work, largely from our lab, on individual differences in ERP studies of language comprehension and discuss the promise of work on individual differences, as well as the challenges. In some cases individual differences in ERPs manifest themselves quantitatively (i.e., systematic differences in effect amplitudes), but in other more complex cases, qualitatively (i.e., different types of effects in different individuals). We will describe work on individual differences in morphosyntactic and semantic processing in both native and nonnative language processing, as

Psychology of Learning and Motivation, Volume 68
ISSN 0079-7421
https://doi.org/10.1016/bs.plm.2018.08.007

well as multi-modal communication and higher-order pragmatic inferencing. In this last vein we will describe some nascent work done in our lab using unsupervised machine learning algorithms to better understand underlying patterns of qualitative individual differences in the processing of scalar implicatures. We conclude by laying out some challenges and suggestions for future work.

1. INTRODUCTION

Language comprehension is rapid and dynamic, necessitating the incorporation of information from sensory representations (e.g., auditory and/or visual information) with information about grammar, word meaning, and world knowledge stored in long term memory, as well as speaker intent, perspective, and other pragmatic inferences. Event-related brain-potentials (ERPs) provide an excellent measure of these processes as they unfold in real time, as ERPs can provide millisecond-level temporal resolution of cortical electrical activity and can be recorded non-invasively at the scalp. However, traditional ERP visualization in publications uses an average-of-averages method, focusing on effects in grand mean waveforms, and statistical analyses typically use parametric tests for differences in central tendency (e.g., mean amplitude). In very many cases, this averaging process has desirable effects. Indeed, a key facet of ERPs' physiological definition is that they reflect neural electrical signal that is both time- and phase-locked to stimuli of interest (e.g., particular words in sentences, the focus of this chapter), and averaging can extract this signal. Averaging is generally carried out in a two-stage process. First, electrical signals are averaged across trials within experimental conditions within participants to derive subject-level ERPs for each condition. Grand mean ERP waveforms for visualization purposes are derived through the second level of averaging in which the subject-level, condition-specific ERPs are then averaged again across participants. The logic of the averaging process is that it should extract the ERP signal (that is, the electrophysiological activity that is consistently elicited by stimuli in the condition of interest) while the noise (that is, random electrophysiological activity) should approximate zero as the number of stimuli and participants increases. The grand mean waveforms thus depict the central tendency of brain responses across both trials and participants.

In many cases this averaging process has the desired consequences described above of increasing the signal-to-noise ratio and revealing cortical electrical activity tied to processing, and can be very revealing about how language is comprehended in real time. However, there are at least two cases in which the averaging process can miss systematic neural activity observable at the scalp that is tied to dynamic information processing. The first case, which we will not go in depth into here, involves changes in oscillatory neural signals that are time-, but not phase-locked to the eliciting stimulus. In these cases, event-related power changes in various frequency bands in the electroencephalogram (EEG) may be obscured in the averaged ERP because of phase mis-alignment, though they would be detectable in spectrographic analyses of single trial dynamics. Here it is the first, trial-level stage of averaging that would lead to systematic distortion of the signal available in the EEG recording (see, e.g., Bastiaansen, Mazaheri, & Jensen, 2012). The second case involves scenarios where the second stage (between-participant) averaging process leads to inappropriate inferences about processing in the general population. These bogus inferences can arise either when ERP effects are present in grand mean waveforms and reliable in statistical analyses, even when they were not present in most (or even any) of the individuals' ERP waveforms (through topographic or temporal distortion), or when ERP effects are found in some (or many) individuals, but are not reflected in the grand mean or corresponding statistical tests. That is, in some cases, there are meaningful between-participant differences, either quantitative or qualitative, in event-related neural activity that traditional ERP visualization and quantification assume to be noise, but which may in fact be meaningful signal. These individual differences in language processing, with a specific focus on EPRs, are the focus of the present chapter.

Individual differences have not generally been the focus of much the work using ERPs to study language processing. The reasons for this are multi-fold. Although there have always been strong counterpoints, the dominant paradigms in theoretical linguistics in the second half of the 20th century focused on invariant traits of the human language faculty and focused on commonalities among languages and language users, to the exclusion of variation, variability, and individual differences in grammatical and semantic knowledge. This focus in theoretical linguistics led to a similar narrow focus in more experimental domains of linguistic inquiry. In experimental and cognitive psychology, the dominant experimental paradigms have generally used within-subjects (repeated measures) designs. While repeated measures designs do not necessarily assume total homogeneity of the population being sampled, the focus of this sort of experimental design is not to identify how and/or why individuals differ. Rather, the rationale is to see whether, on average, participant performance (or brain activity) differs between experimenter-manipulated conditions. That is, individual differences have most frequently been relegated to the error term of an ANOVA or another parametric statistical test in most published experimental psycholinguistic investigations, including ERP studies.

Note that this is a common trait that ERP research shares with much work in experimental psychology. Research frequently involves identifying whether an experimental manipulation has an effect on the central tendency of some outcome measure (mean reaction time, mean ERP amplitude, etc.). Inputs to statistical analyses in ERP research, like many other methods, involve either subject-level condition averages or trial-level observations, with parametric tests giving inferences about whether any differences observed in central tendency are statistically reliable (in a traditional null hypothesis testing approach). Grand mean waveforms in ERP research are thus visualizations of the central tendency of amplitude measurements in each condition, as a function of time, much in the same way that tables of descriptive statistics and bar charts provide information about the central tendency of reaction times across conditions in behavioral research. In this way, the outcomes of statistical analyses serve as a confirmation check about whether or not differences seen between grand mean waveforms (that is, central tendency) are reliable. An important way that ERPs differ from measures like reaction time, however, is in their multidimensionality. ERPs are measured in polarity (positive or negative voltage), time, and scalp topography. Distortions due to averaging can impact any combination of these dimensions, making interpretation of effects problematic in some cases. This can be especially tricky in ERP research, as ERP components are generally defined by a combination of these three dimensions. For example, the N400 effect, which is a centro-parietally distributed negativity around 300–500 ms poststimulus, is distinguished from the left anterior negativity (LAN) based solely on scalp topography: the LAN is also a negativity generally occurring between 300 and 500 ms poststimulus, but primarily over left hemisphere and/or anterior electrodes. As we will describe below, we have shown in several datasets that apparent LAN effects in traditional analyses can be the result superposing a true centro-parietal N400 effect in some participants and a right hemisphere positivity in others (i.e., topographic distortion due to polarity differences across individuals).

Moreover, a common implicit assumption is that the grand mean ERP waveform to a word in a sentence (or more typically a comparison of the ERP elicited by an expected or well-formed word to that elicited by an unexpected or in some way anomalous word) reflects the normative brain response in the population under investigation, be that monolinguals, bilinguals, children, adult second language learners, etc. However, as we will show below, there are many instances where this assumption clearly does not hold. In some cases we find marked variation when studying populations where we would expect marked inter-individual differences (e.g., adult second language learners and bilinguals processing their second language), but we also see variation when studying populations that we previously had assumed to be relatively homogenous in their language processing profiles (i.e., functionally monolingual, literate, right-handed young adults).

As we will explore below, the field has begun to accept that individual differences in the neural dynamics of real-time language comprehension are quite real and are deserving of study. In some cases these individual differences map neatly on to existing hypothesized quantitative individual differences in cognitive or experiential traits relevant to language use, such as verbal working memory capacity (e.g., Kim, Oines, & Miyake, 2018; King & Kutas, 1995; Nakano, Saron, & Swaab, 2010; cf. ; Just & Carpenter, 1992; King & Just, 1991) or language experience, proficiency, and reading skill (e.g., Mehravari, Emmorey, Prat, Klarman, & Osterhout, 2017; Pakulak & Neville, 2010; Tanner, McLaughlin, Herschensohn, & Osterhout, 2013). Notably, and more interestingly, in many cases in which we have observed clear qualitative individual differences, we have also failed to find clear predictors of ERP response type from among this set of usual candidates, such that there is an immense amount of work left for the future.

The growth of interest in individual differences also comes at a time when the fields of psycholinguistics and language science more broadly are accepting that homogenous monolingualism is not the norm globally, that perfectly balanced biliterate two-natives-in-one bilinguals essentially do not exist, that language use and processing changes with both normal aging and neurocognitive pathology, and that Chomsky's (1965) idealized speaker-listener is an abstraction that modern psycholinguistics need not (or even cannot) concern itself with. This rising awareness also coincides with an increase in power and availability of computational and statistical tools for understanding variation. Statistical models which previously took hours, days, or even weeks to fit can now be solved within seconds, so an increasing number of models are now within reach of us 'mere mortals.' These include advanced linear models and their variants such as hierarchical/mixed effects linear or generalized linear mixed effects models, and unsupervised learning algorithms that can find hidden structure in datasets. More recently, generalized additive mixed models (GAMMs) have allowed us to relax the assumption of strict linearity in our measures, which is a welcome advance, since ERPs themselves are highly nonlinear (they show positive and negative voltage changes over time), and ERPs (or other measures of processing) may have non-linear associations with individual difference measures of interest across either time or space. Some nascent work has begun to apply GAMMs to ERP individual difference problems (e.g., Meulman, Wieling, Sprenger, Stowe, & Schmid, 2015; Tremblay & Newman, 2015), though the computational challenges involved in applying GAMMs to ERPs' multiple dimensions and participant-level individual differences simultaneously remain considerable. Note also that the fact that individual differences, and in particular qualitative individual differences, in ERPs have only recently begun to be systematically explored means that we have only just begun to understand the dimensions in which variation in ERPs manifests itself, and how to quantify variation across these dimensions. For these reasons the scope of this chapter will necessarily be limited, and it is intended as a starting point for future work on this issue.

In the next section we will briefly review some of the most commonly studied ERP components related to language processing and highlight a few of the foundational studies of individual differences that have inspired much of our own work. In the subsequent sections we will review some work on individual differences using ERPs, some from other groups and some from our own, some older, and some new and emerging, which looks at individual differences in ERPs in multiple domains: non-native processing and bilingualism, morphosyntactic and semantic processing in monolingualism, multi-modal communication and pragmatic inferencing, and some nascent approaches to unsupervised machine learning to classify sub-groups of participants based on brain responses in an unbiased way. We will finish with some concluding remarks about challenges and suggestions for the future.

2. LANGUAGE-RELATED ERP COMPONENTS

By far, the two most commonly studied ERP components in studies of sentence and discourse comprehension are the N400 and the P600, with the N400 being the better-understood component of the two. Since first being reported in 1980 (Kutas & Hillyard, 1980; see Kutas & Federmeier, 2011, for an extensive overview) the N400 has been understood to broadly reflect semantic processing, with many interpretations holding that the amplitude of the N400 indexes degree of difficulty accessing meaning from long term semantic memory (Kutas & Federmeier, 2000, 2011; Lau, Phillips, & Poeppel, 2008; though see ; Hagoort, Baggio, & Willems, 2009): larger amplitude N400s reflect more difficulty accessing a word's semantics due to the word being unexpected or anomalous, and smaller amplitude N400s reflect facilitated semantic access due to strong contextual support, priming, and/or (partial) preactivation of a word or aspects of a word's meaning.

The P600, on the other hand, was first discovered in the context of syntactic anomalies: dis-preferred garden path continuations, subcategorization violations, and morphosyntactic (agreement) anomalies (Hagoort, Brown, & Groothusen, 1993; Neville, Nicol, Barss, Forster, & Garrett, 1991; Osterhout & Holcomb, 1992, 1993). Later work found that the P600 could be elicited in other syntactically well-formed, but difficult to process, constructions, such as WH-dependencies, where small P600 effects were elicited at the gap position (Kaan, Harris, Gibson, & Holcomb, 2000; Phillips, Kazanina, & Abada, 2005), making the interpretation of the P600 as an index of syntactic anomaly difficult. Later in the 2000s P600s were reported in sentences that were grammatically simple and well-formed, but which violated some semantic constraint (e.g., verb-argument animacy constraints), or contained thematic role reversals, or simply contained very strong implausibilities (as opposed to mild implausibilities where only N400 effects were elicited; Hoeks, Stowe, & Doedens, 2004; Kim & Osterhout, 2005; Kolk, Chwilla, van Herten, & Oor, 2003; Kuperberg, Caplan, Sitnikova, Eddy, & Holcomb, 2006; Kuperberg, Kreher, Sitnikova, Caplan, & Holcomb, 2007; Kuperberg, Sitnikova, Caplan, & Holcomb, 2003; van de Meerendonk, Kolk, Vissers, & Chwilla, 2010; van Herten, Chwilla, & Kolk, 2006). Additional studies have reported that the P600 could be reduced or eliminated when the sentence contained misleading cues about morphosyntactic relations that made unambiguous syntactic violations less salient and/or when words were presented at rates faster than those typically used in ERP studies

(Tanner, Grey, & Van Hell, 2017; Tanner, Nicol, & Brehm, 2014). Still others have found that anomalies involving structured non-linguistic representations (e.g., music and visual events) can also elicit P600 effects (Patel, Gibson, Ratner, Besson, & Holcomb, 1998; Sitnikova, Holcomb, Kiyonaga, & Kuperberg, 2008). This disparate set of eliciting conditions has challenged the notion that the P600 reflects primarily syntactic processing or syntactic reanalysis, and a number of non-mutually-exclusive views of the P600's functional significance have been put forth over the last decade (see e.g., Bornkessel-Schlesewsky & Schlesewsky, 2008; Brouwer, Fitz, & Hoeks, 2012; Kolk & Chwilla, 2007; Kuperberg, 2007; Paczynski & Kuperberg, 2012). Nonetheless, the N400/P600 dichotomy has played a pivotal role in framing much psycholinguistic ERP work in both monolingual and bilingual contexts over the last three decades, and as such, it will play a key role in the present review as well.

Although the N400 and P600 have occupied focal roles in ERP work on sentence processing, one additional component deserves mention here. In some reports, morphosyntactic violations elicit a negative-going wave that sometimes has a left and/or anterior scalp topography preceding the P600: the left anterior negativity, or LAN effect. The LAN surfaces in the same time window as the N400 effect (around 300–500 ms poststimulus), but is distinguished based on its scalp topography: the N400 typically has a broad central or centro-parietal topography, whereas the LAN has the aforementioned left and/or anterior distribution. The LAN features prominently in some neurocognitive theories of morphosyntactic processing as an index of initial detection of anomalies (which in some theories is additionally referred to as an automatic, implicit process), where later controlled reanalysis is indexed by the P600 effect (e.g., Batterink & Neville, 2013; Molinaro, Barber, & Carreiras, 2011; Molinaro, Barber, Caffarra, & Carreiras, 2015; Ullman, 2004), though there are numerous conceptual and empirical issues with these claims, some of which we will discuss below (see also Osterhout, McLaughlin, Kim, Greewald, & Inoue, 2004; Tanner & Van Hell, 2014a, for further discussion).

It is important to note here that the N400, P600, and LAN are not language-specific, nor are they the only ERP components studied in language-related electrophysiological research. However, they are large, and, at least in the case of the N400 and P600, easily elicited by sentence-embedded linguistic manipulations. They have therefore been the focus of the majority of research on native and non-native sentence comprehension. See work by Swaab, Ledoux, Camblin, and Boudewyn (2012) and Payne, Ng, Shantz, and Federmeier (2018) for recent reviews of ERP research on language.

3. SOME QUANTITATIVE INDIVIDUAL DIFFERENCES

The most straightforward way in which individual differences in ERPs may manifest themselves is quantitatively. When we say "quantitatively," we mean that the individual difference 'signal' (that is, the inter-individual variability that is systematic and quantifiable) is found in some component's effect amplitude, e.g., the N400 or P600, and we can find some interesting regressor that predicts the magnitude of the effect across individuals. Here effect amplitude will typically refer to the difference in amplitude between some ill-formed and well-formed condition (e.g., P600 effect amplitude will be the mean amplitude difference in the P600 time window over posterior electrodes from the ungrammatical minus grammatical difference wave in a morphosyntactic violation paradigm).

One notable example that challenged the notion that monolinguals are homogenous is reported by Pakulak and Neville (2010). They investigated syntactic processing in monolingual English speakers, but used a relatively large sample size compared to most language ERP studies (N = 72) and sampled not just university-enrolled subject pool participants that are typically tested in ERP studies, but also tested participants with a range of literacy and socioeconomic status levels. They recorded ERPs to auditorily presented sentences that were either well-formed or included phrase structure violations (*Timmy can ride the horse at his farm./*Timmy can ride the horse at my his farm.*), and also administered a very difficult English proficiency test (the Test of Adolescent and Adult Language, third Edition, TOAL-3). ERPs elicited by the well-formed and anomalous determiners (underlined, above) showed an early LAN and P600 in the grand mean analysis; however, they also found individual differences that varied with participants' proficiency. In particular they found that the LAN was more bilateral and the P600 of smaller amplitude in lower proficiency participants, after controlling for working memory capacity. This finding is notable for two reasons. First, as mentioned above, it provided a clear counterpoint to the notion that all native speakers are equally proficient in their native language and that processing dynamics are homogenous across monolingual speaker populations. Second, it showed quite clearly that between-subjects variation in ERP measures can be systematic, such that – especially in the case of the LAN topography – traditional omnibus statistical analyses of mean amplitudes (as would be depicted in grand mean waveforms) may not provide an accurate depiction of ERP effects in many of the individuals in the sample under investigation. More generally, this study set the stage for the many psycholinguistic ERP studies on individual differences in the subsequent years.

With Pakulak and Neville having shown that language proficiency, over and above working memory, can account for quantitative differences in ERP metrics of syntactic processing, we asked whether a similar observation might hold for semantic processing and the N400. Some have argued that working memory skill is particularly important in semantic processing, as managing competing demands of accessing words and building message level meaning can be cognitively taxing (e.g., Boudewyn, Long, & Swaab, 2013). Others have argued specifically that working memory is less important in sentence comprehension, and that semantic activation and retrieval is mediated by "lexical quality," which is a function of language experience and proficiency (e.g., Van Dyke, Johns, & Kukona, 2014; cf. ; Perfetti, 2007). In a preliminary report, we evaluated these claims in a large cohort of monolingual English speakers (N = 64; Tanner, Bulkes, Shantz, Armstrong, & Reyes, 2016) by recording ERPs while participants read sentences that were well-formed or contained a word that was semantically unexpected and anomalous (e.g., *The doctor diagnosed the tumor/*package while it was still treatable; The delivery man left the package/*tumor on the front porch*; the underlined word marks the time-locking point for ERP averaging and analysis). We also administered a battery of behavioral tasks tapping language experience and lexical knowledge, as well as working memory. The semantically anomalous words elicited much larger amplitude N400 effects than their corresponding well-formed counterparts (the N400 effect), as predicted. However, our question was whether the individual differences related to working memory and/or language experience would predict the N400 effect amplitudes. Because we administered a battery of tasks, we used principal components analysis (PCA) to extract component scores for two latent components; the component loadings showed that principal component (PC) 1 correlated strongly with the language experience tasks and PC 2 correlated strongly with the working memory tasks. We then regressed these latent component scores onto N400 effect amplitudes simultaneously. Scatterplots of the zero-order correlations are depicted in Fig. 1. There was no correlation between the latent language experience

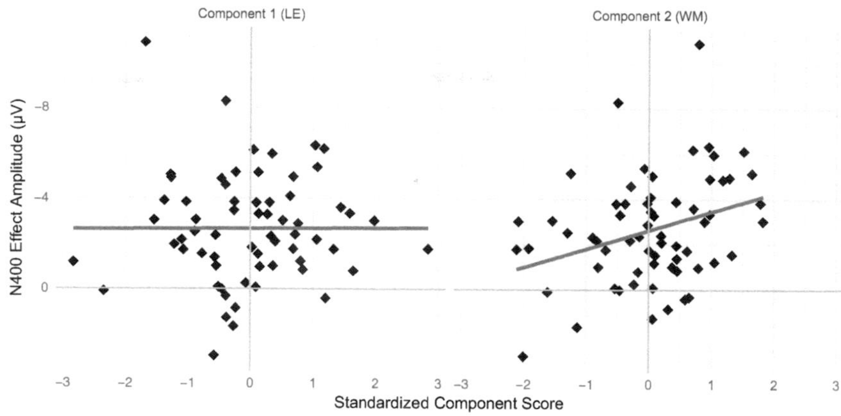

Figure 1 Scatterplot depicting relationship between latent component scores for language experience (left) and working memory (right) and N400 effect amplitude (y-axis) for 64 monolingual English speakers, measured across eight centro-parietal electrodes.

component and N400 amplitude ($r = -0.006$), but there was a reliable correlation between participants' latent working memory component scores and N400 effect amplitudes ($r = .325$, $p = .009$).

These findings show that, among cognitive skills commonly implicated in language processing, there may be some kind of dissociation between individual differences supporting semantic and syntactic processing. Pakulak and Neville (2010) reported that a speaker's proficiency in their native language, over and above working memory, was predictive of P600 amplitudes in syntactic violation contexts; we found that speakers' working memory capacities, over and above native language proficiency, were more predictive of N400 effects in semantic violation contexts. Note that firm conclusions cannot be drawn from this comparison, since the modalities of the two studies were different (auditory versus written presentation), the types of participants sampled were different (a broad spectrum of literacy levels versus literate young adults, who nonetheless varied in their native language proficiency), and the metrics used to assess working memory and language knowledge differed across the two studies. A clear test of this hypothesis should use a within-subjects design, and this remains an active area of work within our lab. Nonetheless, these two studies demonstrate that grand mean analyses can obscure some systematic variation between individuals in ERP responses. In the present scenarios the mappings between ERP responses and behavioral predictors was relatively straight-forward: ERP effect amplitudes in an expected elicited condition (N400 for semantic violations and P600 for syntactic violations) showed linear associations with some well-established behavioral measure. However, as we will see in the coming sections, such straight-forward mappings are not ubiquitous, and the relationship between expected and observed ERPs can be quite complex.

4. QUALITATIVE INDIVIDUAL DIFFERENCES IN NON-NATIVE AND BILINGUAL MORPHOSYNTACTIC PROCESSING

Our initial work on qualitative individual differences in ERP responses was inspired by two lines of research laid out by Lee Osterhout and colleagues. First, in a series of review papers, Osterhout, Judith McLaughlin, and colleagues reported findings from a longitudinal study of second language (L2) learning, which tracked native (L1) English speakers as they progressed through their first year of classroom L2 French instruction (McLaughlin et al., 2010; Osterhout, McLaughlin, Pitkänen, Frenck-Mestre, & Molinaro, 2006; Osterhout et al., 2004, 2008). Most relevant for us, they reported qualitative changes in participants' ERP responses to morphosyntactic subject-verb agreement violations over the course of the year: violations, which elicited P600 effects in native French speakers, initially elicited N400 effects in the L2 group, but the L2 participants' brain responses changed over the course of instruction to approximate natives' P600 by the end of the year. That is, the L2 learners initially processed the agreement violations as though they were lexical or semantic anomalies, and their brain responses changed qualitatively. Second, a separate line of work by Osterhout (1997) in monolingual English speakers demonstrated that a biphasic LAN-like negativity followed by a P600 in grand mean ERPs elicited by garden path disambiguations was not representative of most participants' brain responses. Rather, most participants showed either an N400-like negativity or a P600, but not both, and that the scalp topography of the negativity in the grand mean showed an anterior focus only because the posterior-dominant P600 effect cancelled out the posterior portion of the broad N400 component due to component overlap.

These two threads helped us better understand a perplexing pattern of data found in native English speakers with low proficiency in L2 German, processing German morphosyntactic agreement violations

(Tanner et al., 2013). Unlike native and intermediate proficiency L2 German speakers who showed P600 effects, grand mean ERP waveforms in the group of 20 first year German students showed a biphasic N400-P600 response (Fig. 2). However, inspection of individuals' ERPs showed that this pattern was not representative of most participants' brain responses. The distribution of individuals' brain responses is shown in Fig. 2B. Individuals' N400 effect amplitudes are the y-axis (larger N400 effects are up on the graph) and P600 effect amplitudes are on the x-axis (larger P600 effects are to the right). Across participants, the two ERP responses showed a negative association: individuals who showed large N400 effects tended to show small P600 effects, and vice versa. The dashed line indicates the axis of equal effect sizes. Individuals above/to the left of the line show N400-dominant ERP responses and individuals below/to the right of the line show P600-dominant ERP responses. Averaged waveforms within the N400- and P600-dominant groups (Fig. 2C and D) showed clear monophasic responses of the respective type. Subsequent

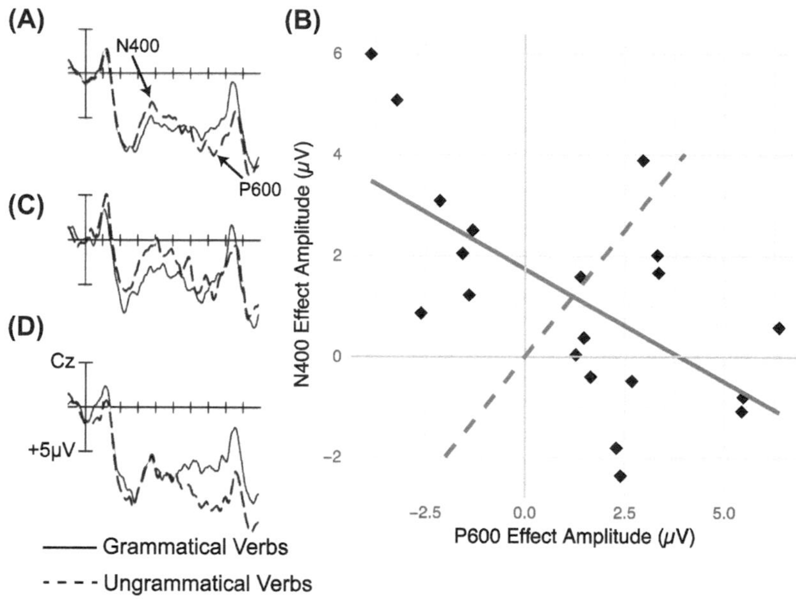

Figure 2 Partial results from Tanner et al. (2013). (A) Grand mean ERPs from 20 L1 English speakers enrolled in first year L2 German courses to well-formed control (solid) and subject-verb agreement violating verbs (dashed). Negative voltage is plotted up in these and all subsequent waveforms. (B) Distribution of the first year participants' N400 and P600 ERP effect amplitudes. (C) ERP responses in the negativity-dominant group (n = 9). (D) ERP responses in the positivity-dominant group (n = 11).

correlation analyses showed that participants' behavioral discrimination of the well-formed and ill-formed sentences correlated negatively with N400 effect amplitudes and positively with P600 effect amplitudes, suggesting that individual differences in L2 proficiency were partly responsible for the pattern of qualitative individual differences in ERP responses. In many ways this finding is a cross-sectional replication of Osterhout and McLaughlin's original longitudinal study, but with the qualitative change in ERP responses associated with proficiency being captured in a single "snapshot." Importantly, this finding shows that the grand mean waveforms and corresponding omnibus statistical analysis misrepresented the true nature of many participants' brain responses. Even though the first-year learners had relatively similar and homogenous classroom exposure to L2 German, their processing differed from one another in marked, yet systematic ways.

In a subsequent study (Tanner, Inoue, & Osterhout, 2014) we investigated morphosyntactic processing in very high proficiency late L1 Spanish-L2 English bilinguals and found a similar distribution of brain responses between N400- and P600-dominant in this group in response to subject-verb agreement violations in English. This was actually quite surprising to us, since previous work had found that ERP responses in non-native language processing could approximate natives' quickly for shared morphosyntactic features, and we specifically targeted high proficiency bilinguals for this study. Nonetheless, English subject-verb agreement violations elicited N400 effects in some, biphasic responses in some, and P600 effects in others. To better understand what factors were related to these qualitative individual differences, we created two new measures for the types of variation we found. The Response Dominance Index (RDI) is a metric of how N400- or P600-dominant a participant's brain response is. Negative values reflect N400-dominance, positive values reflect P600-dominance, and values near zero reflect dominance of neither one response nor the other. However, the RDI does not directly index the *size* of the response, only the relative size of the polarity dominance. In principle, participants with an RDI of zero could show a large, but equisized biphasic response. To complement the RDI, we therefore also proposed the Response Magnitude Index (RMI), which provides an index of the total "amount" of brain response, measured from the anomalous minus control difference wave, regardless of brain response type. RMI values near zero reflect little to no difference in brain responses between the anomaly and control conditions, and large values reflect greater neural sensitivity to the anomalies, but without indicating the identity of the ERP component elicited.

To identify the antecedents of the individual variation in ERP responses among the bilinguals we tested, we computed two multiple regression models, with the RDI and RMI as the respective dependent measures. Predictors included age of arrival in the United States, length of residence in the US, frequency of use of English in daily life, English proficiency (as measured via an abbreviated version of the Michigan Examination for the Certificate of Proficiency in English), and self-rated motivation to speak English like a native speaker. Results showed that RDI values were predicted by age of arrival and motivation, with more P600 bias found among those who arrived earlier and who reported higher motivation to speak like a native speaker. RMI values were predicted by proficiency: higher proficiency correlated uniquely with greater neural sensitivity to subject-verb agreement violations (but not a specific ERP component). For the age of arrival effect, note that all of the participants in this group were late, post-puberty arrivals, putting their arrival in the US after any putative critical period cut-off. The correlation between age and ERP response quality therefore does not lend itself to a straightforward Critical Period Hypothesis explanation, but is consistent with age effects on L2 learning and processing, broadly defined (cf. Birdsong & Molis, 2001; Birdsong, 2006).

The fact that substantial individual variation existed among the highly proficient bilinguals tested in this second study on the one hand may not be surprising, since the circumstances of bilingualism and L2 learning are highly heterogeneous. However, there are several proposals about L2 processing holding that neural markers of non-native processing should converge on "native-like" ERP responses at high L2 proficiency (i.e., a P600 effect to subject-verb agreement violations; see Morgan-Short, 2014; Steinhauer, 2014, for reviews). One might infer, then, that the participants who showed N400 effects in response to English subject-verb agreement violations were poorer learners or processors, since they showed non-native-like ERPs, which were previously shown to be associated with low proficiency L2 learners. However, this inference rests upon the assumption that monolinguals show homogeneous brain responses in their L1 and that no proficient monolinguals show N400 effects. A cursory reading of the literature would lead one to believe that this is the case; however, as we discuss below, careful study shows that even proficient, literate monolinguals can differ qualitatively from one another in their ERP responses, and a substantial minority indeed show N400 effects to morphosyntactic violations.

5. MONOLINGUAL MORPHOSYNTACTIC, SEMANTIC, AND THEMATIC PROCESSING AND SOME PSYCHOLINGUISTIC CONUNDRUMS

In Tanner and Van Hell (2014a) we sought to identify whether similar qualitative individual variation exists among proficient, literate monolingual English speakers processing morphosyntactic relations, and in particular, whether individual variability could explain at least some of the unreliability of the LAN component across studies. Although the LAN features prominently in some ERP-based neurocognitive theories of language processing, including some arguments that it reflects automatic syntactic processing or anomaly detection, a substantial proportion of ERP studies fail to find a LAN in response to morphosyntactic violations (meaning, by definition, that the effect can therefore not be strictly automatic). Based on findings from Osterhout (1997) and our work on individual differences in L2 processing, we hypothesized that at least some instances of biphasic LAN-P600 effects in the reported literature were artifacts of qualitative individual differences, wherein some participants showed N400 effects and some showed P600 effects. Indeed, this is what we found. In response to morphosyntactic violations, grand mean waveforms and traditional omnibus analyses showed a statistically robust left hemisphere (LAN-like) negativity followed by a P600 effect. In this case the LAN was found to be statistically reliable via a grammaticality by hemisphere interaction over lateral electrode sites from both hemispheres in the 300—500 ms time window, i.e., without isolating left hemisphere electrodes alone as a region of interest. However, further investigation revealed a continuous distribution of brain responses between N400- and P600-dominant, very similar to that seen in our previous work on non-native processing. Analyses of the semi-homogeneous sub-groups (negativity- and positivity-dominant groups, respectively) showed that the negativity was not left lateralized, but instead showed a broad, N400-like centro-parietal distribution. The P600 in the other group, on the other hand, showed a right-posterior dominance, with an onset during the 300—500 ms time window over right hemisphere electrodes. That is, the LAN that was detected in the grand mean was a consequence of component overlap and individual differences: the left hemisphere topography was a result of the intersection of a broadly distributed N400 in some individuals and a temporally co-occurring right hemisphere positivity in others, with the residual N400 being visible only over left hemisphere electrodes in the grand mean. This gave the illusion of a LAN effect. This negativity was still statistically significant in omnibus tests of central tendency, despite being present only in a minority of participants. This suggests that at least some of the variation across studies in the LAN might be explained by individual differences and sampling variability; i.e., the presence or absence of a LAN seen in grand mean analyses would be at least partly a function of the number of N400-dominant individuals randomly sampled for a study, as well as the degree of spatiotemporal overlap between the P600 seen in most individuals and the N400 seen in others.

On the one hand, this result suggests that the N400 effects seen in some non-native speakers cannot be clearly interpreted as evidence of non-native-like processing or poor L2 learning, since a substantial proportion of native speakers also showed N400 effects in similar morphosyntactic situations when the data were more closely inspected. However, given that L2-related regressors predicted some proportion of variance in brain response dominance in the previous work (acceptability judgment accuracy as a proxy for L2 proficiency, age of arrival, and motivation to speak in a native-like way), the question remains what individual-level metrics might predict brain response dominance in native speakers. In the Tanner and Van Hell (2014a) study, familial sinistrality accounted for a small proportion of the variance in a way predicted by Bever and colleagues: right-handed participants with left-handed family members showed, on average, more negative-going ERP responses than right-handers without left-handed family members (Hancock & Bever, 2013; Townsend, Carrithers, & Bever, 2001; see also Lee & Federmeier, 2015). A follow-up on this study with left-handed individuals corroborated this finding by showing that left-handers and right-handers with left-handed family members patterned similarly, and both groups had more variation in the response dominance continuum than the right-hander group without familial sinistrality (Grey, Tanner, & Van Hell, 2017). Our lab has also recently completed a large-scale (N = 114) study of individual differences in morphosyntactic processing in English monolinguals where we administered a battery of language experience and working memory measures, in combination with latent variable analyses of language experience and working memory, in an effort to identify cognitive antecedents of variation in the N400-P600 response dominance continuum (Tanner, 2018). This study also addressed some criticisms leveled by Molinaro and colleagues (Molinaro et al., 2015) about the syntactic complexity of the materials used in the Tanner and Van Hell study. Using structurally simple sentences, traditional analyses showed a similar LAN-P600 pattern to subject-verb agreement violations, but again, there were marked individual differences in the N400-P600 continuum, both for lexical verbs where agreement is marked with an overt morpheme (*grow/grows*) and for auxiliary verbs where agreement is marked by lexical alternations (*is/are*). Again, the apparent LAN in the grand mean was a consequence of component overlap between an N400 effect in some individuals and right hemisphere P600 effects in others. However, latent variables related to neither working memory nor language experience were correlated with individuals' ERP effect amplitudes. This is despite findings of robust effects of language experience/proficiency on the P600 effect in our L2 work and in monolingual work by Pakulak and Neville (2010), and despite findings of working memory effects in predicting ERP response dominance in contexts of animacy violations (Kim et al., 2018; Nakano et al., 2010). Thus, although the finding of N400/P600 individual differences in monolinguals is robust and replicable, we do not yet have a clear picture regarding what individual-level traits might predict these differences.

As individual differences helped explain a conundrum regarding variability in the LAN, we asked whether they could help explain conflicting findings in the literature regarding the role of semantic associations between words and their interaction with animacy constraints in thematic role processing. Some have reported that verb-argument animacy violations elicit P600 effects regardless of the verb's semantic association with its subject argument (e.g., Kuperberg et al., 2006, 2007). Others have reported that semantic associations are evaluated independently of combinatorial constraints like animacy and morphosyntax, such that animacy violations involving semantically associated verbs and arguments elicit P600 effects (so-called "semantic attraction"), while animacy violations with unrelated verbs and arguments elicit N400 effects (e.g., Kim & Osterhout, 2005; Kim et al., 2018). We assessed the impacts of morphosyntactic well-formedness and semantic association during thematic role processing in a cohort of 58 monolingual English speakers through an individual differences lens (Tanner & Van Hell, in prep; Tanner & Van Hell, 2014b). The control condition included passive sentences with inanimate subject noun phrases, which were highly semantically related to the verb

(*The broken television was repaired* ...). Experimental sentences included a syntactic violation condition (*The broken television was repairs* ...), a condition with animacy violations indicated by progressive verb morphology and where there was no semantic association between the subject and verb (semantic violation condition: *The hearty meal was repairing* ...), and an animacy violation with strong semantic attraction between the subject and verb (the attraction violation condition: *The broken television was repairing* ...).

Results are depicted in Fig. 3. Grand mean results in the syntactic violation condition showed a large, biphasic N400-P600 effect, a clear N400 effect in the semantic violation condition, and a small, marginal biphasic N400-P600 in the attraction violation condition. However, as with the above studies, interpreting the ERP effects seen in the grand mean waveforms — and confirmed via parametric statistical tests — as representing the normative brain responses in the general population would have led to the wrong conclusions. Instead, the biphasic responses were the result of individual differences in N400/P600 response dominance. Additionally, the monophasic N400 effect in the semantic (no-attraction animacy) violation condition misrepresented the fact that some individuals showed monophasic positive responses in this condition (bottom right quadrants of scatterplots in Fig. 3B). Although there were individual differences in all three conditions, the central tendency of the distribution of differences for the syntactic and semantic violation conditions was moved in the direction predicted by the "classic" take on ERP responses to syntactic and semantic violations: the RDI showed a positive mean in the syntactic condition, indicating a preponderance of P600 effects, and a negative mean in the semantic condition, indicating a preponderance of N400 effects. In the attraction violation condition, there was no clear bias toward positive or negative effects, with the distribution approximately equally distributed between the two effect dominances (Fig. 3C). Participants' RDI values across the three experimental conditions were positively correlated ($rs > .5$), indicating some consistency in response type (N400 versus P600 dominant) within individuals across the various anomaly types. Also interesting to note is that, even though the ERP response visible in the grand mean to attraction violations was small and only marginally significant, the mean of individuals' overall response magnitudes was similar to that seen in the semantic violation condition (Fig. 3C). The effect amplitudes in measures of central tendency were attenuated due to cross-participant cancellation between co-temporal negativities and

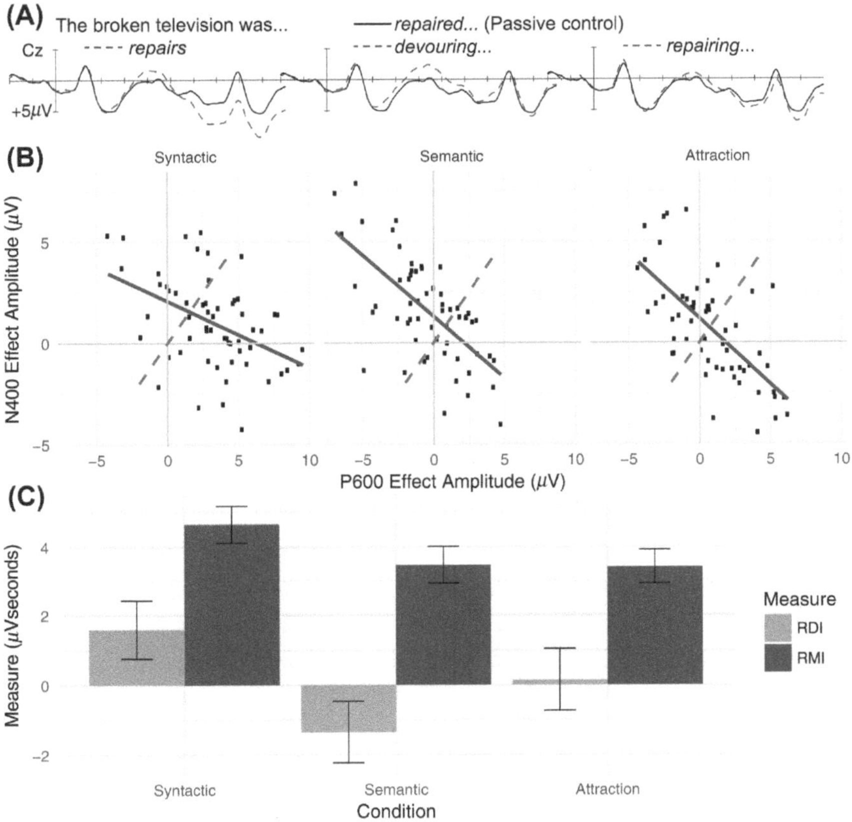

Figure 3 Results from Tanner and Van Hell (2014b). (A) Grand mean ERP results (N = 58) in the three experimental conditions, relative to passive control sentences. (B) Individual differences in N400/P600 effect amplitudes. (C) Measures of ERP response dominance (RDI) and response magnitude (RMI) in the three experimental conditions. RDI measured as positive minus negative integration areas under individuals' difference waves between 200 and 1000 ms poststimulus; RMI measured as rectified area under difference wave. Error bars show 95% CI of the mean.

positivities: the P600 showed an early onset in some individuals and cancelled out the N400 seen in others, and the N400 showed a long time course in some individuals, effectively cancelling out the P600 seen in others.

This last study shows that, although ERP responses to a range of violations (morphosyntactic, semantic, thematic) show individual differences in the N400/P600 continuum, the distribution of these differences is sensitive to the amount and type of information signaling the anomaly in a way that reflects traditional interpretations of the N400 and P600 effects. ERPs were most positive-going when morphosyntactic information signaled an anomaly, most negative-going when violations involved incoherent semantic relations, and there was no clear dominance (equal distribution between N400 and P600 effects) when both semantic and morphosyntactic information were available to signal violations (i.e., world knowledge about what broken televisions can do, and morphological marking on the verb signaling thematic role assignment).

6. HIGHER-ORDER INFERENCING AND MULTI-MODAL COMMUNICATION

Beyond communicating simple morphosyntactic, semantic or thematic relations, an interlocuter must also consider the speaker's intent, and must sometimes compute semantic interpretations that go beyond the literal meaning of the words used in the utterance. These phenomena, of which there are many, can be broadly classified as "pragmatic inferences." Pragmatics, with its rich theoretical foundation, has turned increasingly often to experimental methods within the past two decades (see e.g., Noveck & Sperber, 2004) in order to better understand the time course and processing of these inferences. In this section we discuss two ERP studies that explore individual differences regarding two types of pragmatic phenomena: scalar implicature and multi-modal irony.

Scalar implicatures are generated by interlocutors when they hear a scalar term like *some* and they have reasons (e.g. context, previous experience with this term) to assume the speaker chose *some* over a stronger term (like *all*) due to the stronger term being false or not applicable. When a speaker says *I ate some cookies*, an interlocutor would often interpret that to mean that the speaker did not eat *all* cookies (in contexts where *all* is a relevant alternative). Underinformative *some* statements like *Some people have lungs* have often been used to investigate whether the literal meaning of *some* ("and possibly all") is generated prior to the pragmatically enriched meaning ("but not all"). A common finding is that participants divide into two camps (of varying sizes, depending on the experimental design, the alternative quantifiers used in the experiment, and more): those who judge underinformative *some* items as true, and those who judge them as false (Bott & Noveck, 2004; Degen & Tanenhaus, 2016; Grodner, Klein, Carbary, & Tanenhaus, 2010; Huang & Snedeker, 2009; Papafragou & Musolino, 2003). Several studies have attempted to characterize the two types of responders, the more 'logical' ones, described as relying on the literal meaning, and the more 'pragmatic' ones who rely on the pragmatically enriched meaning. Noveck and Posada (2003) measured truth-value judgments and ERP responses to items like *Some sentences have words* and found no difference in ERP response for the two types of responders, though there were reaction time differences between the two groups.

More recently, others have focused on whether autistic-like traits might underlie differences in processing. Some previous studies have shown participants with autism spectrum disorder (ASD) to be worse than neurotypical populations at understanding metaphor, irony, sarcasm and jokes (Happé, 1993, 1994; Martin & Mcdonald, 2004). With respect to scalar implicatures, ASD participants have been shown to have difficulties in attributing mental states to others (Baron-Cohen, 2000), which directly relates to interpreting communicative intentions (Sperber & Wilson, 1986) and with making inferences related to world-knowledge (Jolliffe & Baron-Cohen, 1999, 2000). For example, Nieuwland, Ditman, and Kuperberg (2010) correlated individuals' Autism Quotient (AQ; Baron-Cohen, Wheelwright, Skinner, Martin, & Clubley, 2001) scores with their ERP responses elicited by temporarily underinformative scalar implicature sentences, compared to sentence with informative scalar implicatures (e.g., *Some people have <u>pets/*lungs</u>, which require good care.*). They found that individuals with more autistic-like traits showed reduced N400 effects to underinformative compared to informative *some* items, whereas participants with less autistic-like traits showed larger N400 effects to the same items. However, their design did not incorporate truth value judgments, so it is not clear how these ERP differences might map onto differences in interpretation. Some more recent behavioral work has found that judgments of underinformative *some* items did not correlate with participants' AQ scores (Antoniou, Cummins, & Katsos, 2016). In another ERP study with a visual verification task, Spychalska, Kontinen, and Werning (2016) found that participants who judged underinformative items as false (pragmatic responders) showed a biphasic N400-P600 effect to underinformative compared to true items, whereas those who judged underinformative items as true (logical responders) showed no ERP differences to the underinformative versus true conditions.

The difference in the methodologies and stimuli used in these studies makes it difficult to determine whether people differ in how they process underinformative *some* items, whether these differences map onto different judgments, and what traits (e.g., ASD-like traits) modulate the differences in judgments and ERP responses. In an attempt to triangulate the different hypotheses about individual differences in the processing and judgment of underinformative *some* items within a single design, we have consolidated some of the previous studies' methodologies and stimuli into one paradigm. Sixty native English speakers read sentences of the form *Some x are y* that were either true (*Some animals are mammals*), false (*Some reptiles are mammals*), or underinformative (*Some cats are mammals*), followed by a true/false judgment. After the ERP experiment, participants completed the AQ-39 questionnaire (Lau et al., 2013) to measure their degree of autistic-like traits (higher score = more autistic-like traits). Results (Fig. 4A) showed that, relative to final words in true items, final words in false items elicited larger N400 amplitudes, and final words in underinformative items elicited a reduced N400 followed by a later posterior P600-like positivity. However, these responses were not reliably mediated by participants' AQ scores, as had been reported by Nieuwland et al. (2010; see Fig. 4B and C). Correlations between participants' AQ scores and N400 effect amplitudes (following Nieuwland et al.) were in the same direction as in Nieuwland et al.'s study, but did not reach significance, despite a larger sample size ($r = -0.16$, $p = .22$). Although our items and task differed from Nieuwland et al.'s, our results suggest that the relationship between autistic traits and ERP effects elicited by underinformative items may not be as strong or as systematic as some studies may imply. Note also that we found very little variation in participants' truth value judgments, with nearly all participants providing "logical"-type responses. We therefore could not directly investigate the relationship between truth value judgments and ERP responses or autistic traits. However, we nonetheless found a difference in ERPs to underinformative versus true items among our predominantly logical-responding participants, in contrast to Spychalska et al.

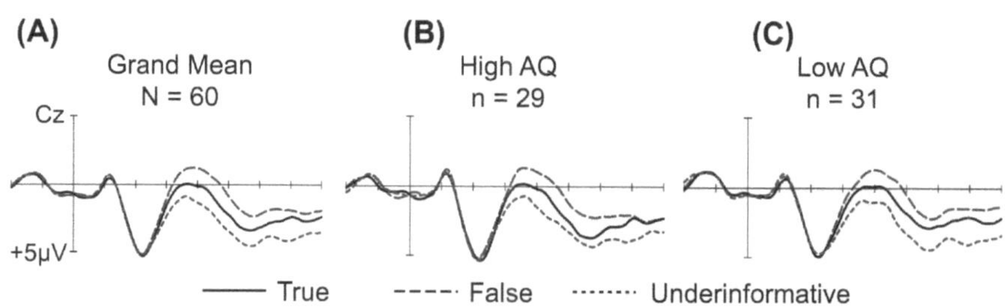

Figure 4 (A) Grand mean ERPs. (B) and (C) Sub-group averages based on median split for AQ: high AQ (B) and low AQ (C). Higher AQ scores reflect more autistic-like traits.

Although we could not establish a clear connection between participants' ERPs and degree of autistic-like traits, we nonetheless observed that individuals' responses to the target conditions did differ qualitatively. It seemed possible, then, that the grand mean waveforms and omnibus analysis obscured variation that was not accounted for by the AQ variable. In order to better understand this variation in a principled, bottom-up way, we have conducted a series of exploratory analyses using unsupervised clustering algorithms to identify whether there are semi-homogeneous sub-groups. One of these analyses used the k-means algorithm to identify clusters for both the false and underinformative conditions, relative to the true (control) condition. K-means clustering is an algorithm that partitions observations into an assigned number of clusters (k). In each run of the algorithm, observations are assigned to a cluster with the nearest centroid mean. This is done iteratively until cluster assignments are stable. We tested models for both the false and underinformative conditions with between two and ten clusters. Model inspection showed that either two or three clusters provided optimally parsimonious solutions for both conditions; however, for brevity we report only the two-cluster solutions here.

Clustering results for the false and underinformative conditions (relative to true) are shown in Figs. 5 and 6, respectively. For the false/true contrast, one cluster showed the clear enhanced N400 effect for false items relative to true seen in the grand mean. The negativity showed a long time course (beyond the typical 300–500 ms time window), and became more frontally

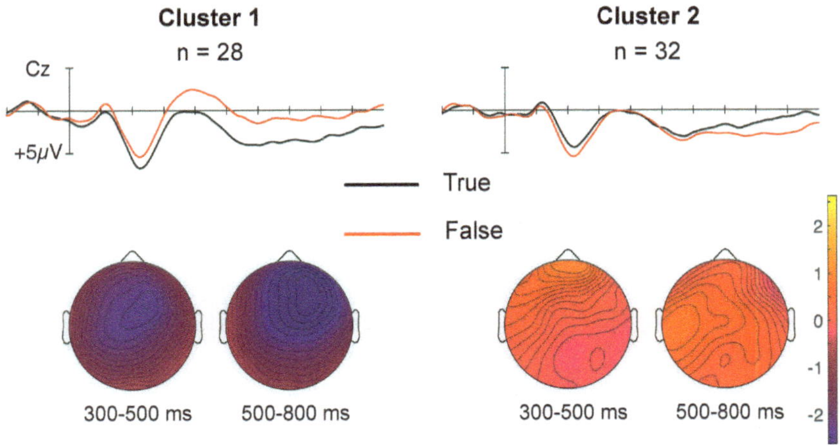

Figure 5 K-means cluster results for the false minus true condition.

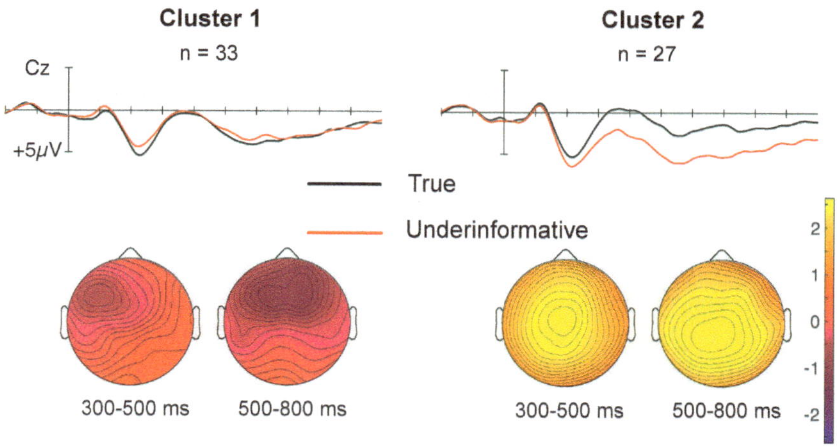

Figure 6 K-means cluster results for the underinformative minus true condition.

distributed in the later 500–800 ms time window. The second cluster showed no clear differences between conditions in the N400 time window, but instead a small positivity with a left hemisphere maximum. For the underinformative/true contrast, the larger cluster showed no clear effects over midline central electrode Cz (the depicted waveforms), but the topographic plots show evidence of a frontally-distributed negative-going component. Only the smaller cluster (cluster 2) showed evidence of the reduced N400 effect and extended positivity for the underinformative relative to true items. An important point to note here is that AQ scores and true/false judgments did not differ between any of the clusters, indicating that the variation we see is not clearly a function of autistic-like traits or judgment behavior. Importantly, as with the morphosyntactic and thematic role processing results reported above, these results further underscore the fact that even highly statistically reliable effects seen in grand mean waveforms and analyses do not always depict the normative brain response in the general population under investigation. Moreover, although the variables we investigated (i.e., AQ, truth value judgments) did not modulate these individual differences, using unsupervised clustering algorithms allowed us to observe these neural differences. This opens up a hypothesis space for further work.

Another classic example of communication beyond that which is literally said is irony. Precise definitions of irony have been debated by theoreticians (e.g., Giora, 1997; Kreuz & Glucksberg, 1989; Wilson & Sperber, 1992), but it can be swiftly summarized as when an utterance is used to communicate the opposite of its literal meaning. Uttering *Living with him would be dreadful* may indeed mean living with him is a horrifying thought, but with certain cues (e.g., prosody, situational context, background knowledge, facial expression, emojis), the utterance could be intended to mean that living with him would be wonderful. Heavy reliance on nonliteral cues means irony interpretation is rife with individual differences (Ivanko, Pexman, & Olineck, 2004; LaMarre, Landreville, & Beam, 2009).

One way to indicate irony in text-based communication is the wink emoticon/emoji, linked with irony in both intent and uptake (Filik et al., 2015; Skovholt, Grønning, & Kankaanranta, 2014; Thompson & Filik, 2016). Used in sentence-final position, as in Fig. 7C, this wink can serve as a marker of illocutionary force (Dresner & Herring, 2010; Searle, 1969, 1979). Here, the wink can mark irony, signaling to the interlocutor that the opposite of the literal meaning is being communicated. Emojis, however, are notably ambiguous (Miller, Kluver, Thebault-Spieker, Terveen, & Hecht, 2017; Miller et al., 2016), and even this conventional ironic wink may not be interpreted as ironic by all individuals nor in all contexts. Interpretation aside, these text with emoji utterances are examples of multimodal communication, in which multiple modalities are used in conjunction in a single utterance (e.g., Cohn, 2016): sentences like those in Fig. 7 are not treated as a sentence with an emoji appended to the end, but rather as a systematic multimodal utterance with a single communicative purpose.

Previous ERP studies of verbal irony (Filik, Leuthold, Wallington, & Page, 2014; Regel, Coulson, & Gunter, 2010; Regel, Gunter, & Friederici, 2011; Spotorno, Cheylus, Van Der Henst, & Noveck, 2013) used stimuli wherein the final word would be either ironic or non-ironic based on a preceding context. Time-locking to that final word, these studies found a P200

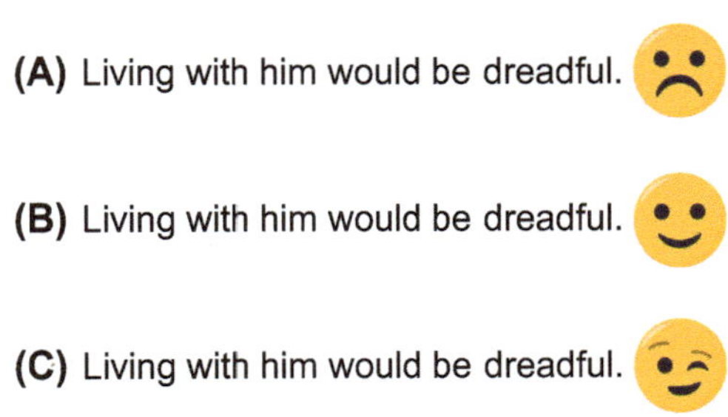

Figure 7 Example sentence followed by a congruent emoji (A), incongruent emoji (B), and ironic (sarcastic) emoji (C).

effect and P600 effect to ironic compared to literal words (i.e. when that same sentence is preceded by context that leads to a non-ironic interpretation of the sentence). The P200 is typically linked to attention (e.g., Carretie, Mercado, & Tapia, 2001; Luck & Hillyard, 1994), and this P600 finding supports re-processing/re-analysis theories of the P600 (e.g., Brouwer et al., 2012; Kolk & Chwilla, 2007) as opposed to purely syntactic ones.

We recently completed a three-experiment ERP study (Weissman & Tanner, 2018) in order to study this multimodal way of communicating irony and the individual differences inherent in such a phenomenon. Sentences such as *Living with him would be dreadful* could be followed by either a frown, smile, or wink emoji, representing conditions that either match, mismatch, or render ironic the sentence, respectively (Fig. 7). ERPs were time-locked to the sentence-final emoji. After some of the sentences, participants answered a yes/no comprehension question (e.g., *Would living with him be bad?*) designed to determine whether a participant had a literal or nonliteral interpretation of the sentence. If participants say "no" after seeing *Living with him would be dreadful <wink>* (i.e., Fig. 7C), they are using the wink as a marker of irony to override the literal meaning of the sentence; a "yes" response indicates the participant has not integrated the emoji as a marker of irony and instead adopts the literal interpretation of the sentence.

In the grand average of all 35 participants in Experiment 1, there was a significant P200 effect but no P600 effect. However, by investigating the behavioral responses, we saw a clear and relevant split among participants. Most participants (n = 25) almost never gave nonliteral answers and instead answered according to the literal semantic content of the sentence. A smaller subset of participants (n = 10) gave nonliteral answers over half the time, indicating that they understood the wink emoji as a marker of irony and allowed it to override the literal meaning. Treating these two sets as separate groups paints a different picture. Fig. 8A and B show the literal and nonliteral groups, respectively. While the P200 effect surfaces for both groups (most strongly over frontal sites), the re-processing P600 effect expectedly showed up only in those participants who actually re-processed and reinterpreted the literal meaning of the sentence. This is a finding that was lost in the overall grand average.

Experiment 2 aimed to guide participants towards understanding the sentences nonliterally by mentioning in the instructions that "some of the sentences will be sarcastic." A new set of 35 monolingual English right-

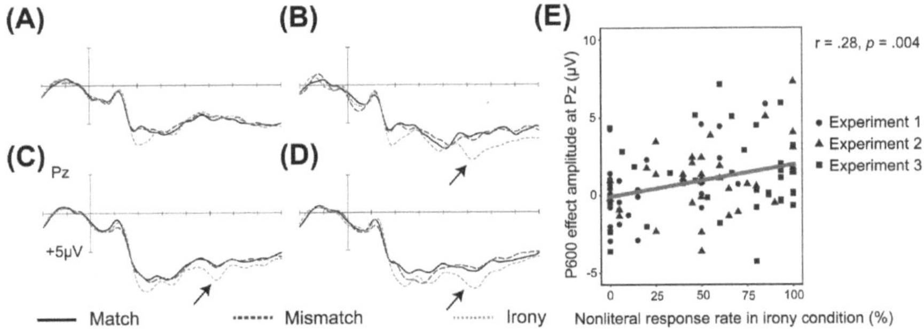

Figure 8 Grand average ERP waveforms at Pz elicited by sentence-final emoji from (A) Experiment 1, literal group (n = 25), (B) Experiment 1, nonliteral group (n = 10), (C) Experiment 2 (N = 35), and (D) Experiment 3 (N = 36). Arrows indicate irony-related P600 effects. (E) Correlation across all three experiments between non-literal response rate and P600 effect amplitude.

handers read the same sentences from Experiment 1 with this additional instruction. The mean nonliteral response rate in irony condition rose from 21.3% in Experiment 1 to 55.0% in Experiment 2. There was no longer a clear split between response groups, and both a P200 and P600 were evident in the grand mean (Fig. 8C). Experiment 3 (N = 36) altered the stimulus make-up from Experiments 1 and 2 to rule out a confound in condition-to-emoji mapping, and still found the P200-P600 complex to ironic sentences (Fig. 8D).

Fig. 8E shows the correlation between nonliteral response rate and P600 effect amplitude at Pz across all three experiments ($r = .28$, $p = .004$); across all 106 participants, nonliteral response rate to the yes/no comprehension questions was a significant predictor of P600 effect amplitude. Interestingly, such was not the case for the P200 effect amplitude at Fz ($r = .03$, $p = .76$). This investigation into individual differences allows for a better understanding of these components. As only the P600 correlates reliably with comprehension question responses, we can determine that this component is linked to the higher-level processing and interpretation of the stimuli, whereas the P200 effect, present no matter the interpretation, is reflecting a different process (e.g., attention) distinct from these pragmatic interpretations. Unlike morphosyntax, there are not strict rules about the "right" and "wrong" interpretations of irony, emojis, or their intersection. Even so, ERPs have proven to be robustly sensitive to the differing interpretations individuals may have when confronted with potentially-ironic emojis.

7. CONCLUSIONS AND CHALLENGES FOR THE FUTURE

As we have demonstrated, although grand mean waveforms and analyses of ERPs can provide meaningful characterizations of the central tendency of brain responses across participants and items, interpreting the grand mean waveform as representative of all (or even most) participants' brain responses in a study can lead to incorrect conclusions. This is especially the case when strong claims are made about multi-phasic ERP effects in grand means indicating a multi-stage processing architecture in the general population. We have clearly seen that these sorts of conclusions were unwarranted based on biphasic N400-P600 or LAN-P600 responses across a broad range of language user types (novice L2 learners, proficient bilinguals, and literate monolinguals): some participants showed primarily negative-going ERPs, some primarily positive-going ERPs, and some a mix of the two. Thus, even statistical reliability in grand mean analyses does not mean that the observed components generalize to all members of the broader population, and surface component scalp topographies and effect amplitudes can be distorted by component overlap. These findings also extend even to cases where there is a clear monophasic response evident in the grand mean: not all individuals show this response, with some even showing a response of the opposite polarity. Thus, focusing exclusively on parametric analyses of central tendency in ERP measures may lead to inappropriate conclusions about processing in the general population. As we have shown here, some of the 'noise' in ERP measures can indeed be structured and interesting; acknowledging this variation and incorporating it into analyses can lead to more robust and appropriate conclusions. This is similar to how some behavioral researchers are beginning to glean important information about processing by examining not simply mean reaction times, but also distributional properties of reaction times (i.e., ex-Gaussian distributional analyses, e.g., Balota & Yap, 2011; Payne & Federmeier, 2017; Staub, 2010).

A clear consequence of these observations is that extreme care must be taken when interpreting ERP effects in populations that might be expected to differ in some substantive way from proficient, literate, right-handed young adult monolinguals (e.g., children, individuals with language impairments, non-native speakers, bilinguals, older adults, etc). For example, in the field of L2 psycholinguistics, ERPs are often used to identify whether or not L2 processing can be "native-like," where native-like-ness is defined as finding qualitatively similar ERPs in the L2 population as seen in grand mean ERPs for native speakers; any deviations from this are frequently interpreted as indicating aberrant processing in some way, or as evidence of critical period effects in language learning (e.g., N400 effects elicited by morphosyntactic anomalies, or a lack of LAN effects). However, this sort of conclusion is only warranted if no native speakers show this sort of "non-native-like" ERP response. Since we have seen that grand mean ERPs can be problematic to interpret in this way, careful consideration of the range and types of variability in the control population need to be considered. Even for young adult monolinguals, we are just beginning to understand these issues, so broad claims about processing in other populations may be premature. Moreover, the absence of LAN-like components in some populations (like L2 users and bilinguals) cannot be interpreted as indicating the lack of some crucial process, since, as we have seen, many young adult monolinguals do not show LAN effects, and LAN-like components in grand mean analyses can be artifacts of individual differences and component overlap.

The approach we have taken to individual differences in the studies we reviewed here is clearly only one way to approach the problem and begin to understand how variation manifests in language-related ERPs. Clearly there is more variation in the data than our analyses capture, and likely in more dimensions than just the N400/P600 continuum we have highlighted. A challenge for the future, then, is to develop more powerful ways to capture variation in all of the dimensions that ERPs have to offer (e.g., effect timing, polarity, topography, and single-trial dynamics). Note that component overlap makes this necessarily difficult. Here we have documented ways in which components overlap across individuals (e.g., P600s in some cancel out part or all of an N400 in others), but the same issue applies for ERPs computed within individuals, and even electrophysiological effects elicited in single trials: the surface topography of an effect will always reflect the spatiotemporal summation of all of the simultaneously active neural sources of the electrophysiological signal. Because of this, simultaneously active sources within individuals (and trials) can make the nature of the underlying latent components difficult to identify (e.g., even if a LAN effect is found within a single individual or a single trial, it will be difficult or impossible to distinguish it from an N400 effect if there is a simultaneous positivity that obscures part of the negativity, even if that positivity is not clearly evident in the corresponding ERP). This is even the case with blind source separation techniques like independent components analysis, which can misallocate variance for spatiotemporally correlated components (Makeig, Jung, Ghahremani, & Sejnowski, 2000).

Nonetheless, we feel that the studies we have described here provide an important starting point for understanding neurocognitive variation for a range of language phenomena and in a number of populations. We are optimistic that future research on individual differences in language processing will help us better understand the mechanisms and dynamics underlying successful language comprehension, learning, and use very broadly.

REFERENCES

Antoniou, K., Cummins, C., & Katsos, N. (2016). Why only some adults reject under-informative utterances. *Journal of Pragmatics, 99*, 78–95. https://doi.org/10.1016/j.pragma.2016.05.001.

Balota, D. A., & Yap, M. J. (2011). Moving beyond the mean in studies of mental chronometry: The power of response time distributional analyses. *Current Directions in Psychological Science, 20*, 160–166. https://doi.org/10.1177/0963721411408885.

Baron-Cohen, S. (2000). Theory of mind and autism: A fifteen year review. In S. Baron-Cohen, H. Tager-Flusberg, & D. J. Cohen (Eds.), *Understanding other minds: Perspectives from developmental cognitive neuroscience* (pp. 3–20). New York: Oxford University Press. https://doi.org/10.1300/J108v05n03.

Baron-Cohen, S., Wheelwright, S., Skinner, R., Martin, J., & Clubley, E. (2001). The Autism Spectrum Quotient: Evidence from Asperger syndrome/high functioning autism, males and females, scientists and mathematicians. *Journal of Autism and Developmental Disorders, 31*, 5–17. https://doi.org/10.1023/A:1005653411471.

Bastiaansen, M., Mazaheri, A., & Jensen, O. (2012). Beyond ERPs: Oscillatory neuronal dynamics. In S. J. Luck, & E. S. Kappenman (Eds.), *The Oxford handbook of event-related potential components* (pp. 31–50). New York: Oxford University Press.

Batterink, L., & Neville, H. J. (2013). The human brain processes syntax in the absence of conscious awareness. *Journal of Neuroscience, 33*, 8528–8533. https://doi.org/10.1523/JNEUROSCI.0618-13.2013.

Birdsong, D. (2006). Age and second language acquisition and processing: A selective overview. *Language Learning, 56*, 9–49. https://doi.org/10.1111/j.1467-9922.2006.00353.x.

Birdsong, D., & Molis, M. (2001). On the evidence for maturational constraints in second-language acquisition. *Journal of Memory and Language, 44*, 235–249.

Bornkessel-Schlesewsky, I., & Schlesewsky, M. (2008). An alternative perspective on "semantic P600" effects in language comprehension. *Brain Research Reviews, 59*, 55–73. https://doi.org/10.1016/j.brainresrev.2008.05.003.

Bott, L., & Noveck, I. A. (2004). Some utterances are underinformative: The onset and time course of scalar inferences. *Journal of Memory and Language, 51*, 437–457. https://doi.org/10.1016/j.jml.2004.05.006.

Boudewyn, M. A., Long, D. L., & Swaab, T. Y. (2013). Effects of working memory span on processing of lexical associations and congruence in spoken discourse. *Frontiers in Psychology, 4*, 1–16. https://doi.org/10.3389/fpsyg.2013.00060.

Brouwer, H., Fitz, H., & Hoeks, J. (2012). Getting real about semantic illusions: Rethinking the functional role of the P600 in language comprehension. *Brain Research, 1446*, 127–143. https://doi.org/10.1016/j.brainres.2012.01.055.

Carretie, L., Mercado, F., & Tapia, M. (2001). Emotion, attention, and the 'negativity bias', studied through event-related potentials. *International Journal of Psychophysiology*, 75–85.

Chomsky, N. (1965). *Aspects of the theory of syntax*. Cambridge, MA: MIT Press.

Cohn, N. (2016). A multimodal parallel architecture: A cognitive framework for multimodal interactions. *Cognition, 146*, 304–323. https://doi.org/10.1016/j.cognition.2015.10.007.

Degen, J., & Tanenhaus, M. K. (2016). Availability of alternatives and the processing of scalar implicatures: A visual world eye-tracking study. *Cognitive Science, 40*, 172–201. https://doi.org/10.1111/cogs.12227.

Dresner, E., & Herring, S. C. (2010). Functions of the nonverbal in CMC: Emoticons and illocutionary force. *Communication Theory, 20*, 249–268. https://doi.org/10.1111/j.1468-2885.2010.01362.x.

Filik, R., Leuthold, H., Wallington, K., & Page, J. (2014). Testing theories of irony processing using eye-tracking and ERPs. *Journal of Experimental Psychology: Learning, Memory, and Cognition, 40*, 811–828. https://doi.org/10.1037/a0035658.

Filik, R., Turcan, A., Thompson, D., Harvey, N., Davies, H., & Turner, A. (2015). Sarcasm and emoticons: Comprehension and emotional impact. *Quarterly Journal of Experimental Psychology, 0218*, 1–17. https://doi.org/10.1080/17470218.2015.1106566.

Giora, R. (1997). Understanding figurative and literal language: The graded salience hypothesis. *Cognitive Linguistics, 8*, 183–206. https://doi.org/10.1515/cogl.1997.8.3.183.

Grey, S., Tanner, D., & Van Hell, J. G. (2017). How right is left? Handedness modulates neural responses during morphosyntactic processing. *Brain Research, 1669*, 27–43. https://doi.org/10.1016/j.brainres.2017.05.024.

Grodner, D. J., Klein, N. M., Carbary, K. M., & Tanenhaus, M. K. (2010). "Some," and possibly all, scalar inferences are not delayed: Evidence for immediate pragmatic enrichment. *Cognition, 116*, 42–55. https://doi.org/10.1016/j.cognition.2010.03.014.

Hagoort, P., Baggio, G., & Willems, R. (2009). Semantic unification. In M. S. Gazzaniga (Ed.), *The cognitive neurosciences* (4th ed., pp. 819–836). Cambridge, MA: MIT Press.

Hagoort, P., Brown, C. M., & Groothusen, J. (1993). The syntactic positive shift as an ERP measure of syntactic processing. *Language & Cognitive Processes, 8*, 439–484.

Hancock, R., & Bever, T. G. (2013). Genetic factors and normal variation in the organization of language. *Biolinguistics, 7*, 75–95.

Happé, F. G. E. (1993). Communicative competence and theory of mind in autism: A test of relevance theory. *Cognition, 48*, 101–119.

Happé, F. G. E. (1994). An advanced test of theory of mind: Understanding of story characters' thoughts and feelings by able autistic, mentally handicapped, and normal children and adults. *Journal of Autism and Developmental Disorders, 24*, 129–154. https://doi.org/10.1007/BF02172093.

Hoeks, J. C. J., Stowe, L. A., & Doedens, G. (2004). Seeing words in context: The interaction of lexical and sentence level information during reading. *Cognitive Brain Research, 19*, 59–73. https://doi.org/10.1016/j.cogbrainres.2003.10.022.

Huang, Y. T., & Snedeker, J. (2009). Online interpretation of scalar quantifiers: Insight into the semantics-pragmatics interface. *Cognitive Psychology, 58*, 376–415. https://doi.org/10.1016/j.cogpsych.2008.09.001.

Ivanko, S., Pexman, P. M., & Olineck, K. (2004). How sarcastic are you?: Individual differences and verbal irony. *Journal of Language and Social Psychology, 23*, 244–271. https://doi.org/10.1177/0261927X04266809.

Jolliffe, T., & Baron-Cohen, S. (1999). A test of central coherence theory: Linguistic processing in high-functioning adults with autism or Asperger syndrome: Is local coherence impaired? *Cognition, 71*, 149–185. https://doi.org/10.1016/S0010-0277(99)00022-0.

Jolliffe, T., & Baron-Cohen, S. (2000). Linguistic processing in high-functioning adults with autism or Asperger's syndrome. Is global coherence impaired? *Psychological Medicine, 30*, 1169–1187.

Just, M. A., & Carpenter, P. A. (1992). A capacity theory of comprehension: Individual differences in working memory. *Psychological Review, 99*, 122–149.

Kaan, E., Harris, A., Gibson, E., & Holcomb, P. J. (2000). The P600 as an index of syntactic integration difficulty. *Language & Cognitive Processes, 15*, 159–201.

Kim, A., Oines, L., & Miyake, A. (2018). Individual differences in verbal working memory underlie a tradeoff between semantic and structural processing difficulty during language comprehension: An ERP investigation. *Journal of Experimental Psychology: Learning, Memory, and Cognition, 44*, 406–420. https://doi.org/10.1037/xlm0000457.

Kim, A., & Osterhout, L. (2005). The independence of combinatory semantic processing: Evidence from event-related potentials. *Journal of Memory and Language, 52*, 205–225. https://doi.org/10.1016/j.jml.2004.10.002.

King, J. W., & Just, M. A. (1991). Individual differences in syntactic processing: The role of working memory. *Journal of Memory and Language, 30*, 580–602.

King, J. W., & Kutas, M. (1995). Who did what and when? Using word- and clause-level ERPs to monitor working memory usage in reading. *Journal of Cognitive Neuroscience, 7*, 376–395.

Kolk, H. H. J., & Chwilla, D. (2007). Late positivities in unusual situations. *Brain and Language, 100*, 257–261. https://doi.org/10.1016/j.bandl.2006.07.006.

Kolk, H. H. J., Chwilla, D. J., van Herten, M., & Oor, P. J. W. (2003). Structure and limited capacity in verbal working memory: A study with event-related potentials. *Brain and Language, 85*, 1–36. https://doi.org/10.1016/S0093-934X(02)00548-5.

Kreuz, R. J., & Glucksberg, S. (1989). How to be sarcastic: The echoic reminder theory of verbal irony. *Journal of Experimental Psychology: General, 118*, 374–386.

Kuperberg, G. R. (2007). Neural mechanisms of language comprehension: Challenges to syntax. *Brain Research, 1146*, 23–49. https://doi.org/10.1016/j.brainres.2006.12.063.

Kuperberg, G. R., Caplan, D., Sitnikova, T., Eddy, M., & Holcomb, P. J. (2006). Neural correlates of processing syntactic, semantic, and thematic relationships in sentences. *Language & Cognitive Processes, 21*, 489–530. https://doi.org/10.1080/01690960500094279.

Kuperberg, G. R., Kreher, D. A., Sitnikova, T., Caplan, D., & Holcomb, P. J. (2007). The role of animacy and thematic relationships in processing active English sentences: Evidence from event-related potentials. *Brain and Language, 100*, 223–237. https://doi.org/10.1016/j.bandl.2005.12.006.

Kuperberg, G. R., Sitnikova, T., Caplan, D., & Holcomb, P. J. (2003). Electrophysiological distinctions in processing conceptual relationships within simple sentences. *Cognitive Brain Research, 17*, 117–129. https://doi.org/10.1016/S0926-6410(03)00086-7.

Kutas, M., & Federmeier, K. D. (2000). Electrophysiology reveals semantic memory use in language comprehension. *Trends in Cognitive Sciences, 4*, 463–470.

Kutas, M., & Federmeier, K. D. (2011). Thirty years and counting: Finding meaning in the N400 component of the event-related brain potential (ERP). *Annual Review of Psychology, 62*, 621–647. https://doi.org/10.1146/annurev.psych.093008.131123.

Kutas, M., & Hillyard, S. A. (1980). Reading senseless sentences: Brain potentials reflect semantic incongruity. *Science, 207*, 203–205.

LaMarre, H. L., Landreville, K. D., & Beam, M. A. (2009). The irony of satire. *The International Journal of Press/Politics, 14*, 212–231. https://doi.org/10.1177/1940161208330904.

Lau, W. Y.-P. P., Gau, S. S.-F. F., Chiu, Y.-N. N., Wu, Y.-Y. Y., Chou, W.-J. J., Liu, S.-K. K., & Chou, M.-C. C. (2013). Psychometric properties of the Chinese version of the autism spectrum quotient (AQ). *Research in Developmental Disabilities, 34*, 294–305. https://doi.org/10.1016/j.ridd.2012.08.005.

Lau, E. F., Phillips, C., & Poeppel, D. (2008). A cortical network for semantics: (de)constructing the N400. *Nature Reviews Neuroscience, 9*, 920–933. https://doi.org/10.1038/nrn2532.

Lee, C., & Federmeier, K. D. (2015). It's all in the family: Brain asymmetry and syntactic processing of word class. *Psychological Science, 26*, 997–1005. https://doi.org/10.1177/0956797615575743.

Luck, S. J., & Hillyard, S. A. (1994). Electrophysiological correlates of feature analysis during visual search. *Psychophysiology*. https://doi.org/10.1111/j.1469-8986.1994.tb02218.x.

Makeig, S., Jung, T.-P., Ghahremani, D., & Sejnowski, T. J. (2000). Independent component analysis of simulated ERP data. In T. Nakada (Ed.), *Integrated human brain science* (pp. 123–146). Amsterdam: Elsevier Science.

Martin, I., & Mcdonald, S. (2004). An exploration of causes of non-literal language problems in individuals with Asperger Syndrome. *Journal of Autism and Developmental Disorders, 34*, 311–328. https://doi.org/10.1023/B:JADD.0000029553.52889.15.

McLaughlin, J., Tanner, D., Pitkänen, I., Frenck-Mestre, C., Inoue, K., Valentine, G., & Osterhout, L. (2010). Brain potentials reveal discrete stages of L2 grammatical learning. *Language Learning, 60*, 123–150. https://doi.org/10.1111/j.1467-9922.2010.00604.x.

Mehravari, A. S., Emmorey, K., Prat, C. S., Klarman, L., & Osterhout, L. (2017). Brain-based individual difference measures of reading skill in deaf and hearing adults. *Neuropsychologia, 101*, 153–168. https://doi.org/10.1016/j.neuropsychologia.2017.05.004.

Meulman, N., Wieling, M., Sprenger, S. A., Stowe, L. A., & Schmid, M. S. (2015). Age effects in L2 grammar processing as revealed by ERPs and how (not) to study them. *PLoS One, 10*. https://doi.org/10.1371/journal.pone.0143328. e0143328.

Miller, H., Kluver, D., Thebault-Spieker, J., Terveen, L., & Hecht, B. (2017). Understanding emoji ambiguity in context: The role of text in emoji-related miscommunication. *Proceedings of the Eleventh International AAAI Conference on Web and Social Media*, 152–161.

Miller, H., Thebault-Spieker, J., Chang, S., Johnson, I., Terveen, L., & Hecht, B. (2016). "Blissfully happy" or "ready to fight": Varying interpretations of emoji. *International AAAI Conference on Web and Social Media*, 259–268. https://doi.org/10.1089/cyber.2011.0179.

Molinaro, N., Barber, H. A., Caffarra, S., & Carreiras, M. (2015). On the left anterior negativity (LAN): The case of morphosyntactic agreement: A reply to Tanner et al. *Cortex, 66*, 156–159. https://doi.org/10.1016/j.cortex.2014.06.009.

Molinaro, N., Barber, H., & Carreiras, M. (2011). Grammatical agreement processing in reading: ERP findings and future directions. *Cortex, 47*, 908–930. https://doi.org/10.1016/j.cortex.2011.02.019.

Morgan-Short, K. (2014). Electrophysiological approaches to understanding second language acquisition: A field reaching its potential. *Annual Review of Applied Linguistics, 34*, 15–36. https://doi.org/10.1017/S026719051400004X.

Nakano, H., Saron, C., & Swaab, T. Y. (2010). Speech and span: Working memory capacity impacts the use of animacy but not of world knowledge during spoken sentence comprehension. *Journal of Cognitive Neuroscience, 22*, 2886–2898. https://doi.org/10.1162/jocn.2009.21400.

Neville, H. J., Nicol, J., Barss, A., Forster, K. I., & Garrett, M. F. (1991). Syntactically based sentence processing classes: Evidence from event-related brain potentials. *Journal of Cognitive Neuroscience, 3*, 151–165.

Nieuwland, M. S., Ditman, T., & Kuperberg, G. R. (2010). On the incrementality of pragmatic processing: An ERP investigation of informativeness and pragmatic abilities. *Journal of Memory and Language, 63*, 324–346. https://doi.org/10.1016/j.jml.2010.06.005.

Noveck, I. A., & Posada, A. (2003). Characterizing the time course of an implicature: An evoked potentials study. *Brain and Language, 85*, 203–210.

Noveck, I., & Sperber, D. (Eds.). (2004). *Experimental pragmatics*. Basingstoke: Palgrave MacMillan. https://doi.org/10.1057/9780230524125.

Osterhout, L. (1997). On the brain response to syntactic anomalies: Manipulations of word position and word class reveal individual differences. *Brain and Language, 59*, 494–522.

Osterhout, L., & Holcomb, P. J. (1992). Event-related brain potentials elicited by syntactic anomaly. *Journal of Memory and Language, 31*, 785–806.

Osterhout, L., & Holcomb, P. J. (1993). Event-related potentials and syntactic anomaly: Evidence of anomaly detection during the perception of continuous speech. *Language & Cognitive Processes, 8*, 413–437.

Osterhout, L., McLaughlin, J., Kim, A., Greewald, R., & Inoue, K. (2004). Sentences in the brain: Event-related potentials as real-time reflections of sentence comprehension and language learning. In M. Carreiras, & C. Clifton (Eds.), *The on-line study of sentence comprehension: Eyetracking-ERPs, and beyond* (pp. 271–308). New York: Psychology Press.

Osterhout, L., McLaughlin, J., Pitkänen, I., Frenck-Mestre, C., & Molinaro, N. (2006). Novice learners, longitudinal designs, and event-related potentials: A means for exploring the neurocognition of second language processing. *Language Learning, 56*, 199–230.

Osterhout, L., Poliakov, A., Inoue, K., McLaughlin, J., Valentine, G., Pitkänen, I., ... Herschensohn, J. (2008). Second-language learning and changes in the brain. *Journal of Neurolinguistics, 21*, 509–521.

Paczynski, M., & Kuperberg, G. R. (2012). Multiple influences of semantic memory on sentence processing: Distinct effects of semantic relatedness on violations of real-world event/state knowledge and animacy selection restrictions. *Journal of Memory and Language, 67*, 426–448. https://doi.org/10.1016/j.jml.2012.07.003.

Pakulak, E., & Neville, H. J. (2010). Proficiency differences in syntactic processing of monolingual native speakers indexed by event-related potentials. *Journal of Cognitive Neuroscience, 22*, 2728–2744.

Papafragou, A., & Musolino, J. (2003). Scalar implicatures: {E}xperiments at the semantics-pragmatics interface. *Cognition, 86*, 253–282.

Patel, A. D., Gibson, E., Ratner, J., Besson, M., & Holcomb, P. (1998). Processing syntactic relations in language and music: An event-related potential study. *Journal of Cognitive Neuroscience, 10*, 717–733.

Payne, B. R., & Federmeier, K. D. (2017). Pace yourself: Intraindividual variability in context use revealed by self-paced event-related brain potentials. *Journal of Cognitive Neuroscience, 25*, 837–854. https://doi.org/10.1162/jocn_a_01090.

Payne, B. R., Ng, S., Shantz, K., & Federmeier, K. D. (2018). Event-related brain potentials in language processing: The N's and the P's. In A. B. Cieślicka, & R. R. Heredia (Eds.), *The bilingual brain unwrapped*. Psychology Press (in press).

Perfetti, C. (2007). Reading ability: Lexical quality to comprehension. *Scientific Studies of Reading, 11*, 357–383. https://doi.org/10.1080/10888430701530730.

Phillips, C., Kazanina, N., & Abada, S. H. (2005). ERP effects of the processing of syntactic long-distance dependencies. *Cognitive Brain Research, 22*, 407–428. https://doi.org/10.1016/j.cogbrainres.2004.09.012.

Regel, S., Coulson, S., & Gunter, T. C. (2010). The communicative style of a speaker can affect language comprehension? ERP evidence from the comprehension of irony. *Brain Research, 1311*, 121–135. https://doi.org/10.1016/j.brainres.2009.10.077.

Regel, S., Gunter, T. C., & Friederici, A. D. (2011). Isn't it ironic? An electrophysiological exploration of figurative language processing. *Journal of Cognitive Neuroscience, 23*, 277–293. https://doi.org/10.1162/jocn.2010.21411.

Searle, J. (1969). *Speech acts: An essay in the philosophy of language*. Cambridge: Cambridge University Press.

Searle, J. R. (1979). *Expression and meaning: Studies in the theory of speech acts*. Cambridge: Cambridge University Press. https://doi.org/10.2307/2184707.

Sitnikova, T., Holcomb, P. J., Kiyonaga, K. a, & Kuperberg, G. R. (2008). Two neurocognitive mechanisms of semantic integration during the comprehension of visual real-world events. *Journal of Cognitive Neuroscience, 20*, 2037–2057. https://doi.org/10.1162/jocn.2008.20143.

Skovholt, K., Grønning, A., & Kankaanranta, A. (2014). The communicative functions of emoticons in workplace e-mails::-). *Journal of Computer-Mediated Communication, 19*, 780–797. https://doi.org/10.1111/jcc4.12063.

Sperber, D., & Wilson, D. (1986). *Relevance: Communication and cognition*. Oxford: Blackwell.

Spotorno, N., Cheylus, A., Van Der Henst, J. B., & Noveck, I. A. (2013). What's behind a P600? Integration operations during irony processing. *PLoS One, 8*, 1–13. https://doi.org/10.1371/journal.pone.0066839.

Spychalska, M., Kontinen, J., & Werning, M. (2016). Investigating scalar implicatures in a truth-value judgement task: Evidence from event-related brain potentials. *Language, Cognition and Neuroscience, 31*, 817–840. https://doi.org/10.1080/23273798.2016.1161806.

Staub, A. (2010). Response time distributional evidence for distinct varieties of number attraction. *Cognition, 114*, 447–454.

Steinhauer, K. (2014). Event-related potentials (ERPs) in second language research: A brief introduction to the technique, a selected review, and an invitation to reconsider critical periods in L2. *Applied Linguistics*. https://doi.org/10.1093/applin/amu028.

Swaab, T. Y., Ledoux, K., Camblin, C. C., & Boudewyn, M. A. (2012). Language-related ERP components. In S. J. Luck, & E. S. Kappenman (Eds.), *The Oxford handbook of event-related potential components* (pp. 397–439). Oxford: Oxford University Press.

Tanner, D. (2018). *Robust neurocognitive individual differences in morphosyntactic processing: A latent variable approach*. https://doi.org/10.17605/OSF.IO/NJGT6. Manuscript submitted for publication.

Tanner, D., Bulkes, N. Z., Shantz, K., Armstrong, A., & Reyes, A. (2016). Working memory, not language experience, predicts N400 effects to semantically anomalous words during reading. *Poster presented at the neurobiology of language conference, London, UK.*

Tanner, D., Grey, S., & Van Hell, J. G. (2017). Dissociating retrieval interference and reanalysis in the P600 during sentence comprehension. *Psychophysiology, 54*, 248–259. https://doi.org/10.1111/psyp.12788.

Tanner, D., Inoue, K., & Osterhout, L. (2014). Brain-based individual differences in online L2 grammatical comprehension. *Bilingualism: Language and Cognition, 17*, 277–293. https://doi.org/10.1017/S1366728913000370.

Tanner, D., McLaughlin, J., Herschensohn, J., & Osterhout, L. (2013). Individual differences reveal stages of L2 grammatical acquisition: ERP evidence. *Bilingualism: Language and Cognition, 16*, 367–382. https://doi.org/10.1017/S1366728912000302.

Tanner, D., Nicol, J., & Brehm, L. (2014). The time-course of feature interference in agreement comprehension: Multiple mechanisms and asymmetrical attraction. *Journal of Memory and Language, 76*, 195–215. https://doi.org/10.1016/j.jml.2014.07.003.

Tanner, D., & Van Hell, J. G. (2014a). ERPs reveal individual differences in morphosyntactic processing. *Neuropsychologia, 56*, 289–301. https://doi.org/10.1016/j.neuropsychologia.2014.02.002.

Tanner, D., & Van Hell, J. G. (2014b). Gradience and individual differences in processing syntactic, semantic, and thematic information: ERP evidence. *Paper presented at the 2014 CUNY conference on human sentence processing, columbus, OH, USA.*

Thompson, D., & Filik, R. (2016). Sarcasm in written communication: Emoticons are efficient markers of intention. *Journal of Computer-Mediated Communication, 21*, 105–120. https://doi.org/10.1111/jcc4.12156.

Townsend, D. J., Carrithers, C., & Bever, T. G. (2001). Familial handedness and access to words, meaning, and syntax during sentence comprehension. *Brain and Language, 78*, 308–331. https://doi.org/10.1006/brln.2001.2469.

Tremblay, A., & Newman, A. J. (2015). Modeling nonlinear relationships in ERP data using mixed-effects regression with R examples. *Psychophysiology, 52*, 124–139. https://doi.org/10.1111/psyp.12299.

Ullman, M. T. (2004). Contributions of memory circuits to language: The declarative/procedural model. *Cognition, 92*, 231–270.

van de Meerendonk, N., Kolk, H. H. J., Vissers, C. T. W. M., & Chwilla, D. J. (2010). Monitoring in language perception: Mild and strong conflicts elicit different ERP patterns. *Journal of Cognitive Neuroscience, 22*, 67–82.

Van Dyke, J. A., Johns, C. L., & Kukona, A. (2014). Low working memory capacity is only spuriously related to poor reading comprehension. *Cognition, 131*, 373–403. https://doi.org/10.1016/j.cognition.2014.01.007.

van Herten, M., Chwilla, D. J., & Kolk, H. H. J. (2006). When heuristics clash with parsing routines: ERP evidence for conflict monitoring in sentence perception. *Journal of Cognitive Neuroscience, 18*, 1181–1197. https://doi.org/10.1162/jocn.2006.18.7.1181.

Weissman, B., & Tanner, D. (2018). A strong wink between verbal and emoji-based irony: How the brain processes ironic emojis during language comprehension. *PLoS One, 3*(8), e0201727. https://doi.org/10.1371/journal.pone.0201727.

Wilson, D., & Sperber, D. (1992). On verbal irony. *Lingua, 87*, 53–76.

CHAPTER ELEVEN

He Gave My Nose a Kick or *He Kicked My Nose?* Argument Structure Alternations and Event Construal

Eva Wittenberg
Department of Linguistics, University of California, San Diego, CA, United States
E-mail: ewittenberg@ucsd.edu

Contents

1. The Linking Structures Between Grammar and Event Construal 339
2. Case Study I: Grammar and Event Roles, and What Happens When They don't Align 342
3. Case Study II: From Grammar to Time 347
4. Loose Ends: Processing, Reference, and Acquisition 351
5. Conclusion: Linguistics Is Worth Your while 355
References 356

Abstract

The most fundamental function of language is to enable people to share mental models of their worlds. For a comprehender, the given mental model she is building will be shaped by the lexical items, and also by the syntactic structures, that a speaker is using. In this chapter, I review literature that unearths the mental models formed by comprehenders, based on the grammatical structure they encounter, as mutually informative for both linguistic theory and event and object cognition. This chapter uses the well-studied case of light verb constructions and reviews data from a range of experimental studies that investigated how linguistic structure shapes core aspects of mental models: the conceptualization of event participants, and temporal structure in events.

I snapped my chin down onto some guy's fist

and hit another one in the knee with my nose!
Woody Allen, "Play it Again, Sam" (1972)

Why is the quote above funny? Any linguist can ruin the joke for you by overexplaining its mechanism: Woody Allen, in order to impress his love

Psychology of Learning and Motivation, Volume 68
ISSN 0079-7421
https://doi.org/10.1016/bs.plm.2018.08.008

interest, describes the *goals* of a violent act ("my chin", "my nose"), as instruments, not goals, of the action — leading her (and us) to draw the wrong inferences about volitionality and agentivity. Allen uses the structure of language to effectively play with our event representations, for just a tiny moment.

Most linguistic choices are subtler than this, but the ramifications they can have on comprehenders are no less interesting. In this chapter, I argue that event representations created by a listener's mind are always, not only in comedy or to dramatic effect, modulated by choice of grammatical construction — and not in any random way, but in a way that is predicted by linguistic theory. To do this, I will take you on a tour de force through the world of light verb constructions: everyday phrases that reveal a lot about how we transverse the space from a syntactic structure to imagining an event, and which we can use to test theories about the architecture of our cognitive capacity for language.

One recurring thread in this chapter will be the subtleness of linguistic choices. Linguists are sometimes scolded by other disciplines about the amount of jargon they use. I agree with the fact that we *do* use a lot of jargon, which makes our work less accessible to non-linguists, and we should try to avoid that. But I hope to convince everyone that if we ignored, downplayed, or simplified the subtleties of linguistic differences, we would not be doing ourselves a favor investigating the connection between language and the mind: No vision scientist decides to ignore color vision just because it may be harder to study (and explain) than edge detection.

In what follows, I will review two cases and a few loose ends on how a linguistic form can influence how people think about events. Events, as percepts and concepts of human cognition, have long held an interest for linguistics, philosophy, and lately also cognitive science. Linguists traditionally focus on the level of syntax, semantics and the lexicon, forming hypotheses about what the mental model would look like, but they leave the investigation of mental models to cognitive scientists, who, in turn, often are less concerned with the underlying linguistic structure. However, much of the mapping between these levels of representation hinges on the particular linguistic architecture one assumes. I argue that in order to learn about each of these domains of inquiry, we are well-advised to take seriously a broad range of linguistic theories — because when we do, they make testable predictions about human behavior.

1. THE LINKING STRUCTURES BETWEEN GRAMMAR AND EVENT CONSTRUAL

One fundamental assumption in many a linguistic theory is that syntax and semantics are in a homomorphic relationship – that is, each noun phrase maps onto a participant in a scene, and the relationships between these participants is determined by the predicate. For instance, "Jeremy is making coffee" is a string of words ("Jeremy", "is", "making", "coffee"). These words stand in a hierarchical, syntactic relationship to each other (sometimes called the *surface form*).

This syntactic form consists of a verb ("make") that denotes the kind of event one is talking about – in this case, creating a hot beverage. The object noun ("coffee") stands in a *thematic relationship* to the verb (Dowty, 1991). Thematic relationships map out the functions of the entities described by the nouns in relation to the action described by the verb. For instance, for every "making" action, there needs to be a maker; this role is called the Agent. The thing being made would be the Patient ("coffee" in "make a coffee"). Similarly, for verbs like "hit", "kiss", or "eat", the syntactic object would be also a Patient of the action, but for transfer verbs such as "give", "steal", "take", or "donate", the syntactic objects and the subject occupy roles of Theme, Source, and Goal: In "Jeremy threw the ball to Julius", Jeremy is the Source (where the ball came from), the ball is a Theme (the entity undergoing the action, but with its state unchanged), and Julius is the Goal (where the ball ends up).

Building from the semantic level, which tells us about the linguistically relevant meaning components, such as Agent and Patient, we can infer a situation model: The spatial structure of the event, maybe its duration, and, depending on the nature of the event, particular perceptual, emotional, and memory associations.

For instance, in the sentence "Jeremy is making coffee", the syntax is a simple transitive sentence (Fig. 1). A proper noun ("Jeremy") is the syntactic subject, mapping onto an Agent in the semantic structure. The noun "coffee" is the syntactic object, and it maps onto a Patient — albeit into a complicated one, since "making coffee" is a process which turns an object (ground beans) into the means of extracting the final product also called "coffee", with the help of another substance (water). The bare-bones information the sentence "Jeremy is making coffee" is giving us is the Agent (JEREMY), the action (MAKE), and the result of the action (COFFEE).

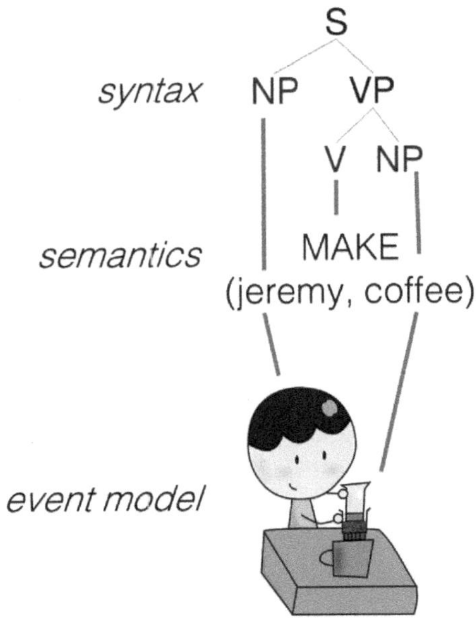

Figure 1 Mapping from syntax to semantics to event structure.

Thus, the grammar of the sentence gives us some helpful hints about the event structure beyond who does what to whom: The progressive tense is used, indicating that the action is ongoing (e.g., Dowty, 1977; Parsons, 1989); "coffee" is a mass noun, which often correlates with being a substance in the world (e.g. Gillon, 1999). Apart from the confusion that the linguistic system of English does not distinguish between coffee, the ingredient, and coffee, the product (Srinivasan & Rabagliati, 2015), "making coffee" is also a hugely complex action composed of a multitude of hierarchically structured subevents (see Jackendoff, 2007a, for a tree-like diagram of the hierarchical actions and subactions involved).

There is evidence that we imagine something like the full event structure upon hearing such a sentence, even without the lexical input explicitly mapping on those subevents. For instance, a recent computational model explains how people are more surprised to hear, in the context of a restaurant, "Anna asked for the menu and left" than "Anna asked for the bill and left", even though the temporal order and co-occurrence frequency of the linguistic items are matched between the sentences. Yet, when people imagine someone going to a restaurant, they know that between asking for the menu and leaving, there are several crucial steps, and leaving them out in the situation description influences how people process the sentence

(Venhuizen, Crocker, & Brouwer, 2018). "Going to a restaurant" or "making coffee" evoke in us as comprehenders a complex conceptual structure with dozens of scripted subevents with the use of just a simple, short verb phrase.

The above example illustrated how language can compress the mess of event cognition by squeezing a group of complex conceptual structures into a single clause; and how, even though the levels of representation in language and cognition are distinct, they are tightly linked. We can probe linguistic constructions to jointly inform our theories on both the architecture of the language system and the theory of event cognition, and how both of those interact with each other. Unlike in the coffee-making example above, however, I will focus less on the implicit and associative knowledge implied by linguistic descriptions, and more on how people imagine events based on grammatical form.

One interesting case helping these investigations are light verb constructions such as "he gave a kick to my nose". Light verb constructions of the kind discussed in the chapter are combinations of verbs and nouns, such as "giving a hug", "taking a shower", or "having a run." They have now been studied from a multitude of angles, and they are very frequent both between and within languages (Butt, 2003; 2010).

What makes these constructions special is the way that their meaning is distributed among its parts. As we have seen in the example above, a verb phrase (such as "make a coffee") consists of a verb ("make") that denotes the kind of event one is talking about, and orders event participants hierarchically. However, in light verb constructions, the relationships between verb and noun are different. For instance, "taking a shower" does not convey the physical removal of a shower. Instead, it means the same thing as "showering". In English and many other languages, light verb constructions have such a simple verb counterpart ("hugging", "showering", "running"; Butt, 2003; Butt & Lahiri, 2013). The verb and object in light verb constructions form a complex predicate — and one of the open questions is how the thematic relationships between light verb, light noun, and the rest of the sentence are structured.

But light verb constructions are also interesting for other reasons. From a syntax/semantics point of view, the way that the eventive nouns are formed and inserted into a sentence matters for other aspects of event structure — temporal duration, for instance. From a discourse processing point of view, the fact that the eventive noun is semantically integrated into the predicate should influence its referential accessibility.

And light verb constructions also pose interesting questions about learnability. In what follows, I will review what we know about each of these domains.

2. CASE STUDY I: GRAMMAR AND EVENT ROLES, AND WHAT HAPPENS WHEN THEY DON'T ALIGN

The connection between syntactic structure and semantic nuances is obviously not a new insight – it is a foundational question of linguistics, one that is best exemplified by syntactic alternations. For instance, when we look at the active—passive alternation ("Tom hit Bill" vs. "Bill was hit by Tom"), some theories (mainly transformational grammars) assume that even though these sentences have a different surface form, they both refer to the same "logical form", that is, to the same underlying mental model of the event (Hornstein, 1995). However, there is both theoretical and now also psycholinguistic evidence that this is not how people use these alternations. For instance, when people read an event description in active voice, they assign more responsibility to the Agent than when they read a description of the same event in passive voice (Fausey & Boroditsky, 2010). These results are more compatible with a theoretical model that views not only lexical items, but also syntactic structures, as linguistic devices that carry meaning (Goldberg, 1995).

Another example hinting at how differences in linguistic structure may have ramifications for mental models is the locative alternation, as in "Jack sprayed paint on the wall" vs. "Jack sprayed the wall with paint." For example, you might draw different inferences about the area covered in paint in each of these examples" (e.g., Gropen, Pinker, Hollander, & Goldberg, 1991). Similarly, the dative alternation ("Bob sent Betty a book" vs. "Bob sent a book to Betty" may lead people to make different inferences about whether Betty actually received the book (e.g., Gropen, Pinker, Hollander, Goldberg, & Wilson, 1989).

In the above examples, the main component of meaning affected by these alternations is the degree to which an event participant is involved or changed, or to which degree an event is completed. However, neither example calls into question an assumption so central to linguistic architectures across the theoretical spectrum that it is often taken as an axiom: The assumption of homomorphism between semantic roles and syntactic arguments, which, for instance, is underlying the Theta-Criterion. The Theta-Criterion states that under no circumstances can it be allowed that a noun gets more than one role (Baker, 1988; Chomsky, 1981/1993; but see; Hornstein, 1999).

Light verb constructions offer a challenge to this axiom, since one question that arises is what *kind* of event one is really imagining when listening to an event described by such a construction. That is, people have strong intuitions that there is a difference between "giving a kiss" and "kissing" – as a colleague said, "when my wife gives me a kiss, I definitely get *something*!" In other words, what people are questioning in these constructions is the notion that the verb ("give") contributes no meaning, and therefore, no semantic roles either.

We took these intuitions seriously and asked how different theories would predict the configuration of semantic roles in light verb constructions, and consequently, which semantic roles people would actually imagine listening or reading an event description. Below is a simplified schema of these predictions (Fig. 2).

The first possibility views light verb constructions as atomic units (Fig. 2, "treelet view"), similar to idioms like "kick the bucket" or "cut some slack," constructions in which the meaning cannot be construed from their parts. In this view, "Douglas gave a kiss to Mary" simply maps "Douglas" to Agent, and "Mary" to Patient, with no semantic role occupied by "kiss": "give a kiss" as a whole is taken as a semantic-role launching predicate – it is an

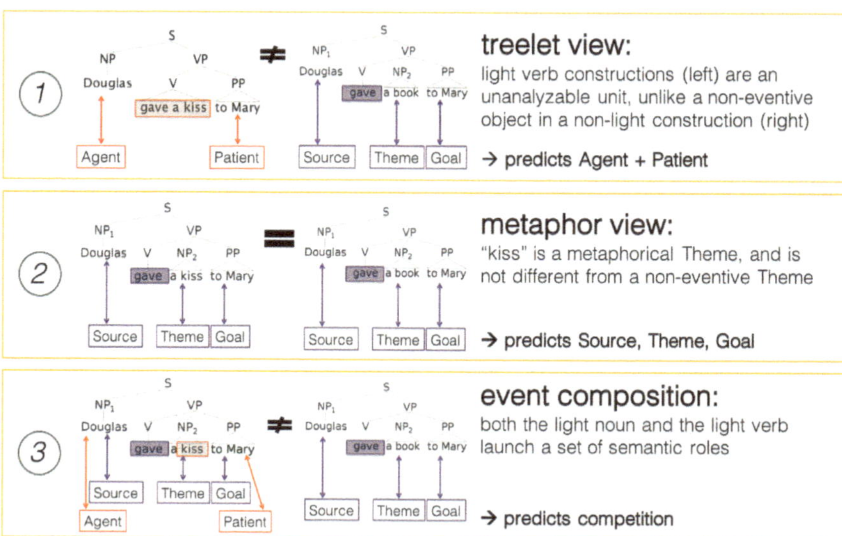

Figure 2 Schema of predictions: How many semantic roles do people imagine in light verb constructions (left) and non-light constructions (right)?

atomic construction. This is an approach taken both by a Construction Grammarian view and has also been argued for idioms by proponents of the Minimalist Framework, using so-called "treelets" (e.g., Family, 2014; Goldberg, 1995; Nediger, 2017). (Obviously, these frameworks differ in most other aspects of their view on linguistic architecture, but they happen to make the same prediction in this case).

The second view argues that there is nothing special about light verb constructions like "give a kiss" – "kiss" is a metaphorical Theme, and "give" is a metaphorical transfer action, just like "send a text message" does not involve the physical transfer of one text message from one phone to the other (rather, each phone ends up with a digital copy). In this view, people would not imagine "giving a kiss" of having anything but three event roles (Fig. 2, "metaphor view"; e.g., Newman, 2010).

Finally, there is an analysis of light verb constructions consisting of two co-events: In this view, the core meaning, and argument structure, of "give" coexists with the meaning and argument structure of "kiss:" "giving a kiss" is both "giving" and "kissing." There are different proposals about the exact mechanism, often called 'argument sharing'; but the general prediction is that people would categorize a light verb construction's thematic roles as hybrids (Fig. 2, "event composition"; e.g., Ahmed, Butt, Hautli, & Sulger, 2012; Culicover & Jackendoff, 2005; Müller, 2010; Pantcheva, 2009).

We tested these different predictions in two behavioral studies (Wittenberg & Snedeker, 2014; Wittenberg, Khan, & Snedeker, 2017). In both studies, we taught participants to categorize events according to the number of the semantic roles they encode or depict; the dependent variable was whether events described by light verb constructions ("Douglas gave a kiss to Mary") were categorized like base-verb sentences ("Douglas kissed Mary"), or three-role sentences ("Douglas gave a book to Mary").

In Wittenberg and Snedeker (2014), we explicitly taught people to categorize events according to semantic roles, by telling them to sort cards with drawings on them according to the number of event roles they depicted (see Fig. 3). Once they succeeded with non-verbal stimuli in the training phase, they also sorted sentences. We found that while one-, two-, and three-role sentences were reliably categorized into their respective piles, light verb constructions were sorted partly into the two-role-pile, and partly into the three-role-pile, with significant differences to both two-role and three-role sentences and pictures. This effect was neither driven by particular items or participants, nor by orthogonal strategies: Several control conditions showed that extraneous strategies, like counting the number of syntactic

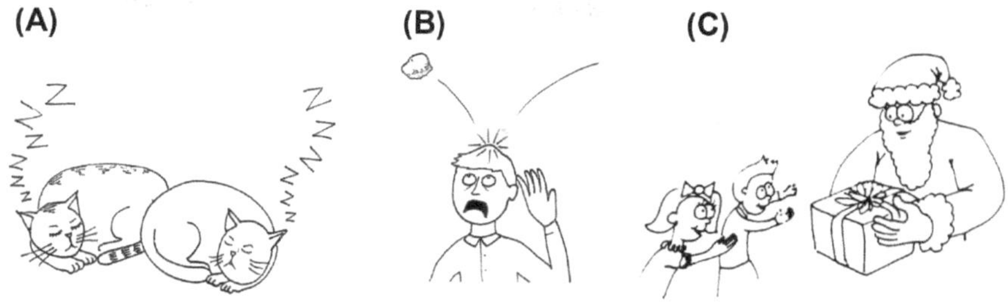

Figure 3 In Wittenberg and Snedeker (2014), participants were trained to categorize nonverbal scenes according to number of semantic roles: Picture (A) is a one-role event (showing two Agents sleeping), Picture (B) is a two-role event (showing the rock as Agent and the boy as Patient of hitting), and Picture (C) is a three-role event (showing Santa Claus as Agent/Source, the children as Patient/Recipients, and the present as Theme of giving).

constituents, or paying attention to abstractness of the event participants, or to animacy, played no discernible role in how people approached the task. We therefore interpreted these results as evidence against the treelet and metaphor view, and in favor of the co-event hypothesis.

However, one issue could have been that we explicitly taught people the concept of semantic roles (albeit concealed under a cover story), and our participants' attention may have been directed to the object of study. In a follow-up experiment (Wittenberg et al., 2017), we avoided this problem by implicitly training participants to associate one side of a computer screen with two-role events, and the other, with three-role events.

Adapting a method used by Rohde and Horton (2014), we had people listen to 100 maximally heterogeneous sentences, while they watched a ball rolling into a Y-shaped tube. Unbeknownst to them, the ball always exited on one side when it contained two semantic roles, and on the other, when it contained three semantic roles. Participants had to click on the ball as soon as it came out, but for the critical conditions, the ball appeared stuck, and so participants had to indicate where they thought it would come out by clicking on one of the two potential landing sites. Thus, we had two dependent measures: mouse-clicks and eye-movements.

Equivalent to the categorization results obtained by Wittenberg and Snedeker (2014), the mouse-click results showed that while people learned to categorize homomorphic two- and three-role sentences, they were uncertain how to classify light verb constructions. The eye-movement data confirmed this pattern, and in addition, they showed that when participants listened to light verb constructions, they first tended to gaze toward the three-role side, and their eye-movements only diverged late in the two sentence types. We interpreted this as evidence for the dominating role of syntactic argument structure.

In sum, both studies obtained the same pattern of results: When people implicitly or explicitly categorize light verb constructions according to the number of semantic roles, they are able to do this for homomorphic constructions, but not for constructions violating the principle of homomorphism. These results would not have been predicted by a model of language that preserves homomorphism by analyzing these constructions as unanalyzable atoms, or metaphors (see Fig. 2).

Thus, what these studies show, together with a small but growing body of similar literature connecting syntactic structure and event representations (for instance, Fausey & Boroditsky, 2010; Gropen et al., 1991; Johnson & Goldberg, 2013; Majid, Sanford, & Pickering, 2007), is not only that comprehenders of language follow their syntactic and semantic cues very prudently, which in turn influences the way they think about an event, but also that we can get to the bottom of these intuitions by pitting different models of linguistic representation against each other.

But the results I summarized above also make a prediction about other aspects of cognition: They predict what people keep in their attentional working memory focus, and what aspects they shift attention away from — that is, for which kinds of structural representations neural activation remains high, and for which kinds of structural representation neural activation decreases.

We tested whether people keep a syntactic structure in their working memory more when the structure is homomorphic with the semantics by using a structural priming paradigm. Priming paradigms use the human preference to reuse structures they just processed, and are still in working memory. Usually, this method is used to find structural similarities between constructions (e.g., Bock, 1986; for overviews and discussions,; Pickering & Ferreira, 2008; Branigan & Pickering, 2017), but we already know that "Anna gave a ball to Bill" and "Anna gave a kiss to Bill" has the same structure, and these sentences will prime each other. However, what is different between ball-giving and kiss-giving are the mappings between syntax and semantics: ball-giving is homomorphic, and kiss-giving is non-homomorphic. Our hypothesis is that, structure being equal, homomorphic sentences should be better primes than non-homomorphic sentences, because you do not need to retain the syntax of "giving a kiss" in working memory in order to create an event representation. In fact, if you try to remember that Anna kissed Bill, keeping a ditransitive syntactic structure in working memory may be counterproductive, because the syntactic structure does not match the semantics.

In fact, using this paradigm, we found more priming between structures that were homomorphic, such as a prime like "Anna gave a ball to Bill", than between a non-homomorphic structure, such as "Anna gave a kiss to Bill", or "Anna gave a pink slip to Bill", and a target picture like Picture (c) in Fig. 3, which could be described with a homomorphic structure (Ziegler, Snedeker, & Wittenberg, 2018). That is, when the syntactic structure supports the event structure, it seems like both syntax and semantics remain active in working memory, leading to larger priming effects. But when the syntactic structure does not support the event structure, such as in light verb constructions and idioms, people's working memory will discard the useless syntactic structure, leading to lower priming rates.

In sum, the case study of semantic roles in light verb constructions can help us understand how people comprehend language, and how the intuition that the verb does contribute to the semantics can be predicted by linguistic theory. First, we know that people do not simply understand the words in a sentence and somehow infer the event; instead, they diligently pay attention to the syntactic structure — a strategy that often pays off, since the syntactic structure usually reliably predicts the semantics. Second, the end goal of language comprehension is not to build a syntactic tree; that structure is simply a means to an end, that is, it helps to understand the meaning of a sentence. But when it does not help, it is discarded quickly, as the priming data show.

In the next section, we will take another set of intuitions seriously: That some light verb constructions really mean something different from their simple verb counterparts. What I want to emphasize, though, is that these intuitions are not based on some idiosyncratic, unstructured and unprincipled blip of the lexicon, but that these intuitions can be explained by theoretical linguistic insights that go back decades.

3. CASE STUDY II: FROM GRAMMAR TO TIME

Our choice of linguistic expression has ramifications for lots of aspects of event construal. For instance, there is a sense of a difference between "he kicked my nose" or "he gave my nose a kick" — this difference likely has to do with force, distance, intention, resulting pain level, and likely many other factors. These particular shades of meaning are likely specific to the action of nose-kicking; but other aspects may be more systematic. In the previous section, we have seen an instance for a systematic divergence, namely that between constructions like "kicking" and constructions like "giving a kick" (at least when two animate characters are described), people construe semantic roles differently. In this section, I will review studies on how people construe another dimension of event structure, based on linguistic cues: time and aspect.

One crucial dimension in which events are different from objects is that they take place over time (Casati & Varzi, 2008). Information about this temporal dimension — how long an event took, and whether it is ongoing, repeated, or concluded, for instance – is usually either encoded lexically ("She played piano for an hour", "He already finished washing the dishes") or by morphosyntactic markers on and around the verb ("He was kicking my nose" vs. "He has kicked my nose"), in most languages of the world (Dahl, 1985; Dahl & Velupillai, 2013). And indeed, several studies have shown that people take these morphosyntactic cues into account when they try to understand whether an event is completed or not (Foppolo et al., under review; Foppolo et al., 2017). That is, there is a very strong link between the temporal characteristics of an event, and the form of a verb encoding that event.

However, we have recently shown that speakers of a language use any kind of cue they can get in order to make completeness inferences. For instance, completeness of certain kinds of events can be marked at two places in a split-ergativity language like Hindi: At the subject, which usually occurs in the very beginning of a sentence, and at the verb (e.g., Narasimhan, 1998). For instance, "Mohan" in "Mohan has peeled an orange" receives a special suffix (a so-called ergative case-marker) when the event is completed and in past tense (such as in the upper half of examples in Fig. 4). But when the event is ongoing and in the present (such as in the lower half of examples in Fig. 4), the noun "Mohan" remains unmarked. That is, when a Hindi native speaker hears an ergative marker on a subject, she can draw inferences about whether an event is completed.

We asked Hindi native speakers to quickly match a picture to a given auditory sentence which either exhibited split ergativity or not, and in which an endpoint was either inherent in the event or not (Wittenberg & Vaidya, 2018, Fig. 4). The pictures either showed incomplete actions (an orange being peeled, or an orange being dropped), or complete actions (a peeled orange, a dropped orange). Given the choice between an

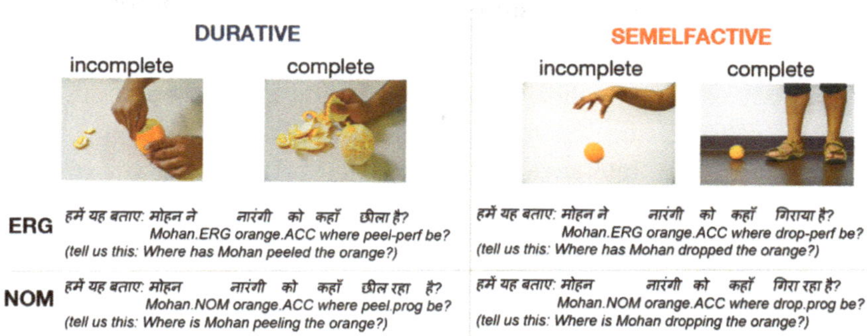

Figure 4 Split ergativity (ERG above) gives early cues as to whether an event is completed or not. In the case of durative verbs like "peeling", the inference that an event is completed is only derivable from the case-marking on the subject, and the tense morphology (has peeled/is peeling), while verbs like "dropping" have an internal endpoint, which already should bias people to pick the complete event.

incomplete action and a complete action, a native speaker should be able to make a completeness inference as soon as she hears either the marked, ergative subject noun (Mohan-ne), or the unmarked, nominative subject noun (Mohan). This is strikingly different from languages like English or Italian, where any inference about completeness is only possible once the verb is heard. Reaction time and accuracy data indicate that people indeed take these cues into account when understanding a sentence. This is in line with Foppolo et al.'s (2018) findings, but unexpected in light of how subtle these cues are, and how many other functions the ergative marker has in Hindi (Mohanan, 1994).

And another factor, apart from verb type and verbal morphology, plays a role when people make inferences about event completion: The kind of object an event is combined with. In philosophy and linguistic theory, the connection between verb and object has long been an object of linguistic and philosophical inquiry; both from the perspective of ontological commonalities, and how people integrate the meaning of a verb and its syntactic object (e.g., Bach, 1986; Gennari & Poeppel, 2003; Husband, Stockall, & Beretta, 2010; Ramchand, 2008, 2018).

In terms of ontological commonalities, it has long been noted that objects and event-denoting verbs show parallels: For instance, Vendler (1967) classified verbs into atelic (unbounded) and telic (bounded) verbs: unbounded verbs would be, for instance, "talking" — there is no inherent end to talking, as opposed to, say, "kicking." Objects are not bound in time, but they can be bound in space: When they are, they usually appear in count syntax ("an apple", "two apples"; exceptions are aggregates like "furniture"; Jackendoff, 1991), and when they are not, they usually appear in mass syntax — in English, that means without a definite or indefinite determiner ("applesauce" as opposed to *"an applesauce").

However, the reference properties of syntactic objects (that is, whether they denote mass or count objects) affects the reference properties of the whole predicate (Krifka, 1992). For instance, "eating an apple" is telic, but "eating apples" is not. At least for English, research has shown that people are consistently aware of this and similar distinctions in their event construal. For instance, when Proctor and Rips (2006) had participants judge scenarios to be either bounded or unbounded, they judged activity-type destruction events (such as "eating") as bounded when the object was a count noun (like "a chocolate bar"), but as unbounded when it was a mass noun (like "chocolate sauce"). Similarly, Barner, Wagner, and Snedeker (2008) found that count syntax ("to do climb<u>s</u>"), but not mass syntax ("to do climbing"), affected how people individuate and quantify events, especially when those events are bounded themselves.

In a series of six studies, we used the contrast between light verb constructions such as "He gave a kick to my nose" and their non-light counterparts (i.e., "He kicked my nose") to investigate how the use of one of these alternatives systematically affects the construal of event similarity and duration in a way that is systematically predictable from the interaction of mass/count syntax and verb semantics (Wittenberg & Levy, 2017).

We found that whenever events were described with count syntax ("give <u>a</u> kiss", "give <u>a</u> talk"), people thought that there were fewer events than when they were described as simple verbs ("kiss", "talk"). When events like "kiss" and "advise" were used in a ditranstitive frame with "give", people judged them to take a shorter amount of time than when they occurred as simple verbs. Finally, when unbounded events, like "talk", were described with count syntax ("give a talk"), they underwent a semantic shift, similar to uncountable objects occurring in count syntax (like the etymological shift from "iron", the material, to "an iron", the instrument; or the conventional portion sizes from "beer" to "a beer", vs "schnapps" to "a schnapps"). That is, "giving a talk" and "talking" were judged as less similar to each other than, say, "giving a kick" to "kicking." Finally, we could also show that people's estimates for how long an event took depends both on the semantics of the verb, and count and mass syntax.

4. LOOSE ENDS: PROCESSING, REFERENCE, AND ACQUISITION

Above, I discussed two dimensions of event construal: mapping from grammatical entities onto event participants, and temporal characteristics of events. I argued that these dimensions are construed differently in light verb constructions than in homomorphic constructions. If this is true, the difference should be observable in real-time language processing: Instead of deriving semantic or event roles directly from the syntactic structure, one has to go through the semantics of the noun. That is, instead of inferring that "He gave my nose a kick" involves a Source, Goal, and Theme, one has to use the lexical meaning of "kick" to map "he" and "my nose" to Agent and Patient of the kicking action. This should be a different, and perhaps costly, process.

Indeed, studies measuring brain activity to light verb constructions and other complex verb constructions in German have found evidence for different cortical processes for homomorphic, compared to non-homomorphic verbs (Briem et al., 2009; Wittenberg, Paczynski, Wiese, Jackendoff, & Kuperberg, 2014). Interestingly, these cortical processes are distinct from those found for metaphor comprehension, another piece of evidence that the metaphor view of light verb constructions in Fig. 2 may not be on the right track (e.g., Arzouan, Goldstein, & Faust, 2007; Lai & Curran, 2013; Lai, Curran, & Menn, 2009; Obert, Gierski, & Caillies, 2018; Tartter, Gomes, Dubrovsky, Molholm, & Stewart, 2002), and seem to point to increased involvement in working memory and shifts in event construal, such as when people comprehend jokes (e.g., Coulson & Kutas, 2001; Du et al., 2013) — that is, Woody Allen's frame shift of noses hitting knees may involve a similar comprehension process as kicks on the nose concealed by a syntactic frame suggesting a transfer action. It has been argued that this process of frame-shifting involves an increase in demand on working memory.

And in fact, studies in English and German indicate that people's working memory is more taxed when they comprehend light verb constructions, since they are slower at executing a secondary task when they do so, compared to when they listen to non-light constructions (Piñango, Mack, & Jackendoff, 2006; Wittenberg & Piñango, 2011). However, we should be careful to generalize these findings: When we tried to replicate the results for English and German in Hindi, we could not — Hindi native speakers do not slow down when reading light verb constructions, nor do they slow down at a secondary task when listening to light verb constructions (Vaidya & Wittenberg, 2018). This is likely because they're used to it — while English and German are languages with a fair number of light verbs, and light verbs in these familiar languages are used more often with light than non-light objects ("give a kiss" is more frequent than "give a book"), it is still more common to describe an event with a simple verb. However, in Hindi, the reverse is true — there are very few non-light constructions in the language overall, which also means that generally the only way to describe an event is by using a complex verb.

But in English and German, as well as in many other languages, expressions like "giving a kick" and "kicking" alternate. One question is why anyone would use the clunkier light verb construction, as opposed to the simple verb; and like for other syntactic alternations, we can think of several different explanations. First, we have seen above that each construction carries with it a different shade of meaning; second, I want to put forth another explanation of why one would prefer a light verb construction over a simple verb: Reference to a noun is easier than reference to a verb.

In a recent study (Wittenberg, under review), I used five constructions with varying levels of semantic integration (1–5) in a forced-choice production task (Scholten & Aguilar Guevara, 2010). Participants had to pick either a pronoun ("it") or a full definite DP ("the note" for (3); "the church" in (4), etc.) in order to fill in the blank. Scholten and Aguilar Guevara (2010) had found that people dislike reference with a pronoun to weak definites (4) and bare nouns (5) compared to full verbs and objects (3), because the nouns in these constructions seem to be much more semantically integrated in the verb. This is a similar situation as in light verb constructions: One could imagine a verb for church-going, like "churching", just as we have a verb for kiss-giving, namely "kissing".

(1) The teenager kicked his rival. ___ was rather mean. (*simple verb, referent: event*)

(2) The teenager gave a kick to his rival. __was rather mean. (*light verb, referent: event*)

(3) The teenager gave a note to his rival. __was rather mean. (*full verb, referent: object*)

(4) The neighbors go to church every Sunday. __ is always packed. (*bare noun*)

(5) Claire read the newspaper over breakfast. ___ is from last week. (*weak definite*)

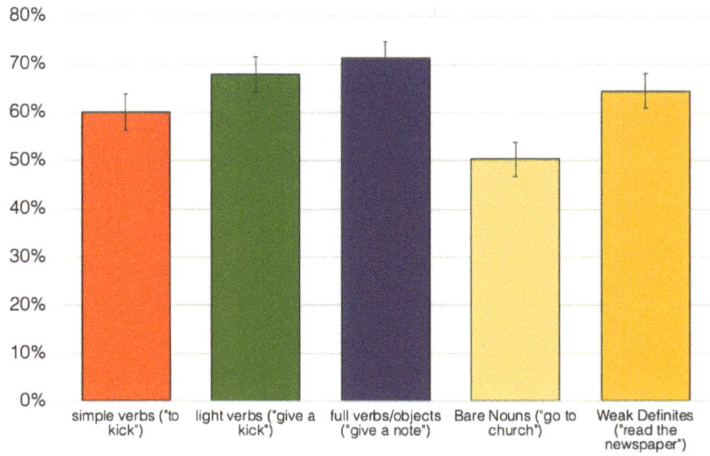

Figure 5 Rate at which people chose to create reference with a pronoun ("it"), as compared to using the full noun phrase, in Wittenberg (under review).

Like in the study by Scholten and Aguilar Guevara (2010), people disliked to refer to weak definites (5) and bare nouns (4) with the simple pronoun "it", compared to fully compositional structures ((3); see Fig. 5). Simple verbs (1) patterned like bare nouns. The rate of pronominal reference for light verbs (2), which 'unpack' the event into a noun phrase, was significantly higher compared to simple verbs, but it was only numerically less than for full verbs. What this means is that light verb constructions help people because they make the event itself readily available for future reference in the discourse.

A third factor why light verb constructions may not be dying out soon could be the issue of learnability (He & Wittenberg, 2018). Children should have a really hard time learning light verb constructions, based on two reasons: First, we know that kids use the usual homomorphism in order to draw inferences for word learning – either going from syntax to semantics or vice versa. Gleitman (1990, p. 29) observes on children's verb learning, "[...] if a verb has been discovered to mean 'give', then it will appear in three-argument structures such as *John gives the book to Mary*. This is because the logic of 'give' implies one who gives, one who is given, and that which is given, and each of these entities requires a noun phrase to express." Obviously, light verb constructions do not allow this inference process.

Second, another factor that should make light verb constructions difficult to learn is a different kind of homomorphism violation: The expectation that verbs convey events, and nouns, objects. In light verb constructions, the noun itself conveys an event, and it has been shown that children have difficulties learning the meaning of eventive nouns, even if they use them (Nelson, Hampson, & Shaw, 1993): instead, they seem to use these eventive nouns to denote "what happens in association with object X", as Arunachalam and He (2018) reported. In their study, toddlers had to match an event noun like "nap" to a picture. Children correctly picked out a picture of a child sleeping in a bed, when the alternative was a picture of a bed alone, without a sleeper in it. But when the alternatives were a picture of someone sleeping on the floor, and a picture of a bed without a sleeper, children were at chance. That means that children may use event nouns like "nap" thinking that it may denote "something with a bed", when they use them at all — most nouns that children produce early on adhere to the homomorphism between noun and object (Barner, 2001).

So where do all the light verb constructions come from? It has been argued — especially for languages that have more complicated verbal morphology than English — that light verb constructions are actually easier to learn than their base verbs: Instead of having to learn the verbal paradigm of "kick", "kiss", "talk", etc., people can use "give" in combination with any of these verbs in their nominal form — and likewise, they can combine the same noun with different verbs to express the desired shade of meaning. For instance, Wiese (2009) and Kotsinas (1988) have shown that in multilingual contexts, light verb constructions are used as a productive device in order to express verbal meaning for a range of nouns: "make red traffic light" denotes "cross while the light is red", "make knife" denotes "to knife someone"; "make song" denotes "to sing". Perhaps children use this strategy as well.

It would tie in with what Braine (1963) called the 'pivot schema' stage of language acquisition — when children use a restricted number of words in combination with a larger variety of other words (also see Diessel, 2013; Tomasello, 2003): for instance, Tomasello (1992), calling these 'constructional islands' reports a productive schema with "find-it X" ("find-it funny", "find-it bird", "find-it ball" etc.) or "X gone" ("raisins gone", "doo—doo gone" etc.). Do children use light verb constructions as pivot schemas? While the jury is still out for this specific question, it seems like light verb constructions appear late in children's production (Barner, 2001; Oshima-Takane, Barner, Elsabbagh, & Guerriero, 2001; He & Wittenberg, 2018). In short, the trajectory of acquiring light verb construction remains an open problem to be investigated; but we can tentatively conclude that these constructions do not only come with a cost for adults, but they are also a thorny issue in acquisition.

5. CONCLUSION: LINGUISTICS IS WORTH YOUR WHILE

In this chapter, I have given a short overview of my own and other's work on what I think is one the most fascinating aspects of human cognition: The ability to construe a representation of an event, simply based on a few sound waves and armed with the ability to learn and remember (linguistic) structures. I showed how the study of the relationship between grammar and cognition is fruitful on both sides, since different linguistic theories can provide testable hypotheses for how people imagine event participants and temporal characteristics; what they keep in working memory, and what they discard; and in turn, this may inform our story on how people actually *do* language comprehension, and how they acquire all the intricacies of each language's ornate and puzzling array of grammatical possibilities.

One of these possibilities in many languages of the world is using light verb constructions — combinations of event-denoting nouns and verbs, which taken together form a predicate. As we have seen, these constructions carry with them shades of meaning, from the configuration of event participants to the temporal structure of events, that are not conveyed by simple verbs; however, the downside of these constructions is that they seem to require more working memory, and they may or may not be harder to learn.

If you are a non-linguist and have read this far, you may be asking yourself: Why should we care about these subtle linguistic distinctions? My answer would be that ever since economics became interested in behavior, researchers have been asking people about their behavior using linguistic frames (e.g., Kahneman & Tversky, 1977; Kruger & Evans, 2004; Roy, Christenfeld, & McKenzie, 2005). Showing that subtle linguistic argument structure alternations can reliably and predictably influence how people imagine events (and likely anything else) is not only (for some of us) fascinating and interesting in and of itself; it also delivers prerequisite data for other cognitive scientists, and anyone interested in human behavior. For instance, if you were a social scientist interested in how people perceive benefit from an action, you may want to think about whether you frame your scenario with a transfer-implying light verb construction ("Jeremy gave Julius a hug"), as opposed to an agent-patient event ("Jeremy hugged

Julius"), because, as we have seen, even if people know that these are the same events, they may tend to feel that Julius received *something* that made him feel good in the hug-giving scenario; but less so in the hugging scenario, where something was done to him. Many shades of meaning remain to be empirically probed.

My aim for this chapter was, consequently, to convince the small percentage of people not yet fascinated by the knottiness of grammar that linguistics is worth every social scientist's while. The reason I stress this is that it has transpired that until very recently, linguistics as a field did not enjoy the best reputation in neighboring fields. Apparently, a famous neuroscientist at the University of Chicago used to discourage his students to do research on higher-level language processing, because, as he put it, "linguistics is what happens to science unconstrained by data." This assessment did not seem entirely unfair, given the history of the field: Some linguists were arguing that language is *especially* special among the very special cognitive capacities humans possess — so special that the science of language should not be constrained by what we know about cognition. This led, for instance, to the unfortunate uncoupling of 'competence' and 'performance': Human behavior not predicted by the grammatical model of one's choice was explained by factors outside of the grammatical system, and thus outside of the linguist's research interest (see e.g., Jackendoff, 2007b, for a discussion).

This bifurcation between the fields of linguistics and cognitive psychologists has thankfully narrowed, and in this paper, I hope to have convinced the non-linguists of the probably obvious point that the human capacity for language can be a window into the study of the human mind, and how humans understand the world (Pinker, 2007). But to open this window, we need to study all its aspects, and what people do with them. Even light verb constructions.

REFERENCES

Ahmed, T., Butt, M., Hautli, A., & Sulger, S. (2012). A reference dependency bank for analyzing complex predicates. In *Proceedings of the eight international conference on language resources and evaluation (LREC'12).*

Arunachalam, S., & He, A. X. (2018). Children's acquisition of nouns that denote events. In A. B. Bertolini, & M. J. Kaplan (Eds.), *Proceedings of the 42nd annual Boston University conference on language development.* Somerville, MA: Cascadilla Press.

Arzouan, Y., Goldstein, A., & Faust, M. (2007). Brainwaves are stethoscopes: ERP correlates of novel metaphor comprehension. *Brain Research, 1160*, 69–81.

Bach, E. (1986). The algebra of events. *Linguistics and Philosophy, 9*(1), 5–16.

Baker, M. C. (1988). *Incorporation: A theory of grammatical function changing*. Chicago, IL: The University of Chicago Press.
Barner, D. (2001). *Light verbs and the flexible use of words as noun and verb in early language learning* (MSc thesis). Montréal, Canada: McGill University.
Barner, D., Wagner, L., & Snedeker, J. (2008). Events and the ontology of individuals: Verbs as a source of individuating mass and count nouns. *Cognition, 106*, 805–832.
Bock, J. K. (1986). Syntactic persistence in language production. *Cognitive Psychology, 18*(3), 355–387.
Braine, M. (1963). The ontogeny of English phrase structure: The first phase. *Language, 39*, 11–14.
Branigan, H. P., & Pickering, M. J. (2017). Structural priming and the representation of language. *Behavioral and Brain Sciences, 40*.
Briem, D., Balliel, B., Rockstroh, B., Butt, M., Walde, S. S., & Assadollahi, R. (2009). Distinct processing of function verb categories in the human brain. *Brain Research, 1249*, 173–180.
Butt, M. (2003). *The morpheme that wouldn't go away*. Handout. University of Manchester seminar series.
Butt, M. (2010). The light verb jungle: Still hacking away. In M. Amberber, M. Harvey, & B. Baker (Eds.), *Complex predicates in cross-linguistic perspective* (pp. 48–78). Cambridge University Press.
Butt, M., & Lahiri, A. (2013). Diachronic pertinacity of light verbs. *Lingua, 135*, 7–29.
Casati, R., & Varzi, A. C. (2008). Event concepts. In T. F. Shipley, & J. M. Zacks (Eds.), *Understanding events: From perception to action* (pp. 31–53). Oxford University Press.
Chomsky, N. (1993). *Lectures on government and binding: The Pisa lectures (No. 9)*. Walter de Gruyter.
Coulson, S., & Kutas, M. (2001). Getting it: Human event-related brain response to jokes in good and poor comprehenders. *Neuroscience Letters, 316*(2), 71–74.
Culicover, P. W., & Jackendoff, R. (2005). *Simpler syntax*. Oxford University Press.
Dahl, Ö. (1985). *Tense and aspect systems*. Oxford: Blackwell.
Dahl, Ö., & Velupillai, V. (2013). Tense and aspect. In M. S. Dryer, & M. Haspelmath (Eds.), *The world atlas of language structures online*. Leipzig: Max Planck Institute for Evolutionary Anthropology.
Diessel, H. (2013). Construction grammar and first language acquisition. *The Oxford Handbook of Construction Grammar, 347*, 364.
Dowty, D. R. (1977). Toward a semantic analysis of verb aspect and the English 'imperfective' progressive. *Linguistics and Philosophy, 1*(1), 45–77.
Dowty, D. (1991). Thematic proto-roles and argument selection. *Language, 67*(3), 547–619.
Du, X., Qin, Y., Tu, S., Yin, H., Wang, T., Yu, C., & Qiu, J. (2013). Differentiation of stages in joke comprehension: Evidence from an ERP study. *International Journal of Psychology, 48*(2), 149–157.
Family, N. (2014). *Semantic spaces of Persian light verbs: A constructionist account*. Brill.
Fausey, C. M., & Boroditsky, L. (2010). Subtle linguistic cues influence perceived blame and financial liability. *Psychonomic Bulletin & Review, 17*, 644–650.
Foppolo, F., Greco, C., Panzeri, F., & Carminati, M. N. (2018). *The incremental processing of accomplishment predicates* (under review).
Foppolo, F., Santin, M., Danu, J., & van Hout, A. (2017). Telicity cross-linguistically: An eye-tracking study. In *Poster session presented at architectures and mechanisms of language processing, Lancaster, United Kingdom.*
Gennari, S., & Poeppel, D. (2003). Processing correlates of lexical semantic complexity. *Cognition, 89*(1), B27–B41.
Gillon, B. S. (1999). The lexical semantics of English count and mass nouns. In *Breadth and depth of semantic lexicons* (pp. 19–37). Dordrecht: Springer.

Gleitman, L. (1990). The structural sources of verb meanings. *Language Acquisition, 1*(1), 3–55.
Goldberg, A. E. (1995). *Constructions: A construction grammar approach to argument structure.* University of Chicago Press.
Gropen, J., Pinker, S., Hollander, M., & Goldberg, R. (1991). Affectedness and direct objects: The role of lexical semantics in the acquisition of verb argument structure. *Cognition, 41*(1–3), 153–195.
Gropen, J., Pinker, S., Hollander, M., Goldberg, R., & Wilson, R. (1989). The learnability and acquisition of the dative alternation in English. *Language*, 203–257.
He, A., & Wittenberg, E. (2018). *The strengths and limits of syntax-semantics homomorphism in lexical acquisition: Event nominals and light verbs* (under review.).
Hornstein, N. (1995). *Logical form: From GB to minimalism.* Blackwell.
Hornstein, N. (1999). Movement and control. *Linguistic Inquiry, 30*(1), 69–96.
Husband, E. M., Stockall, L., & Beretta, A. (2010). The online composition of events. *Queen Mary's Occasional Papers Advancing Linguistics, 19*, 1–32.
Jackendoff, R. (1991). Parts and boundaries. *Cognition, 41*(1–3), 9–45.
Jackendoff, R. (2007a). *Language, consciousness, culture: Essays on mental structure* (vol. 2007). MIT Press.
Jackendoff, R. (2007b). A parallel architecture perspective on language processing. *Brain Research, 1146*, 2–22.
Johnson, M. A., & Goldberg, A. E. (2013). Evidence for automatic accessing of constructional meaning: Jabberwocky sentences prime associated verbs. *Language and Cognitive Processes, 28*(10), 1439–1452.
Kahneman, D., & Tversky, A. (1977). *Intuitive prediction: Biases and corrective procedures.* Decisions and Designs Inc Mclean Va.
Kotsinas, U. B. (1988). Immigrant children's Swedish—a new variety? *Journal of Multilingual and Multicultural Development, 9*(1–2), 129–140.
Krifka, M. (1992). Thematic relations as links between nominal reference and temporal constitution. In I. Sag, & A. Szabolcsi (Eds.), *Lexical matters* (pp. 29–53). CSLI Publications, Chicago University Press.
Kruger, J., & Evans, M. (2004). If you don't want to be late, enumerate: Unpacking reduces the planning fallacy. *Journal of Experimental Social Psychology, 40*, 586–598.
Lai, V. T., & Curran, T. (2013). ERP evidence for conceptual mappings and comparison processes during the comprehension of conventional and novel metaphors. *Brain and Language, 127*(3), 484–496.
Lai, V. T., Curran, T., & Menn, L. (2009). Comprehending conventional and novel metaphors: An ERP study. *Brain Research, 1284*, 145–155.
Majid, A., Sanford, A. J., & Pickering, M. J. (2007). The linguistic description of minimal social scenarios affects the extent of causal inference making. *Journal of Experimental Social Psychology, 43*(6), 918–932.
Mohanan, T. (1994). *Argument structure in Hindi.* Center for the Study of Language (CSLI).
Müller, S. (2010). Persian complex predicates and the limits of inheritance-based analyses. *Journal of Linguistics, 46*(3), 601–655.
Narasimhan, B. (1998). A lexical semantic explanation for 'quirky' case marking in Hindi. *Studia Linguistica, 52*(1), 48–76.
Nediger, W. (2017). *Unifying structure-building in human language: The minimalist syntax of idioms* (Dissertation). University of Michigan.
Nelson, K., Hampson, J., & Shaw, L. K. (1993). Nouns in early lexicons: Evidence, explanations and implications. *Journal of Child Language, 20*(1), 61–84.
Newman, J. (2010). *Give: A cognitive linguistic study* (vol. 7). Walter de Gruyter.
Obert, A., Gierski, F., & Caillies, S. (2018). He catapulted his words from the dais: An ERP investigation of novel verbal metaphors. *Journal of Neurolinguistics, 47*, 59–70.

Oshima-Takane, Y., Barner, D., Elsabbagh, M., & Guerriero, A. S. (2001). Learning of deverbal nouns. In *Proceedings of the 8th conference of the international association for the study of child language* (pp. 1154–1170).

Pantcheva, M. (2009). First phase syntax of Persian complex predicates: Argument structure and telicity. *Journal of South Asian Linguistics, 2*(1).

Parsons, T. (1989). The progressive in English: Events, states and processes. *Linguistics and Philosophy, 12*(2), 213–241.

Pickering, M. J., & Ferreira, V. S. (2008). Structural priming: A critical review. *Psychological Bulletin, 134*(3), 427.

Piñango, M. M., Mack, J., & Jackendoff, R. (2006). Semantic combinatorial processes in argument structure: Evidence from light-verbs. *Annual Meeting of the Berkeley Linguistics Society, 32*(1), 573–583.

Pinker, S. (2007). *The stuff of thought: Language as a window into human nature.* Penguin.

Proctor, A. S., & Rips, L. J. (2006). Interactions between nouns and verbs in interpreting events. *Proceedings of the Annual Meeting of the Cognitive Science Society, 28*(28).

Ramchand, G. C. (2008). *Verb meaning and the lexicon: A first phase syntax* (vol. 116). Cambridge University Press.

Ramchand, G. C. (2018). *Situations and syntactic structures: Auxiliaries and ordering in English.* MIT Press Monographs.

Rohde, H., & Horton, W. S. (2014). Anticipatory looks reveal expectations about discourse relations. *Cognition, 133*(3), 667–691.

Roy, M. M., Christenfeld, N. J., & McKenzie, C. R. (2005). Underestimating the duration of future events: Memory incorrectly used or memory bias? *Psychological Bulletin, 131*, 738.

Scholten, J., & Aguilar Guevara, A. (2010). Assessing the discourse referential properties of weak definite NPs. *Linguistics in the Netherlands, 27*(1), 115–128.

Srinivasan, M., & Rabagliati, H. (2015). How concepts and conventions structure the lexicon: Cross-linguistic evidence from polysemy. *Lingua, 157*, 124–152.

Tartter, V. C., Gomes, H., Dubrovsky, B., Molholm, S., & Stewart, R. V. (2002). Novel metaphors appear anomalous at least momentarily: Evidence from N400. *Brain and Language, 80*(3), 488–509.

Tomasello, M. (1992). *First Verbs. A case study of early grammatical development.* Cambridge: Cambridge University Press.

Tomasello, M. (2003). *Constructing a language: A usage-based theory of language acquisition.* Harvard University Press.

Vaidya, A., & Wittenberg, E. (2018). *Frequency effects modulate argument sharing effects in Hindi light verb constructions* (in prep.).

Vendler, Z. (1967). *Linguistics in philosophy*. Ithaca. NY: Cornell UP.

Venhuizen, N. J., Crocker, M. W., & Brouwer, H. (2018). Expectation-based Comprehension: Modeling the interaction of world knowledge and linguistic experience. *Discourse Processes*. https://doi.org/10.1080/0163853X.2018.1448677.

Wiese, H. (2009). Grammatical innovation in multiethnic urban Europe: New linguistic practices among adolescents. *Lingua, 119*(5), 782–806.

Wittenberg, E., Khan, M., & Snedeker, J. (2017). Investigating thematic roles through implicit learning: Evidence from light verb constructions. *Frontiers in Psychology, 8*, 1089.

Wittenberg, E., & Levy, R. (2017). If you want a quick kiss, make it count: How choice of syntactic construction affects event construal. *Journal of Memory and Language, 94*, 254–271.

Wittenberg, E., Paczynski, M., Wiese, H., Jackendoff, R., & Kuperberg, G. (2014). The difference between "giving a rose" and "giving a kiss": Sustained neural activity to the light verb construction. *Journal of Memory and Language, 73*, 31–42.

Wittenberg, E., & Piñango, M. M. (2011). Processing light verb constructions. *The Mental Lexicon, 6*(3), 393–413.

Wittenberg, E., & Snedeker, J. (2014). It takes two to kiss, but does it take three to give a kiss? Categorization based on thematic roles. *Language, Cognition and Neuroscience, 29*(5), 635–641.

Wittenberg, E., & Vaidya, A. (2018). *Ergativity in Hindi as cue for completeness inferences* (in prep.).

Ziegler, J., Snedeker, J., & Wittenberg, E. (2018). *Event structures drive semantic structural priming, not thematic roles: Evidence from idioms and light verbs* (under review.).

CHAPTER TWELVE

The Role of Discourse Context in Reference Production and Comprehension: Insights From the Lexical Differentiation Effect

Si On Yoon
Beckman Institute for Advanced Science and Technology, University of Illinois, Urbana, IL, United States
E-mail: syoon10@illinois.edu

Contents

1. Introduction 362
   1.1 Context and Referential Design 362
   1.2 Overspecification in the Current Context 364
   1.3 Discourse History Supports Overspecification 365
2. Lexical Differentiation in Production 367
   2.1 Lexical Differentiation 367
   2.2 Is Lexical Differentiation a Listener-Driven Process? 370
   2.3 Does Repetition of Entrainment Trials Modulate Lexical Differentiation? 372
   2.4 Is the Latency Between Two Referents Crucial to Elicit Lexical Differentiation? 373
3. The Source of Lexical Differentiation 374
   3.1 Pre-emption by Similar Form 374
   3.2 Memory for Past Referents 376
   3.3 The Role of Relevant Context 378
4. Lexical Differentiation in Comprehension 381
   4.1 Lexical Differentiation While Tracking Eye-Movements 382
   4.2 Lexical Differentiation While Measuring Brain Electrical Activity 384
5. Future Directions 385
   5.1 Prosodic Differentiation 385
   5.2 Differentiation in Aging 386
6. Conclusion 387
References 388

Abstract

In conversation, interlocutors design referential expressions with respect to both local and historical contexts. Referential expressions that are more specific than necessary given the local context often can be motivated by the historical discourse context. To explore the breadth of the relevant discourse context in conversation, many studies

Psychology of Learning and Motivation, Volume 68
ISSN 0079-7421
https://doi.org/10.1016/bs.plm.2018.08.009

have examined how past references from the previous discourse context affect current language use. Lexical differentiation is a phenomenon that reflects the contribution of discourse history to referential design. The term refers to speakers' tendency to use two different labels (or modifiers) to name two subsequently presented objects from the same category. In multiple studies, this has been replicated in production, but the size of the effect has been small. I discuss the source of lexical differentiation and identify the factors that can cause the discourse history to exert a larger effect on referential design. Further, I introduce mixed findings regarding lexical differentiation in comprehension: listeners' brain responses reveal clear sensitivity to the discourse history, but this is not observed in behavioral measures (e.g., eye-movements). Lastly, lexical differentiation can reveal the interaction of memory and referential design in cognitive processes, and, thus, I discuss potential future work that could elucidate the mechanisms that support interlocutors' reference use on the basis of multifaceted measures.

1. INTRODUCTION

1.1 Context and Referential Design

In face-to-face conversation, delivering an intended message is essential for efficient communication. For successful communication, the interlocutor needs to design and use appropriate references, that is, to decide how to label the objects and events they wish to talk about. What constitutes a good reference that serves efficient communication? Speakers can choose from wide range of references when they design a reference. An appropriate reference is thought to be one that provides an "optimal" amount of information—be neither under-informative nor over-informative (Grice, 1975)—so that the conversational partner can identify the unique referent among many candidates in a conversational context. The amount of information that interlocutors need varies with respect to the conversational context. For example, a basic level noun (e.g., the rabbit) would suffice to refer to a rabbit in a context where a single rabbit is present. In contrast, in a context like Fig. 1, a modifier is necessary to describe a rabbit on the top right (e.g., the spotted rabbit) in order to distinguish it from the rabbit on the bottom right. Speakers do not typically produce all the attributes that distinguish two competitors; instead, the most salient attribute is selected among many possible candidates (e.g., black and gray, white, facing towards the left/right...). The optimal amount of information that interlocutors need depends on the context, and this information allows interlocutors to uniquely identify the intended message. Under- or

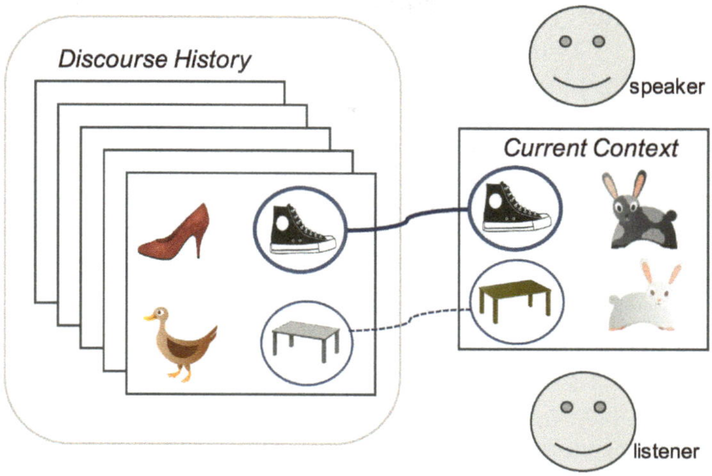

Figure 1 An example of how interlocutors link the current and past discourse context. Linking the same referent across the current and past discourse context may be easier (e.g., lexical entrainment: a phenomenon in which speakers tend to maintain the same reference across the contexts (e.g., "the sneaker")) than linking two similar referents from the same category (e.g., lexical differentiation: a phenomenon in which speakers distinguish the current referent from the past referent (e.g., "the desk" referring the metal desk in the past context vs. "the wooden desk" in the current context)).

over-informative references may make communication ambiguous or redundant (Engelhardt, Bailey, & Ferreira, 2006).

Speakers are known to be good at flexibly tailoring references depending on how much information is required in the relevant discourse context (Brown-Schmidt, Ryskin, & Yoon, 2015; Olson, 1970; Osgood, 1971; Yoon & Brown-Schmidt, under review). However, speakers do not always provide the optimal or "ideal" amount of information (Dale & Viethen, 2010; Gatt & Belz, 2010). Although they are relatively unlikely to underspecify references, speakers do often overspecify expressions, thus providing redundant information that it is not (strictly) necessary (e.g., underspecification 4% vs. overspecification 75% of trials in Pechmann (1989)) (Arts, Maes, Noordman, & Jansen, 2011; Davies & Katsos, 2013; Deutsch & Pechmann, 1982; Engelhardt, et al., 2006; Koolen, Gatt, Goudbeek, & Krahmer, 2011; Tarenskeen, Broersma, & Geurts, 2015).

Importantly, in some cases, overspecification of references in a local context—e.g., saying "spotted rabbit" in a context where a single rabbit is present (termed "exophoric redundancy")—seems to be supported by the discourse history (termed "endophoric redundancy") (Koolen, et al., 2011; Levelt, 1989; Pechmann, 1984). For example, in Pechmann's (1984) study, in a context that included four objects (a black bird, a white bird, a black cup, and a white cup), after the speaker referred to one of the birds (e.g., the black bird), the context changed to exclude the mentioned item—now, there are two cups and one bird. Here, the speaker still needs a color modifier to refer to a cup because the two cups are of different colors. In contrast, although a color modifier is not necessary to refer to the bird, speakers are likely to refer to it as "the white bird," seemingly to contrast it with the previously mentioned (though not physically co-present) item that was recently in focus. In what follows, I will discuss overspecification and its relationship to the discourse model (Section 1.2) and introduce a phenomenon called *lexical differentiation* (Section 1.3), wherein speakers are likely to overspecify references in a local context, but the overspecification is supported by the historical discourse context. Sections 2–4 discuss how the discourse context modulates the effect of lexical differentiation in production (Section 2) and comprehension (Section 4) and examine the source of lexical differentiation (Section 3). Future directions to fully understand the role of discourse context in lexical differentiation are discussed in Section 5.

1.2 Overspecification in the Current Context

An extensive literature shows that speakers often produce redundant information that is not needed to discriminate the target referent from competitors in the current physical context—i.e., exophoric redundancy (Arts, et al., 2011; Davies & Katsos, 2013; Deutsch & Pechmann, 1982; Engelhardt et al., 2006; Koolen et al., 2011; Pechmann, 1989; Sedivy, 2003; Tarenskeen et al., 2015). One of the most common ways to deliver extra information is through the use of modifiers (e.g., the brown duck in Fig. 1 (left)).

How do speakers decide to use modifiers and select one among many? One factor is saliency: the degree to which the attribute of the referent is distinctive and unique in the context compared to other distractors (Koolen et al., 2011; Tarenskeen et al., 2015; Vogels, Krahmer, & Maes, 2015). The salient attribute is not necessarily critical information that interlocutors need to identify the referent; instead, it is an attribute that quickly attracts the interlocutor's attention. For example, when a pink watch is present, the noun phrase "watch" would minimally specify the referent for the listener, but the modified noun phrase "pink watch" is usually produced. Although there are many different types of modifiers (e.g., color, pattern, material), the color adjective is the most widely used because color is an intrinsically salient feature of objects (Arts et al., 2011; Gatt, Krahmer, van Gompel, & van Deemter, 2013; Koolen, Goudbeek, & Krahmer, 2013 Markman & Makin, 1998; Sedivy, 2003, 2004). Moreover, color is perceived more quickly than other features (Gatt, 2007; Pechmann, 1989). Typicality is another factor that determines the saliency of objects. Atypical referents (e.g., a square plate) attract an interlocutor's attention and are more likely to be overspecified than typical referents (Ford & Olson, 1975; Mangold & Pobel, 1988; Rubio-Fernandez, 2016; Van Der Wege, 2009; Westerbeek, Koolen, & Maes, 2015). The accessibility of potential attributes also drives speakers' production of overspecification. Koolen et al. (2011), who manipulated the domain of reference type (e.g., furniture vs. people), found that the rate of overspecification was higher in domains wherein more potential referential possibilities are available (e.g., people) than in domains wherein there are fewer possibilities (e.g., furniture). A recent study showed that goals and task demands also modulate the rate of overspecification (Rubio-Fernandez, 2016).

1.3 Discourse History Supports Overspecification

In addition to the local physical context, what is known as "endophoric reference" considers the discourse history of what has been mentioned during a conversation (Koolen et al., 2011; Levelt, 1989). Even though referring expressions may include information that is unnecessary from the exophoric perspective, the discourse history that interlocutors have shared can explain the source and utility of some types of overspecification. Previous work has shown that exophoric redundancy derives from the speakers' discourse model, in which the discourse history of what has been discussed and shared is recorded and continuously updated as the conversation unfolds (Levelt, 1989). The phenomenon of *lexical entrainment*, wherein speakers converge on the same labels for a given entity across repeated reference, suggests that discourse history strongly guides reference design even if it is locally over-informative (Brennan & Clark, 1996) (Fig. 1). In Brennan and Clark (1996), speakers maintained a subordinate level label that had been previously used (e.g., the sneaker) in a new context, where only a single shoe (e.g., the sneaker-type shoe) was present. In the new context, a specified subordinate noun is redundant, as it provides more information than a basic level noun (e.g., the shoe) (Rosch, 1973). Nevertheless, in the speaker's discourse model, the sneaker is the shared term between interlocutors; maintaining the same term can help listeners identify the (familiar) target referent (Levelt, 1989).

This demonstrates that how speakers referred to an object in the past influences how they will refer to the same object in the current or in future conversation.

In other examples of endophoric reference, a term used in an earlier context can affect the design of the current reference for *a similar referent* (Markman & Makin, 1998; Pechmann, 1984; Van Der Wege, 2009; Yoon & Brown-Schmidt, 2013; Yoon, Benjamin, & Brown-Schmidt, 2016). As described earlier in Section 1.1, speakers in the study by Pechmann (1984) used a color adjective to refer to a single bird (e.g., the white bird) in a new context after they had referred to a different colored bird (e.g., the black bird) in an earlier context. Here, the speaker maintains the focus on the previously mentioned object and introduces an adjective that reflects the difference between the past and current item (Levelt, 1989; Markman & Makin, 1998). In this example described by Pechmann (1984), it is relatively easy to compare the two similar items because they were once presented together. Stronger evidence of this kind of endophoric reference can be seen when speakers produce an overspecified expression to refer to an item (e.g., the wooden table) after they had referred to a similar item in the past (e.g., the table in Figs. 1 and 2), even when those two items were never shown together before. Van Der Wege (2009), who first identified this phenomenon, named it lexical differentiation. In Pechmann's study, speakers compared the two birds in a context and named one of them; consequently, both birds had already been marked with a modifier in

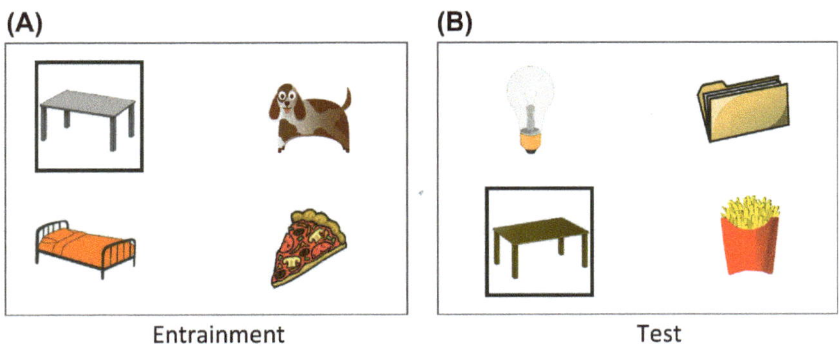

Figure 2 Example display of (A) entrainment and (B) test trial. On the speaker's screen, but not on the listener's screen, the referent was marked with a black box. Participants were exposed to the contrast item (e.g., metal table) in entrainment trials and the target item (e.g., wooden table) in test trials. Two items were from the same category (e.g., table) in the contrast condition, and they were unrelated in the non-contrast condition (e.g., an apple on entrainment).

speaker's discourse model. In contrast to Peckmann's (1984) simultaneous presentation, Van der Wege separately showed her participants two items in two different contexts (e.g., a metal table and a wooden table in Figs. 1 and 2). In her study, speakers overspecified their referring expressions with modifiers (e.g., the wooden table) after a different exemplar from the same category (e.g., the table) has previously been mentioned. The two separate presentations in Van der Wege's design demonstrate that speakers must represent discourse history in their discourse model to compare current and past contexts. Her findings also support the claim that historical context affects referential design in the current discourse (Brennan & Clark, 1996).

2. LEXICAL DIFFERENTIATION IN PRODUCTION

What is the relevant discourse context that interlocutors consider when they communicate? The relevant discourse context has been widely defined to include both the current physical context (e.g., a place where the conversation takes place) and the linguistic information that has been discussed between interlocutors (Brennan & Clark, 1996; Brown-Schmidt, Yoon, & Ryskin, 2015; Levelt, 1989; Yoon & Brown-Schmidt, 2013). Extensive work has shown that speakers are sensitive to the current context and tailor references accordingly (Krahmer & van Deemter, 2012; Olson, 1970; van Deemter, Gatt, van Gompel, & Krahmer, 2012) and maintain the same references with respect to discourse history (Brennan & Clark, 1996; Horton & Gerrig, 2002; Van Der Wege, 2009). However, less explored is how the discourse history affects referential design for a similar item compared to the earlier referent, and this deserves more attention. Here I focus on how speakers represent discourse context both currently and historically, and how this representation affects referential expressions, especially those related to the phenomenon called *lexical differentiation*.

2.1 Lexical Differentiation

When speakers refer to an exemplar from a category in a context wherein a similar competitor is absent, for example as in Fig. 2, they are likely to use a basic-level noun phrase, such as "the table," for both cases in Fig. 2A and B (Rosch, 1973). In a single context, speakers do not need to use a modifier in either case because a single table is present in each local context. Unnecessary modifiers provide extra information that can confuse listeners (Engelhardt et al., 2006). In another case, when speakers are subsequently exposed to the two contexts, they are more likely to produce a modified noun phrase to refer to the second table than they are if they have not previously seen the first table. A speaker's propensity to use two different labels to refer to two similar items from the same category, even though they do not appear in the same immediate context, is called *lexical differentiation*. Lexical differentiation supports the claim that the referent from previous contexts guides subsequent referring expressions (Brennan & Clark, 1996; Pechmann, 1984; Van Der Wege, 2009). If speakers consider only the local context, the same label ("the table") would suffice even though it refers to two different types of table—one from the past and the other in the current context. However, speakers differentiate between two similar entities by using two different labels; this is consistent with the principle of contrast (Clark, 1988), which claims that two linguistic forms imply a semantic distinction between the two.

Delving into lexical differentiation, recent studies have used a natural conversation task to understand how the historical discourse context guides referential design. In a series of studies by Yoon and her colleagues (Yoon et al., 2016; Yoon, Benjamin, & Brown-Schmidt, in preparation; Yoon & Brown-Schmidt, 2013), a pair of participants were recruited and randomly assigned the role of speaker and listener. They sat in front of each screen, which presented the same four pictures in each trial. A black box indicating the current target referent appeared only on the speaker's screen (Fig. 2). The speaker was asked to describe the target picture to the listener, and then the listener identified the target on her screen and clicked it. To test lexical differentiation, critical trials consisted of pairs of entrainments and test trials (Fig. 2). How speakers described a target item (e.g., wooden table) during test trials was examined after they described either a similar item (contrast item; e.g., metal table) or an unrelated item (non-contrast item; e.g., apple) during entrainment trials. During test trials, speakers were more likely to produce a modified noun phrase that referred to the target item if they had already referred to the contrast item in the past (contrast condition) than if they had not previously been exposed to the contrast item (non-contrast condition). The modification rate on test trials was 17.6% in the contrast condition and 10.1% in the non-contrast condition (Fig. 3). This was a smaller effect (7.5%) than Van Der Wege's (2009) differentiation effect. Van Der Wege's (2009) larger effect probably is due to stimulus differences

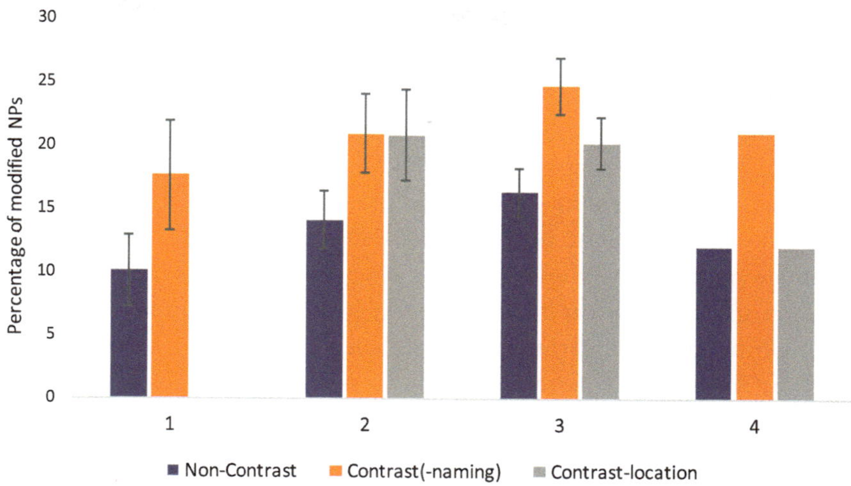

Figure 3 Percentage of noun phrases that were modified across three experiments—1: Experiment 1 in Yoon and Brown-Schmidt (2013), 2: Experiment 1 in Yoon, et al. (2016), and 3: Experiment 3 in Yoon, et al. (2016)—and 4: a predicted outcome by pre-emption by similar form (Van Der Wege, 2009).

(including the use of atypical category exemplars) and other methodological differences.

The effect of lexical differentiation has been replicated across various types of experimental setup, including diverse stimuli (e.g., typical vs. atypical items from a category), manipulating presence of partners (e.g., interacting with a live addressee vs. with an imagined addressee), local context (e.g., the presence or absence of a competitor in entrainment trials), the number of exposures to the contrast item, and so on (Van Der Wege, 2009; Yoon et al., 2016; Yoon & Brown-Schmidt, 2013). What is less clear is why speakers do not always differentiate two exemplars from the same category. The proportion of differentiation measured as the difference in the modification rate between the contrast and the non-contrast condition is not always high (e.g., considering only typical items, 4%–8%). The lexical differentiation effect is robustly consistent, but it is a small effect compared to the effect of discourse history on referential design, such as lexical entrainment (46%–52% in Brennan & Clark, 1996; 38%–66% in Van Der Wege, 2009).

Lexical differentiation may inherently have a smaller effect than other effects that support the influence of discourse history on referential design. Speakers appreciate discourse history in both lexical differentiation and entrainment, but in these two cases the manner in which discourse history guides the current reference can differ (Fig. 1). When speakers design a reference, they need to identify whether the current referent is a past referent or a new one (see Mandler, 1980). If it is a past referent, the linguistic precedent is recalled and reused, which is lexical entrainment. In contrast, if the current referent does not exactly reproduce a past referent but is similar to it, then the speakers first must recall the linguistic precedent set by the similar referent. Next they need to determine how the two items resemble and differ from one another. This process of lexical differentiation may demand more effort and attention from speakers than the process of lexical entrainment, and thus naturally elicit a smaller effect. Alternatively, speakers may use a different strategy to differentiate two subsequently shown items from the same category (see also discussion in Section 5.1). For example, in Yoon and Brown-Schmidt (2013), instead of producing modified noun phrases on test trials (17.6%), different noun phrases (16.4%), such as synonyms (e.g., bunny-rabbit) or subordinate-level nouns (e.g., dog-puppy), were introduced to distinguish the current from the past referents when speakers referred to the second exemplars from the same category.

2.2 Is Lexical Differentiation a Listener-Driven Process?

Speakers often provide extra information if there is a chance of failing to deliver a correct message to the listener—for example, when the listener can be confused and select the wrong referent (Paraboni, van Deemter, & Masthoff, 2007; Van Der Sluis & Krahmer, 2007) or feedback from the listener is not available (Arts, 2004). However, there have been mixed findings regarding listeners' interpretations of overspecified expressions. Some claim that overspecified expressions facilitate listeners' identification of intended referents (Arts et al., 2011; Paraboni, Masthoff, van Deemter, 2006; Nadig & Sedivy, 2002), whereas others argue that unnecessary extra information can disturb comprehension (Engelhardt et al., 2006; Engelhardt, Baris Demiral, & Ferreira, 2011).

Van Der Wege (2009) examined different types of addressees to examine why speakers produce lexically differentiated expressions—in other words, do these expressions benefit the speaker, the listener, or both? Participants in her study performed the task under three different conditions: a real audience, an imagined audience, and no audience. In the real audience condition, a pair of participants interactively engaged in the referential communication task, while in the imagined audience and no audience conditions, a participant performed the task without a partner present. In both the imagined audience and no audience conditions, participants were asked to describe a set of pictures, whereas in the imagined audience condition only, they were told that their descriptions would be recorded for a future participant, who would listen and follow their descriptions. If speakers produce lexically differentiated expressions to facilitate their partner's comprehension, then the magnitude of the effect should be bigger when speakers interact with a live audience rather than with an imagined audience. Similarly, the magnitude should be smallest when speakers talk to themselves (no audience condition). The live feedback of partners should motivate speakers to appropriately design referential expressions with respect to the shared discourse history. Alternatively, if speakers design references for a generic addressee (Dell & Brown, 1991) rather than adjusting references to a specific partner, the magnitude of effect should be similar in the real audience and the imagined audience conditions, but lower in the no audience condition when no audience is believed to be present. Finally, if the speaker is less concerned about whether listeners understand him or her, the effect will be equally sized across the three conditions. Van Der Wege's (2009) results showed that speakers produced lexically differentiated expressions under all three conditions. The effect was largest in the real audience condition and smallest in the no audience condition; the imagined audience condition fell in the middle. The fact that there is a larger effect of lexical differentiation when the speaker interacts with a live addressee than when he or she talks alone (regardless of the belief of an audience in the future) suggests that speakers form references for a specific partner, rather than designing references for a generic addressee (see Dell & Brown, 1991). Further, the finding that speakers showed lexical differentiation in the no audience condition implies that speakers consistently update their discourse model and produce references that track discourse history for both listeners and themselves. Thus, lexical differentiation is derived from a speaker's own discourse model and is encouraged by the listener's needs during communication.

Even when the listener's need to correctly identify referents motivates the speaker's performance of lexical differentiation, the effect can be small because the speaker may fail to appreciate the listener's perspective and needs when he or she designs referential expressions. In Van Der Wege (2009) and Yoon and Brown-Schmidt (2013), speakers maintained their speaking role and never experienced the listener's perspective. This could prevent speakers from aptly gauging their partner's needs because they lack the experience of what listeners are likely to do (see Horton & Gerrig, 2002).

In another study, a pair of participants alternated their roles as speaker and listeners during each block (Yoon et al., 2016). Although the design did not systematically manipulate the alternating roles to examine the effect of lexical differentiation, the magnitude of effect did not change much relative to the previous study in which participants assumed only a speaking role. The results suggest that appreciating a partner's needs does not compel speakers to produce lexically differentiated expressions.

2.3 Does Repetition of Entrainment Trials Modulate Lexical Differentiation?

When interlocutors refer to the same object multiple times during the course of a conversation, they agree to use the same label to refer it in an arrangement called conceptual pacts (Clark & Wilkes-Gibbs, 1986; Wilkes-Gibbs & Clark, 1992). The strength of the conceptual pacts can vary across contexts. For example, the number of repetitions or the agreeableness of labels can determine the strength of the conceptual pacts and the duration of the time during which they are maintained (Brennan & Clark, 1996 Knutsen & Le Bigot, 2014; Horton, 2008). Strong conceptual pacts will be sustained longer and are less likely to be violated (see Ibarra & Tanenhaus, 2016; Kronmüller & Barr, 2015). Likewise, in lexical differentiation, when speakers refer to a contrast item, they establish the conceptual pact with their partners. The strength of the conceptual pact in the case of a contrast item can influence how speakers design a referential expression for the target item and, consequently, modulate the rate of lexical differentiation: a greater the number of repeated entrainment trials can lead to a larger proportion of lexical differentiation.

Contrary to this idea, previous studies showed no relationship between the number of entrainment repetitions and the size of lexical differentiation (Van Der Wege, 2009; Yoon et al., 2016). Entrainments were repeated between one to four times by Van Der Wege (2009)[1] and between one and six times by Yoon et al. (2016). In neither study did entrainment repetition have a significant effect. Speakers produced lexically differentiated expressions to refer to the target item even when they were exposed only once to the contrast item (Yoon et al., 2016). This finding is consistent with the one-bit model (Galati & Brennan, 2010), which suggests that conceptual

[1] Van Der Wege (2009) calls this the *introductory phrase*.

pacts can be developed following one experience. Less clear is whether interlocutors establish deliberate conceptual pacts through a single exposure or make an automatic alignment of conceptualization to refer a series of referents (Pickering & Garrod, 2004, 2013). What is striking is that one single reference is sufficient to drive lexical differentiation and that the number of exposures to the same contrast item did not modulate the effect size of lexical differentiation.

2.4 Is the Latency Between Two Referents Crucial to Elicit Lexical Differentiation?

The repetition of entrainment does not affect the size of lexical differentiation. Another factor that could contribute to the magnitude of lexical differentiation is latency between two referents. Compared to distant discourse history, recent discourse history is easily accessible and retrievable (see Higgins, 1989; Raghubir & Menon, 2005). Thus, adjacent trials should lead to a larger lexical differentiation effect than nonadjacent exposures. Imagine, for example, a wooden table followed by a metal table. Immediately after the contrast, the speaker can retrieve the contrast item (e.g., the metal table) and compare the two more easily than he or she could after interrupting referents have been interposed between them. However, in a post-hoc analysis in Yoon and Brown-Schmidt (2013), they did not find a significant effect of latency between the two critical trials (the last entrainment and the test trials), which ranged from 1 to 15 trials (Fig. 4). Yoon et al. (2016) made another attempt to systematically manipulate latency between the critical pairs of trials. Their latencies were either short or long. Under the short latency condition the target immediately followed the contrast item, whereas under the long latency condition nine trials separated the two critical trials. Yoon et al. (2016) predicted that there would be a bigger differentiation effect in the short latency, but, in fact, the effect of latency was not significant. One explanation is that the

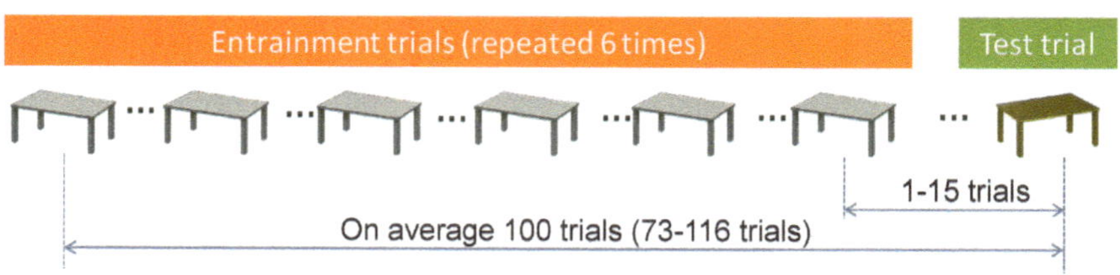

Figure 4 The latency between referents in entrainment and test trials in Yoon and Brown-Schmidt (2013).

differentiation effect was very small (~8%), and, consequently, there was not enough power to detect an interaction.

In Yoon and Brown-Schmidt (2013), they also tested the latency between the first entrainment—when entrainment was repeated six times—and the test trials in order to examine the frequency among entrainment trials affects lexical differentiation (Fig. 4). Again, it did not modulate the size of differentiation.

3. THE SOURCE OF LEXICAL DIFFERENTIATION

Lexical differentiation reveals how discourse history influences current reference design. In a variety of studies, the effect was small, but it was a consistent effect that was replicated many times under different sets of stimuli and manipulations (Van Der Wege, 2009; Yoon et al., 2016; Yoon & Brown-Schmidt, 2013). Interestingly, the rate of lexical differentiation is modulated not by how strongly or recently the discourse history has been established but by different types of audience. When speakers interact with a live partner, they lexically differentiate between two similar items more often than when they differentiate the items for a future partner or to themselves. Less explored is why speakers often, but not always, choose to differentiate between two exemplars. Understanding the source of differentiation is crucial to uncovering the mechanisms that underlie lexical differentiation and to appreciating the role that discourse history plays in referential design. A few potential sources that can lead to lexical differentiation in production have been tested (Yoon et al., 2016; Yoon et al., in preparation). Below I discuss each potential source of lexical differentiation.

3.1 Pre-emption by Similar Form

To lexically differentiate between two similar items, it is important to identify what the two objects have in common and what distinguishes them. When speakers introduce a new referential conceptualization, it should not have the same form as an already-established conceptualization in discourse. For example, when a speaker formulates a reference to name a metal table as "the table," a different type of table, such as a wooden one, cannot be described as "the table" again. The different type requires a different form of referential conceptualization (e.g., "the wooden table"). As suggested by Van Der Wege (2009), this is pre-emption by a similar form (cf. pre-emption of synonymy, Clark and Clark (1979)).

If pre-emption by similar form is a source of lexical differentiation, an already-established referential conceptualization serves as a key that inspires speakers to produce a new referential conceptualization. In other words, if the established referential conceptualization does not exist, there is no need to produce a lexically differentiated referential expression. To test this idea, Yoon et al. (2016) explored the potential source of lexical differentiation, pre-emption by a similar form, and they added a "contrast-location" condition. In this condition, participants were exposed to the contrast item during entrainment trials, but they were asked to describe it using a locative phrase (e.g., "the top left one") and to describe the target item during test trials. The non-contrast and contrast(-naming) conditions also were included. In the contrast-naming and contrast-location conditions, participants were exposed to pairs of contrast and target items, but the contrast items were only labeled in the contrast-naming condition. How speakers describe the target item during test trials was examined. If pre-existing referential conceptualization is required to derive lexical differentiation, then we would expect speakers to produce less modified noun phrases in the contrast-location condition than in the contrast-naming condition (Fig. 3). Even though they referred to the contrast item, speakers did not label it; thus, referential conceptualization had not been formulated in the contrast-location condition. Consistent with previous findings, speakers differentiated locally unique referents with modified noun phrases during test trials significantly more when the similar referent had previously been discussed (contrast-naming) than when it had not (non-contrast) (Fig. 3). Most importantly, there was no difference in the modification rates under the contrast-naming and the contrast-location conditions. This suggests that establishing referential conceptualization by naming the referent is not necessary for lexical differentiation—a finding that is not consistent with the prediction of pre-emption by similar form.

What if another label has been used to refer to the contrast item? For example, if a modified noun phrase has already been used (e.g., the metal table on an entrainment trial), then speakers may not need to produce another modified noun phrase; instead, a simple noun phrase would suffice for both local and historical contexts (e.g., the table on test trial). This scenario is also relevant to the question of whether pre-existing referential conceptualization is required for lexical differentiation. To test this situation, Yoon et al.'s (2016) Experiment 3 includes a context item in the entrainment trial. The context item is another exemplar from the same category in which the contrast and target items are involved (e.g., context item: plastic table; contrast and target: metal and wooden table). In the entrainment trials, both context and contrast items were simultaneously presented. A modified noun phrase is required to uniquely refer to the contrast item during entrainment trials (e.g., the metal table). The question is how speakers will refer to the target item during the test. Because a bare noun phrase (e.g., the table) has not been used, using the bare noun phrase does not violate pre-emption by similar form. Yet contrary to the prediction of pre-emption by similar form, speakers still produced more modified noun phrases to refer the target item in the contrast-naming and contrast-location conditions than in the non-contrast condition (Fig. 3). No difference was observed in the modification rates of the contrast-naming and the contrast-location conditions. Even though bare noun phrases had not been used during entrainment, speakers were less likely to use the bare noun phrases during test when they had seen a pair of items from the same category that the target was involved than when they had not. This finding suggests that lexical differentiation is not a product of a speaker's attempt to avoid giving two referents to the same label. Instead, speakers distinguish current from past referents by modifying noun phrases. An open question is how the mere presence of context items that were never mentioned during entrainments affects the speakers' discourse representations (Fukumura et al., 2010) and the rate of differentiation.

3.2 Memory for Past Referents

Another prerequisite for distinguishing current from past referents is to correctly encode and retrieve past referents when needed. While memory has long been thought to play an essential role in successful communication (Brown-Schmidt & Duff, 2016; Horton & Gerrig, 2005; McKinley, Brown-Schmidt, & Benjamin, 2017), the link between memory for language and language use—how past referents were encoded and retrieved in speakers' discourse model and how these memory representations affect referential design—has not been well explored. Examining the relationship between a speaker's memory for the historical discourse context and the referential form during a conversation allows us to address the question of why the differentiation rate in previous studies was so low. One explanation is that low rates of differentiation result from a poor memory of past contrast objects. Although speakers intend to design expressions with respect to the discourse history, if they fail to remember the past contrast items, they will not be able to differentiate the current target from the past contrast items. Memory for past referents was tested to determine whether it was a cause of low differentiation (Yoon, et al., 2016; Yoon et al., in preparation). If it is, speakers' performance on memory for contrast items would not be high, which would be consistent with the low differentiation rate. Furthermore, correctly recognizing contrast items would be significantly related to the use of modified referential expressions during the test.

A surprise memory test was performed after a conversation task that tested lexical differentiation in two critical conditions (contrast-naming and non-contrast, Yoon et al., 2016). In the memory test, two pictures—one old and one new picture—were presented on the screen. Participants were asked to identify the old picture that they had seen previously during the conversation task (a two-alternative forced choice). The critical pairs of pictures—contrast and target—were tested. The results showed that speakers' memory for past referents was good: the accuracy of contrast and target items was above 90%. When the contrast items were repeated six times during the conversation task, their memory for contrast items even reached ceiling. This was not a surprising result given that recognition memory for pictures is known to be good (Shepard, 1967). The high accuracy of memory for contrast item indicated that memory representation for the contrast item might not be related to modification of the target item during the conversation task. A few attempts were made to avoid memory performance reaching ceiling (Yoon et al., 2016; Yoon et al., in preparation). For example, participants took a 20-min break between the conversation task and the memory test, and during the conversation task the contrast item appeared not six times but once in entrainment trials. A different type of memory test was also tried—a yes/no recognition test. Across all studies, the gist of the results remained the same: in the memory test, speakers' accuracy for the contrast item was high. There was no significant relationship between the correct memory of the contrast item and modification of the target item during the conversation task. These results do not support the claim that speakers do not produce a higher rate of lexical differentiation because they fail to remember past referents. The recognition memory of speakers for discourse referents was, in fact, quite good. Yoon et al. (2016) conclude that remembering past referents is necessary for lexical differentiation, but it is not sufficient for speakers to produce differentiated references.

Although these memory test results allow us to better understand how memory contributes to the referential phenomenon of lexical differentiation, testing memory representations after the conversation task may have a few limitations. One of the limitations is that there are important differences in the memory processes needed to recall past references versus simply recognizing them (see Tversky, 1973). In conversation, when speakers design references, they actively search their memory for potential competitors in the relevant discourse contexts. The ability to recall past referents is more closely related to the speaker's task of probing discourse context than is the ability to recognizing those references. However, the setup in previous studies, in which critical pairs of contrast and target items look similar, made it difficult to test speakers' recall of past referents. Speakers' recall of simple noun phrases (e.g., saying they recalled "a table") would not allow researchers to identify the intended referent, which could be either a contrast or a target item (e.g., the metal or wooden desk).

In sum, speakers recognized past contrast and target items well. This finding is not consistent with the idea that small effect of lexical differentiation is caused by a failure to remember discourse context. Furthermore, no relationship between memory and reference form was observed; the accuracy of memory for the contrast item was not related to modification of the target item during the conversation task. However, it is not necessarily the case that memory for discourse context and reference design do not interact. During the conversation task, the nature of the reference did predict later memory performance. When a referent was described by a modified noun phrase (e.g., the metal table), it was better recognized in the future memory test than when it was described by a bare noun phrase (e.g., the table). Modification helps interlocutors remember the referent better in the future because it provides extra information, which suggests that memory for referents and reference form are closely related to one another. This raises the possibility that the failure to find a relationship between memory for past referents and lexical differentiation might be due to a low rate of modification during interlocutors' references to target objects.

3.3 The Role of Relevant Context

Previous studies showed that speakers lexically differentiate in order to distinguish current from past referents. However, differentiation in these studies was relatively infrequent. The low rate of differentiation could not be explained by failure to remember past referents. Instead, it is possible that speakers did not regard the previous discourse context as relevant. In previous studies, participants saw four pictures and described one picture on each trial. Once they had described a picture, the local context changed. This experimental context might be so discrete that speakers did not link previous discourse context and the current context across a series of trials. The speaker's perception of discourse context might be the key to eliciting a higher rate of lexical differentiation—that is, is the discourse history perceived to be relevant and cohesive?

What would make the context be perceived as relevant? A recurring cue might help interlocutors make links with the past and thus render the current context relevant. The presence of a recurring cue could help interlocutors retrieve information related to referents, labels, and contexts in the discourse history (Horton & Gerrig, 2005). For example, after Ricky talked about his rabbits to Marilyn at the table near the window in a coffee shop, Ricky remembers that conversation when he revisits the same table. The table at the coffee shop is a recurring cue that links the discourse history (e.g., "Do you remember when we talked about my rabbits?"). To test this idea, a recurring context cue was added to the type of experiment already described; the question is whether this might help speakers perceive the discourse history as relevant and, thus, elicit a higher rate of lexical differentiation (Yoon et al., in preparation). In a set of 8 trials, which included a critical pair of a contrast and a target item, Yoon et al. (in preparation) manipulated the coherence of the discourse context on the basis of a recurring background (see Boyce, Pollatsek, & Rayner, 1989). In the coherent condition, the background scenic picture changed across different sets; thus, each set was perceptually distinct. The discourse context remained relevant within a set because of the recurring background picture. In contrast, in the non-coherent condition, a gray background was used across sets. Each set used a 3×5 grid that potentially contained a unique image in each square, but only four pictures were presented during each trial. The position of each picture was manipulated to make the context more or less relevant: pictures in the grid remained in the same positions in the coherent condition, whereas they were randomly presented in the non-coherent condition. As in previous studies, speakers' expressions referring to the target item after they described either the contrast or the unrelated item were analyzed. Surprisingly, the modification rate overall was higher ($\sim$50%) in the current design than in previous studies ($\sim$25%). Although the presence of four pictures during each trial was consistent across the studies, the perceptual complexity in the current study (a grid potentially has a picture in each square) might induce speakers to produce more modification compared to the setup in previous studies (cf. Davies & Arnold, 2018). Further, the effect of lexical differentiation was significantly larger in the coherent condition (~22%) than in the non-coherent condition (~10%). This effect size is comparably substantial compared to previous findings (Yoon et al., 2016; Yoon & Brown-Schmidt, 2013). Independent of the coherence manipulation, speakers linked the past and current context more because there were more potential competitors in the grid. However, the more they regarded the context as relevant, the more modifications they produced to refer to the target item. In other words, when speakers perceive the context as relevant, they access and connect the past and current context and, as a result, the discourse history strongly guides the reference design.

Although a complex context may induce more discourse links, and thus higher rates of lexical differentiation, complex contexts are not necessary for such effects. Supporting this claim, a study that uses a very simple perceptual setup also demonstrated a large effect of lexical differentiation when speakers considered the discourse history relevant (Yoon & Federmeier, 2018). Yoon and Federmeier (2018) tested lexical differentiation using a different measure: the rate of different noun phrases that refer to the target referents. Note that in Yoon and Brown-Schmidt (2013), speakers used both modified noun phrases and different noun phrases to distinguish the current from past referents (e.g., rabbit vs. spotted rabbit or rabbit vs. bunny). In previous studies, it was difficult to compare the rate of different noun phrase use between the contrast and non-contrast conditions because only one noun is relevant in the non-contrast condition; the reference for the contrast item in the non-contrast condition was unrelated to the target category (e.g., apple). Consequently, Yoon and Federmeier (2018) employed a novel design to test the rate of different noun phrases in lexical differentiation. In their study, participants were shown one picture on a screen and asked to describe it using one noun. They were told that their response would be recorded for a future partner, who would listen to their descriptions, similar to the imagined audience condition in Van Der Wege (2009). In a series of picture stimuli, either the same picture was repeated or two similar pictures (e.g., two different types of sofas—sofa1 and sofa2) were presented. Following the picture description task, participants performed a picture identification task, during which they listened to their own description and had to identify the target among four pictures on the screen. A pair of picture description and identification tasks was repeated eight times. Because participants' own descriptions in the picture description task were reused during the

Table 1 Four Conditions and an Example in Each Condition in Yoon and Federmeier (2018)

	Same Picture	Different Picture
Same label	**Repetition** picture: sofa1—sofa1 label: "sofa"—"sofa"	**Under-differentiation** picture: sofa2—sofa1 label: "sofa"—"sofa"
Different label	**Over differentiation** picture: sofa1—sofa1 label:"couch"—"sofa"	**Differentiation** picture: sofa2—sofa1 label: "couch"—"sofa"

identification task, it was clear that the discourse context was relevant. The descriptions for critical pairs (either the same pictures or different pictures) in the picture description task were transcribed and coded into four different categories (Table 1).

When the same picture was repeated, participants mostly used the same labels (93%), and they rarely over-differentiated referents by providing two different labels to refer the same picture (7%). This finding is consistent with previous studies (Brennan & Clark, 1996; Horton & Gerrig, 2002; Van Der Wege, 2009). The critical question was whether speakers would produce two different labels when two similar pictures were shown. If speakers consider only the local context, a different noun phrase is not necessary to describe a single picture on the screen. The results showed that speakers produced under-differentiated expressions for 54% of the trials and appropriately-differentiated expressions for 46% of the trials. Although speakers did not always differentiate two items, they notably showed a higher rate of lexical differentiation than previous studies have found. This finding also supports the claim that speakers are more likely to distinguish two similar referents across the discourse context when they perceive that the past discourse context is tightly linked with the current context.

4. LEXICAL DIFFERENTIATION IN COMPREHENSION

In interactive conversation, interlocutors perform both roles of speaking and listening; The speaker will also be the listener by taking turns. Thus, speakers and listeners are expected to align each level of linguistic representation, such as phonetic, lexical and syntactic representations, to facilitate efficient communication (Pickering & Garrod, 2004, 2013). It is also expected that the representation of conceptual pacts—an agreement on how to refer objects between interlocutors—and the scope of relevant discourse context is aligned between interlocutors. Accordingly, an extensive amount of previous work has supported the idea that not only speakers but also listeners are sensitive to discourse history. Listeners process previously used references quickly (Brennan & Hanna, 2009; Brown-Schmidt, 2009; Kronmüller & Barr, 2007; Metzing, 2003), even when the references are over-informative in the local context (Yoon & Brown-Schmidt, 2018). They expect speakers to retain the labels used in previous contexts, and they delay interpretation when that expectation is violated (Metzing & Brennan, 2003; Yoon & Brown-Schmidt, 2018). This suggests that the salient context for speakers is also aligned to be salient for listeners (Brown-Schmidt & Tanenhaus, 2008; Rubio-Fernandez, 2016), and these coordinated representations of the discourse facilitate understanding (Richardson & Dale, 2005). Less explored is how listeners interpret lexically differentiated expressions that are over-informative in the local context but appropriately differentiated in the discourse history. To examine lexical differentiation in comprehension, two different methodologies (eye-movements and electrical brain activity) have been employed.

4.1 Lexical Differentiation While Tracking Eye-Movements

The measurement of eye movements is known to reflect incremental spoken language processing (Salverda, Brown, & Tanenhaus, 2011; Tanenhaus & Trueswell, 2006). Yoon and Brown-Schmidt (2013) monitored participant's eye movements while they listened to an experimenter's instructions (e.g., "Click on the table."). Participants saw two exemplars from the same category: a contrast item in entrainment trials and a target item in test trials (e.g., a metal and a wooden table in Fig. 2). During entrainment trials, the experimenter described the contrast items with either a basic noun phrase (e.g., the table—historical condition) or a locative phrase (e.g., top right one—ahistorical condition). Following entrainment, either a basic noun phrase (e.g., the table) or a modified noun phrase (e.g., the wooden table) was used to describe the target item. If listeners predict lexically differentiated referents that refer to the target item, modified noun phrases will facilitate listeners' interpretation in the historical but not in the ahistorical condition. In contrast, when they hear basic noun phrases during test trials, comprehension should be faster in the ahistorical condition (after hearing locative phrases) than in the historical condition (after hearing the same noun phrase that refers to a different referent). Surprisingly, however, the results were different from what was predicted. When modified noun phrases were presented during test trials, participants interpreted references with equal speed in both historical and ahistorical conditions, and no significant difference was elicited between them. Further, upon hearing a bare noun phrase during test trials, participants' identification of the target was faster in the historical condition than in the ahistorical condition. That is, participants showed a simple priming of basic level nouns when they had already heard the same label referring to a different item in the historical condition. The same pattern of results was replicated with a cohort-competitor paradigm that used different noun phrases (e.g., pool-poodle/dog-dock) rather than modified noun phrases (Experiment 3 in Yoon & Brown-Schmidt, 2013). The authors tested whether listeners showed a facilitation effect for subordinate level nouns (e.g., poodle) when they had previously heard the basic level noun (e.g., dog) refer to a contrast object from the same category. Consistent with the study of testing modified noun phrases, listeners showed no evidence of expecting subordinate-level expressions. Taken together, the two studies fail to demonstrate that listeners predict lexically differentiated expressions with respect to discourse history.

Why did these studies fail to show lexical differentiation in comprehension? There are a few possible explanations. First, listeners might not be able to use historical context. Compared to speakers, listeners have a weaker memory for what was previously said and described (Yoon et al., 2016). This outcome is consistent with the generation effect described in the memory literature, which has shown that what is generated is remembered better than what is passively processed (e.g., silently-read items, Slamecka & Graf, 1978). When comprehending references, then, listeners' weaker memory from past referents might keep them from utilizing discourse history in their predictions about what reference might be used. However, although the memory of speakers was better than that of listeners, Yoon and her colleagues found that the memory of listeners for past referents was actually fairly good (>80% in Experiment 1; Yoon et al., 2016). An alternative explanation is that listeners might remember past context but be more sensitive to the local context. If so, the relevant contexts for speaking and listening are different. Speakers consider a broader context, including past referents, but listeners attend more to the local context than to the entire discourse history.

Finally, the experimental setup might not offer a fair comparison between comprehension and production. The historical and ahistorical conditions in these comprehension studies are comparable to the contrast-naming and contrast-location conditions in the production studies (see Section 3.1). Note that the modification rate in production also did not differ under the contrast-naming and contrast-location conditions. If production and comprehension are interwoven processes that align multiple levels of representations (Pickering & Garrod, 2013), then the absence of significant difference in eye-movements under historical and ahistorical conditions might not be surprising. Although the ahistorical condition was added to control for the effect of previous exposure (Barr & Keysar, 2002), all three conditions, including the non-contrast condition, should ideally be tested to allow a fair comparison between production and comprehension.

4.2 Lexical Differentiation While Measuring Brain Electrical Activity

In addition to using different conditions to look for listeners' sensitivity to lexical differentiation, it is also useful to examine this question with different measures. In a recent study, electrical brain activity was recorded while listeners saw a series of pictures one at a time and listened to them being described (Yoon & Federmeier, 2018). Two critical factors were manipulated: type of picture (the same picture, repeated vs. a pair of similar, but different pictures from the same category (e.g., sofa1-sofa2)) and type of label (the same label vs. different labels (e.g., sofa-couch)). Critical pairs of labels consist of either basic-subordinate level nouns (e.g., cat-kitten) or synonyms (e.g., sofa-couch). Four different conditions were formulated as in Table 1. Participants listened to the picture description after 2 s of picture presentation. ERPs (event-related potentials) were time locked to the presentation of the second label (either the same or different label). If listeners appropriately use discourse history, they should predict the same label for the same picture that they saw before vs. a different label for a new picture. The results showed that when listeners' prediction was validated in the repetition and differentiation conditions, no significant difference was observed between the two conditions. In contrast, when the prediction was violated in the over-differentiation and the under-differentiation conditions, effects were elicited on sensory components—an increased P2 for under-differentiation and increased N1 for over-differentiation compared to the repetition condition. As N1 and P2 increases are seen under conditions of enhanced selective attention (e.g., Luck & Hillyard, 1994; Luck, Woodman, & Vogel, 2000), it seems that listeners may have attended more to these descriptions that violated their predictions. In addition, enhanced late positivity was elicited in the over-differentiation condition. Similar late positivities have been seen in the context of episodic memory, linked to recollection (Rubin, Van Petten, Glisky, & Newberg, 1999; Van Petten & Senkfor, 1996). Thus, it may be that, under these conditions—unusual in everyday discourse—wherein the same object was given two different labels, participants may have accessed the previous event and retrieve the label that was used earlier. Regardless of the precise mechanisms at work, however, the fact that brain responses differed for both over- and under-differentiated referents, compared to appropriately differentiated ones, makes clear that listeners are sensitive to the historical discourse context. Thus, these data support the claim that not only speakers, but also listeners track the discourse context and predict upcoming referents, even if such predictions do not always manifest in overt behavior.

5. FUTURE DIRECTIONS

5.1 Prosodic Differentiation

Speakers and listeners can use different means to differentiate two similar objects from the same category. Using different words or modifiers can be one way to differentiate, but it may not fully capture the mechanisms of differentiation in both production and comprehension, as differentiation could also be accomplished with paralinguistic features, such as prosody or accentuation. If so, rates of lexical differentiation may be low because differentiation through paralinguistic features is not captured in the modification rates that were the main focus of previous studies.

Speakers use prosody to efficiently indicate utterance components (e.g., words and syntactic constituents) and to convey their intention or meaning (Arnold & Watson, 2015; Snedeker & Trueswell, 2003; Watson, 2010, 2012; Watson & Gibson, 2004). Speakers are inclined to refer to new information using pitch accents (Brown, 1983; Hirschberg, 1993). Pitch accents are also used in a context containing contrastive entities (e.g., John bought a blue hat because Bill has a RED hat (Williams, 1997)). Similarly, distinguishing two similar referents in two local contexts can be accomplished by using pitch accent, which is *prosodic differentiation*. When prosodic differentiation accompanies lexical differentiation (e.g., the desk—the WOODEN desk), differentiation can be more effective in production. What is unknown is whether speakers use prosody when they lexically differentiate two referents that are separately presented.

Previous research on comprehension has also demonstrated that listeners can immediately single out a referent by using pitch accent that provides information about the focus of an utterance (Arnold, 2008; Dahan, Tanenhaus, & Chambers, 2002). Listeners use pitch accent as utterances unfold, assuming that an accented syllable refers to the new entity while an unaccented syllable refers to a given entity. They also rely on contrastive accents of a modifier to predict the upcoming referent during conversation (Ito & Speer, 2008). Contrast accents guide visual search, and felicitous intonation facilitates the interpretation of referent (e.g., "First hang the orange ball. Next, hang the BLUE ball/blue BALL."). Fixations to the referent were faster when a prominent pitch accent was assigned to contrastive information (e.g., the BLUE ball) than when the accent was produced infelicitously on the noun (e.g., the blue BALL). Thus, when prosodic information is available, listeners show sensitivity to lexical differentiation even in behavioral output.

An open question is how prosody works in lexical differentiation for both production and comprehension—whether speakers use prosody to differentiate two similar items from the same category, and if so, whether pitch accent can help listener predict lexically differentiated expressions. Yoon and Brown-Schmidt (2013) showed that speakers differentiate new discourse referents from similar referents discussed in past contexts, whereas listeners did not show any evidence of lexical differentiation expectation in the behavioral measures (e.g., eye movements). While they tried to control for all possible prosodic cues, it remains possible that prosody could have contributed to the observed asymmetry between production and comprehension in behavioral measures.

5.2 Differentiation in Aging

One special characteristic of lexical differentiation is that it represents an intersection of memory and language use. It requires interlocutors to remember past referents and discourse context to successfully design a referential form with respect to the past discourse history. An open question is how contextual information is encoded and used in lexical differentiation.

One way to explore the interaction between memory and language use is to examine how interlocutors—implicitly or explicitly—encode and represent discourse context and referents during conversation by testing lexical differentiation in a various populations. In particular, older adults, who experience more memory problems than young adults in daily life (Cutler & Grams, 1988), can provide informative data. Processing contextual information might be automatic and implicit, contrary to content (target) information (Chun & Jiang, 1998; Davenport & Potter, 2004; Munneke, Brentari, & Peelen, 2013). Research on aging suggests that implicit memory is generally well preserved against the effects of aging compared to explicit memory (Dew & Cabeza, 2011; Light & Singh, 1987; O'Hanlon, Wilcox, & Kemper, 2001). Older adults have more difficulty in learning new information consciously than learning without intending to do so. Examining how older adults take the historical context into account when designing a reference would bridge the fields of memory and language use (Horton & Spieler, 2007; Lysander & Horton, 2012). If contextual information is implicitly encoded without effort or attention, we would predict no deficits in memory for contextual information and lexical differentiation in older adults compared to young adults. Exploration of lexical differentiation in older adults can thus allow us to build up a better understanding of how memory affects referential designs and the consequences of age-related memory decline on referential communication.

6. CONCLUSION

Interlocutors often encounter overspecification in the local context. Some overspecified references are supported by the discourse history. Lexical differentiation, in particular, reflects the breadth of the relevant discourse context in task-based conversation in both language production and comprehension. Research looking at lexical differentiation in production suggests the relevant context is quite broad: speakers distinguish new discourse referents from non-present referents discussed in a previous context. To differentiate between a current target referent and previously-described contrast objects, speakers select modifiers and different nouns (e.g., subordinate level nouns), and this process is influenced by how they perceive the discourse context (e.g., relevant vs. discrete). Even though behavioral measures, listeners' eye movements, have failed to reveal evidence for lexical differentiation in comprehension, listeners also predict appropriately differentiated expressions and show a different response in their electrical brain activity when their prediction regarding the discourse history is violated. Taken together, the breadth of relevant discourse context is aligned across language production and comprehension. Future work should examine the process of prosodic differentiation: whether interlocutors use a paralinguistic cue, such as prosody, to differentiate two exemplars more effectively. Further, lexical differentiation is a good way to examine the interaction between memory and language in cognitive processes. Examining the memory of an interlocutor for both a past referent and its context will provide a more salient picture of how language is connected to memory representations during communication.

REFERENCES

Arnold, J. E. (2008). The BACON not the bacon: How children and adults understand accented and unaccented noun phrases. *Cognition, 108*, 69–99.

Arnold, J. E., & Watson, D. G. (2015). Synthesizing meaning and processing approaches to prosody: Performance matters. *Language, Cognition, and Neuroscience, 30*, 88–102.

Arts, A. (2004). *Overpsecification in instructive texts* (dissertation). Tilburg: Tilburg University.

Arts, A., Maes, A., Noordman, L., & Jansen, C. (2011). Overspecification facilitates object identification. *Journal of Pragmatics, 43*(1), 361–374.

Barr, D. J., & Keysar, B. (2002). Anchoring and comprehension in linguistic precedents. *Journal of Memory and Language, 46*, 391–418.

Boyce, S. J., Pollatsek, A., & Rayner, K. (1989). Effect of background information on object identification. *Journal of Experimental Psychology: Human Perception and Performance, 15*, 556–566.

Brennan, S. E., & Clark, H. H. (1996). Conceptual pacts and lexical choice in conversation. *Journal of Experimental Psychology. Learning, Memory, and Cognition, 22*(6), 1482–1493.

Brennan, S. E., & Hanna, J. E. (2009). Partner-specific adaptation in dialogue. *Topics in Cognitive Science, 1*, 274–291.

Brown, G. (1983). Prosodic structure and the given/new distinction. In A. Cutler, & D. R. Ladd (Eds.), *Prosody, models, and measurements*. New York: Springer-Verlag.

Brown-Schmidt, S. (2009). Partner-specific interpretation of maintained referential precedents during interactive dialog. *Journal of Memory and Language, 61*(2), 171–190.

Brown-Schmidt, S., & Duff, M. C. (2016). Memory and common ground processes in language use. *Topics in Cognitive Science, 8*(4), 722–736.

Brown-Schmidt, S., & Tanenhaus, M. K. (2008). Real-time investigation of referential domains in unscripted conversation: A targeted language game approach. *Cognitive Science, 32*, 643–684.

Brown-Schmidt, S., Yoon, S., & Ryskin, R. A. (2015). People as contexts in conversation. *Psychology of Learning and Motivation, 62*, 59–99.

Chun, M. M., & Jiang, Y. (1998). Contextual cueing: Implicit learning and memory of visual context guides spatial attention. *Cognitive Psychology, 36*, 28–71.

Clark, E. V. (1988). On the logic of contrast. *Journal of Child Language, 15*, 317–335.

Clark, H. H., & Clark, E. V. (1979). When nouns surface as verbs. *Language, 55*, 430–477.

Clark, H. H., & Wilkes-Gibbs, D. (1986). Referring as a collaborative process. *Cognition, 22*(1), 1–39.

Cutler, S. J., & Grams, A. E. (1988). Correlates of self-reported everyday memory problems. *Journal of Gerontology: Social Sciences, 43*, 82–90.

Dale, R., & Viethen, J. (2010). Attribute-centric referring expression generation. In E. Krahmer, & M. Theune (Eds.), *Lecture notes in computer science: Vol. 5790. Empirical methods in natural language generation*. Berlin, Heidelberg: Springer.

Dahan, D., Tanenhaus, M. K., & Chambers, C. G. (2002). Accent and reference resolution in spoken-language comprehension. *Journal of Memory and Language, 47*, 292–314.

Davenport, J. L., & Potter, M. C. (2004). Scene consistency in object and background perception. *Psychological Science, 15*, 559–564.

Davies, K., & Arnold, J. (2018). *Reference and informativeness: How context shapes referential choice* (pp. 1–36) (forthcoming).

Davies, C., & Katsos, N. (2013). Are speakers and listeners "only moderately Gricean"? An empirical response to Engelhardt et al. (2006). *Journal of Pragmatics, 49*(1), 78–106.

Dell, G. S., & Brown, P. M. (1991). Mechanisms for listener-adaptation in language production: Limiting the role of the "model of the listener". In D. Napoli, & J. Kegl (Eds.), *Bridges between psychology and linguistics*. New York: Academic Press.

Deutsch, W., & Pechmann, T. (1982). Social interaction and the development of definite descriptions. *Cognition, 11*, 159–184.

Dew, I. T., & Cabeza, R. (2011). The porous boundaries between explicit and implicit memory: Behavioral and neural evidence. *Annals of the New York Academy of Sciences, 1224*(1), 174–190.

Engelhardt, P., Bailey, K., & Ferreira, F. (2006). Do speakers and listeners observe the Gricean Maxim of quantity? *Journal of Memory and Language, 54*(4), 554–573.

Engelhardt, P. E., Baris Demiral, S. B., & Ferreira, F. (2011). Over-specified referring expression impair comprehension: An ERP study. *Brain and Cognition, 77*, 304–314.

Ford, W., & Olson, D. (1975). The elaboration of the noun phrase in children's descriptions of objects. *Journal of Experimental Child Psychology, 19*, 371–382.

Fukumura, K., van Gompel, R. P. G., & Pickering, M. J. (2010). The use of visual context during the production of referring expressions. *Quarterly Journal of Experimental Psychology, 63*, 1700–1715.

Galati, A., & Brennan, S. (2010). Attenuating information in spoken communication: For the speaker, or for the addressee? *Journal of Memory and Language, 62*, 35–51.

Gatt, A. (2007). *Generating coherent references to multiple entities* (Ph.D. dissertation). University of Aberdeen.

Gatt, A., & Belz, A. (2010). Introducing shared task evaluation to NLG: The TUNA shared task evaluation challenges. In E. Krahmer, & M. Theune (Eds.), *Empirical methods in natural language generation*. Berlin: Springer Verlag.

Gatt, A., Krahmer, E., van Gompel, R. P., & van Deemter, K. (2013). Production of referring expressions: Preference trumps discrimination. In *Proceedings of the 35th annual conference of the cognitive science society, Berlin, Germany*.

Grice, H. P. (1975). Logid and conversation. In P. Core, & J. L. Morgan (Eds.), *Syntax an semantics 3: Speech acts* (pp. 41–58). New York: Academic Press.

Higgins, E. T. (1989). Knowledge accessibility and activation: Subjectivity and suffering from unconscious sources. In J. S. Uleman, & J. A. Bargh (Eds.), *Unintended thought*. New York: Guilford.

Hirschberg, J. (1993). Pitch accent in context: Predicting intonational prominence from text. *Artificial Intelligence, 63*(1/2), 305–340.

Horton, W. S. (2008). A memory-based approach to common ground and audience design. In I. Kecskes (Ed.), *Intention, common ground, and the egocentric speaker-hearer* (pp. 189–222). Berlin: Mouton de Gruyter.

Horton, W. S., & Gerrig, R. J. (2002). Speakers' experiences and audience design: Knowing when and knowing how to adjust utterances to addressees. *Journal of Memory and Language, 47*, 589–606.

Horton, W. S., & Gerrig, R. J. (2005). Conversational common ground and memory processes in language production. *Discourse Processes, 40*(1), 1–35.

Horton, W. S., & Spieler, D. H. (2007). Age-related effects in communication and audience design. *Psychology and Aging, 22*, 281–290.

Ibarra, A., & Tanenhaus, M. K. (2016). The flexibility of conceptual pacts: Referring expressions dynamically shift to accommodate new conceptualizations. *Frontiers in Psychology, 7*, 561.

Ito, K., & Speer, S. (2008). Anticipatory effects of intonation: Eye movements during instructed visual search. *Journal of Memory and Language, 58*, 541–573.

Knutsen, D., & Le Bigot, L. (2014). Capturing egocentric biases in reference reuse during collaborative dialogue. *Psychonomic Bulletin & Review, 21*, 1590–1599.

Koolen, R., Gatt, A., Goudbeek, M., & Krahmer, E. (2011). Factors causing overspecification in definite descriptions. *Journal of Pragmatics, 43*, 3231–3250.

Koolen, R., Goudbeek, M., & Krahmer, E. (2013). The effect of scene variation on the redundant use of color in definite reference. *Cognitive Science, 37*, 395–411.

Krahmer, E., & van Deemter, K. (2012). Computational generation of referring expressions: A survey. *Computational Linguistics, 38*(1), 173–218.

Kronmüller, E., & Barr, D. J. (2007). Perspective-free pragmatics: Broken precedents and the recovery-from-preemption hypothesis. *Journal of Memory and Language, 56*, 436–455.

Kronmüller, E., & Barr, D. J. (2015). Referential precedents in spoken language comprehension: A review and meta-analysis. *Journal of Memory and Language, 83*, 1–19.

Levelt, W. J. M. (1989). *Speaking: From intention to articulation.* Cambridge, MA: MIT Press.

Light, L. L., & Singh, A. (1987). Implicit and explicit memory in young and older adults. *Journal of Experimental Psychology: Learning, Memory, and Cognition, 13*(4), 531.

Luck, S. J., & Hillyard, S. A. (1994). Electrophysiological correlates of feature analysis during visual search. *Psychophysiology, 31*, 291–308.

Luck, S. J., Woodman, G. F., & Vogel, E. K. (2000). Event-related potential studies of attention. *Trends in Cognitive Sciences, 4*, 432–440.

Lysander, K., & Horton, W. S. (2012). Conversational grounding in younger and older adults: The effect of partner visibility and referent abstractness. *Discourse Processes, 49*, 29–60.

Mandler, G. (1980). Recognizing: The judgment of previous occurrence. *Psychological Review, 87*, 252–271.

Mangold, R., & Pobel, R. (1988). Informativeness and instrumentality in referential communication. *Journal of Language and Social Psychology, 7*, 181–191.

Markman, A. B., & Makin, V. S. (1998). Referential communication and category acquisition. *Journal of Experimental Psychology: General, 127*(4), 331–354.

McKinley, G. L., Brown-Schmidt, S., & Benjamin, A. S. (2017). Memory for conversation and the development of common ground. *Memory & Cognition, 45*, 1281–1294.

Metzing, C. (2003). When conceptual pacts are broken: Partner-specific effects on the comprehension of referring expressions. *Journal of Memory and Language, 49*(2), 201–213.

Metzing, C., & Brennan, S. E. (2003). When conceptual pacts are broken: Partner specific effects on the comprehension of referring expressions. *Journal of Memory & Language, 49*, 201–213.

Munneke, J., Brentari, V., & Peelen, M. V. (2013). The influence of scene context on object recognition is independent of attentional focus. *Frontiers in Psychology, 4*, 552. https://doi.org/10.3389/fpsyg.2013.00552.

Nadig, A. S., & Sedivy, J. C. (2002). Evidence of perspective-taking constraints in children's on-line reference resolution. *Psychological Science, 13*, 329–336.

Olson, D. R. (1970). Language and thought: Aspects of a cognitive theory of semantics. *Psychological Review, 77*, 257–273.

Osgood, C. E. (1971). Where do sentences come from? In D. D. Steinberg, & L. A. Jakobovits (Eds.), *Semantics: An interdisciplinary reader in philosophy, linguistics and psychology* (pp. 497–529). Cambridge, MA: Cambridge University Press.

O'Hanlon, L., Wilcox, K. A., & Kemper, S. (2001). Age differences in implicit and explicit associative memory: Exploring elaborative processing effects. *Experimental Aging Research, 27*, 341–359.

Paraboni, I., Masthoff, J., & van Deemter, K. (2007). Making referents easy to identify. *Computational Linguistics, 33*, 229–254.

Paraboni, I., van Deemter, K., & Mastohoff, J. (2006). Overspecified reference in hierarchical domains: Measuring the benefits for readers. In *Proceddings of the 4th international conference on natural language generation, Syndey, Australia.*

Pechmann, T. (1984). Accentuation and redundancy in children's and adult's referential communication. In H. Bouman, & D. Bouwhuis (Eds.), *Vol. X. Attention and performance. Control of language Processees* (pp. 417–431). Hillsdale, NJ: Laurence Erlbaum.

Pechmann, T. (1989). Incremental speech production and referential overspecification. *Linguistics, 27*, 89–110.

Pickering, M. J., & Garrod, S. (2004). Toward a mechanistic psychology of dialogue. *The Behavioral and Brain Sciences, 27*(2), 169–190, 226.

Pickering, M. J., & Garrod, S. (2013). An integrated theory of language production and comprehension. *The Behavioral and Brain Sciences, 36*(4), 329–347.

Raghubir, P., & Menon, G. (2005). When and why is ease of retrieval informative? *Memory & Cognition, 33*, 821–832.

Richardson, D. C., & Dale, R. (2005). Looking to understand: The coupling between speakers' and listeners' eye movements and its relationship to discourse comprehension. *Cognitive Science, 29*, 1045–1060.

Rosch, E. (1973). Natural categories. *Cognitive Psychology, 4*, 328–350.

Rubin, S. R., Van Petten, C., Glisky, E. L., & Newberg, W. M. (1999). Memory Conjunction errors in younger and older adults: Event-related potential and neuropsychological data. *Cognitive Neuropsychology, 16*, 459–488.

Rubio-Fernandez, P. (2016). How redundant are redundant color adjectives? An efficiency-based analysis of color overspecification. *Frontiers in Psychology, 7*, 153.

Salverda, A. P., Brown, M., & Tanenhaus, M. K. (2011). A goal-based perspective on eye movements in visual world studies. *Acta Psychologica, 137*(2), 172–180.

Sedivy, J. C. (2003). Pragmatic versus form-based accounts of referential contrast: Evidence for effects of informativity expectations. *Journal of Psycholinguistic Research, 32*(1), 3–23.

Sedivy, J. C. (2004). Evaluating explanations for referential context effects: Evidence for Gricean mechanisms in online language interpretation. In J. C. Trueswell, & M. K. Tanenhaus (Eds.), *Approaches to studying world-situated language use: Bridging the language-as-product and language-as-action traditions* (pp. 345–364). Cambridge, MA: MIT Press.

Shepard, R. N. (1967). Recognition memory for words, sentences and pictures. *Journal of Verbal Learning and Verbal Behavior, 6*, 156–163.

Slamecka, N. J., & Graf, P. (1978). The generation effect: Delineation of a phenomenon. *Journal of Experimental Psychology: Human Learning & Memory, 4*, 592–604.

Snedeker, J., & Trueswell, J. (2003). Using prosody to avoid ambiguity: Effects of speaker awareness and referential context. *Journal of Memory and Language, 48*, 103–130.

Tanenhaus, M. K., & Trueswell, J. C. (2006). Eye movements and spoken language comprehension. In M. J. Traxler, & M. A. Gernsbacher (Eds.), *Handbook of psycholinguistics* (2nd ed.). New York: Elsevier Academic Press.

Tarenskeen, S., Broersma, M., & Geurts, B. (2015). Overspecification of color, pattern, and size: Salience, absoluteness, and consistency. *Frontiers in Psychology, 6*, 1703.

Tversky, B. (1973). Encoding processes in recognition and recall. *Cognitive Psychology, 5*, 275–287.

van Deemter, K., Gatt, A., van Gompel, R. P. G., & Krahmer, E. (2012). Toward a computational psycholinguistics of reference production. *Topics in Cognitive Science, 4*(2), 166–183.

Van Der Sluis, I., & Krahmer, E. (2007). Generating multimodal references. *Discourse Processes, 44*, 145–174.

Van Der Wege, M. M. (2009). Lexical entrainment and lexical differentiation in reference phrase choice. *Journal of Memory and Language, 60*(4), 448–463.

Van Petten, C., & Senkfor, A. J. (1996). Memory for words and novel visual patterns: Repetition, recognition, and encoding effects in the event-related brain potential. *Psychophysiology, 33*, 291–508.

Vogels, J., Krahmer, E., & Maes, A. (2015). How cognivie load influences speakers' choice of referring expressions. *Cognitive Science, 39*, 1396–1418.

Watson, D. G. (2010). The many roads to prominence: Understanding emphasis in conversation. *The Psychology of Learning and Motivation, 52*, 163–183.

Watson, D. G. (2012). Prosody in production. In H. Pashler (Ed.), *Encyclopedia of mind*. Sage Publications.

Watson, D. G., & Gibson, E. (2004). The relationship between intonational phrasing and syntactic structure in language production. *Language & Cognitive Processes, 19*, 713–755.

Westerbeek, H., Koolen, R., & Maes, A. (2015). Stored object knowledge and the production of referring expressions: The case of color typicality. *Frontiers in Psychology, 6*, 935.

Wilkes-Gibbs, D., & Clark, H. H. (1992). Coordinating beliefs in conversation. *Journal of Memory and Language, 31*(2), 183–194.

Williams, E. (1997). Blocking and anaphora. *Linguistic Inquiry, 28*, 577–628.

Yoon, S., Benjamin, A. S., & Brown-Schmidt, S. (2016). The historical context in conversation: Lexical differentiation and memory for the discourse history. *Cognition, 154*, 102–117.

Yoon, S., Benjamin, A. S., & Brown-Schmidt, S. Referential form and memory for the discourse history (in preparation).

Yoon, S., & Brown-Schmidt, S. (2013). Lexical differentiation in language production and comprehension. *Journal of Memory and Language, 69*(3), 397–416.

Yoon, S., & Brown-Schmidt, S. (2018). The influence of the historical discourse record on language processing in dialogue. *Discourse Processes, 55*, 31–46.

Yoon, S., & Brown-Schmidt, S. Contextual integration in multiparty audience design (under review).

Yoon, S., & Federmeier, K. D. (2018). ERP evidence revisits lexical differentiation in production and comprehension. In *Poster presented at the 31th annual CUNY conference on human sentence processing, Davis, CA*.

FSC
www.fsc.org
MIX
Papier | Fördert
gute Waldnutzung
FSC® C083411

Kolibri 360 GmbH
Ferdinand-Jühlke-Straße 7
99095 Erfurt, Deutschland
produktsicherheit@kolibri360.de